T0185231

Gravitational Radiation, Luminous Black Holes, and Gamma-Ray Burst Supernovae

Black holes and gravitational radiation are two of the most dramatic predictions of general relativity. The quest for rotating black holes – discovered by Roy P. Kerr as exact solutions to the Einstein equations – is one of the most exciting challenges currently facing physicists and astronomers.

Gravitational Radiation, Luminous Black Holes and Gamma-ray Burst Supernovae takes the reader through the theory of gravitational radiation and rotating black holes, and the phenomenology of GRB supernovae. Topics covered include Kerr black holes and the frame-dragging of spacetime, luminous black holes, compact tori around black holes, and black hole–spin interactions. It concludes with a discussion of prospects for gravitational-wave detections of a long-duration burst in gravitational waves as a method of choice for identifying Kerr black holes in the universe.

This book is ideal for a special topics graduate course on gravitational-wave astronomy and as an introduction to those interested in this contemporary development in physics.

MAURICE H. P. M. VAN PUTTEN studied at Delft University of Technology, The Netherlands and received his Ph.D. from the California Institute of Technology. He has held postdoctoral positions at the Institute of Theoretical Physics at the University of California at Santa Barbara, and the Center for Radiophysics and Space Research at Cornell University. He then joined the faculty of the Massachusetts Institute of Technology and became a member of the new Laser Interferometric Gravitational-wave Observatory (MIT-LIGO), where he teaches a special-topic graduate course based on his research.

Professor van Putten's research in theoretical astrophysics has spanned a broad range of topics in relativistic magnetohydrodynamics, hyperbolic formulations of general relativity, and radiation processes around rotating black holes. He has led global collaborations on the theory of gamma-ray burst supernovae from rotating black holes as burst sources of gravitational radiation. His theory describes a unique link between gravitational waves and Kerr black holes, two of the most dramatic predictions of general relativity. Discovery of *triplets* – gamma-ray burst supernovae accompanied by a long-duration gravitational-wave burst – provides a method for calorimetric identification of Kerr black holes in the universe.

Gravitational Radiation, Luminous Black Holes, and Gamma-Ray Burst Supernovae

MAURICE H. P. M. VAN PUTTEN

Massachusetts Institute of Technology

CAMBRIDGE
UNIVERSITY PRESS

CAMBRIDGE UNIVERSITY PRESS
Cambridge, New York, Melbourne, Madrid, Cape Town, Singapore,
São Paulo, Delhi, Dubai, Tokyo

Cambridge University Press
The Edinburgh Building, Cambridge CB2 8RU, UK

Published in the United States of America by Cambridge University Press, New York

www.cambridge.org
Information on this title: www.cambridge.org/9780521143615

First published 2005
This digitally printed version 2010

A catalogue record for this publication is available from the British Library

ISBN 978-0-521-84960-9 Hardback
ISBN 978-0-521-14361-5 Paperback

To my parents Anton and Maria,
and
Michael, Pascal, and Antoinette

Contents

Foreword

General relativity is one of the most elegant and fundamental theories of physics, describing the gravitational force with a most awesome precision. When it was first discovered, by Einstein in 1915, the theory appeared to do little more than provide for minute corrections to the older formalism: Newton's law of gravity. Today, more and more stellar systems are discovered, in the far outreaches of the universe, where extreme conditions are suspected to exist that lead to incredibly strong gravitational forces, and where relativistic effects are no longer a tiny perturbation, but they dominate, yielding totally new phenomena. One of these phenomena is gravitational radiation – gravity then acts in a way very similar to what happens with electric and magnetic fields when they oscillate: they form waves that transmit information and energy.

Only the most violent sources emit gravitational waves that can perhaps be detected from the Earth, and this makes investigating such sources interesting. The physics and mathematics of these sources is highly complex.

Maurice van Putten has great expertise in setting up the required physical models and in solving the complicated equations emerging from them. This book explains his methods in dealing with these equations. Not much time is wasted on philosophical questions or fundamental motivations or justifications. The really relevant physical questions are confronted with direct attacks. Of course, we encounter all sorts of difficulties on our way. Here, we ask for practical ways out, rather than indulging on formalities. Different fields of physics are seen to merge: relativity, quantum mechanics, plasma physics, elementary particle physics, numerical analysis and, of course, astrophysics. A book for those who want to get their hands dirty.

Gerard 't Hooft

xi

Acknowledgments

Epigraph to Chapter 11, reprinted with permission from Oxford University Press from *The Mathematical Theory of Black Holes*, by S. Chandrasekhar (1983).

Epigraph to Chapter 16, reprinted with permission from *Gravitation and Cosmology*, by Stephen Weinberg.

Introduction

Observations of gravitational radiation from black holes and neutron stars promise to dramatically transform our view of the universe. This new topic of gravitational-wave astronomy will be initiated with detections by recently commissioned gravitational-wave detectors. These are notably the Laser Interferometric Gravitational wave Observatory LIGO (US), Virgo (Europe), TAMA (Japan) and GEO (Germany), and various bar detectors in the US and Europe.

This book is intended for graduate students and postdoctoral researchers who are interested in this emerging opportunity. The audience is expected to be familiar with electromagnetism, thermodynamics, classical and quantum mechanics. Given the rapid development in gravitational wave experiments and our understanding of sources of gravitational waves, it is recommended that this book is used in combination with current review articles.

This book developed as a graduate text on general relativity and gravitational radiation in a one-semester special topics graduate course at MIT. It started with an invitation of Gerald E. Brown for a *Physics Reports* on gamma-ray bursts. *Why study gamma-ray bursters?* Because they are there, representing the most energetic and relativistic transients in the sky? Or perhaps because they hold further promise as burst sources of gravitational radiation?

Our focus is on gravitational radiation powered with rotating black holes – the two most fundamental predictions of general relativity for astronomy (other than cosmology). General relativity is a classical field theory, and we believe it applies to all macroscopic bodies. We do not know whether general relativity is valid down to the Planck scale without modifications at intermediate scales, without any extra dimensions or additional internal symmetries.

Observations of neutron star binaries PSR 1913 + 16 and, more recently, PSR 0737-3039, tell us that gravitational waves exist and carry energy. This discovery is a considerable advance beyond the earlier phenomenology of quasi-static spacetimes in general relativity, such as the deflection of light by the Sun and the orbital precession of Mercury.

Observational evidence of black holes is presently limited to compact stellar mass objects as black hole candidates in soft X-ray transients and their super-massive counterparts at centers of galaxies. Particularly striking is the discovery of compact stellar trajectories in SgrA* in our own galaxy, which reveals a supermassive black hole of a few million solar masses.

Rotating black holes are believed to nucleate in core collapse of massive stars. The exact solution of rotating black holes was discovered by Roy P. Kerr[293]. It shows frame-dragging to be the explicit manifestation of curvature induced by angular momentum. It further predicts a large energy reservoir in rotation in the black hole: its energy content may exceed that in a rapidly rotating neutron star by at least an order of magnitude. While in isolation stellar black holes are stable and essentially nonradiating, in interaction with their environment black holes can become luminous upon emitting angular momentum in various radiation channels.

Essential to the interaction of Kerr black holes with the environment is the Rayleigh criterion. Rotating black holes tend to lower their energy by radiating high specific angular momentum to infinity. In isolation, these radiative processes are suppressed by canonical angular momentum barriers, rendering macroscopic black holes stable. Penrose recognized that, in principle, the rotational energy of a black hole can be liberated by splitting surrounding matter into high and low angular momentum particles[416, 417]. Absorption of low-angular momentum and ejection of high-angular momentum with positive energy to infinity is consistent with the Rayleigh criterion and conservation of mass and angular momentum. These processes are restricted to the so-called *ergosphere*. Black hole spin-induced curvature and curvature coupling to spin combined further give rise to spin–orbit coupling – an effective interaction of black hole spin with angular momentum in an *ergotube* along the axis of rotation. Calorimetry on the ensuing radiation energies promises first-principle evidence for Kerr black holes and, consequently, evidence for general relativity in the nonlinear regime.

While currently observed neutron star binary systems provide us with labora-tories to study linearized general relativity, could gamma ray burst supernovae serve a similar role for fully nonlinear general relativity?

Cosmological gamma-ray bursts were accidentally discovered by Vela and Konus satellites in the late 1960s. Their association with supernovae, in its earliest form proposed by Stirling Colgate, has been confirmed by GRB 980425/SN1988bw[224, 536] and GRB030329/SN2003dh[506, 265]. Thus, Type Ib/c supernovae are probably the parent population of long GRBs. It has been appreciated that the observed GRB afterglow emissions repre-sent the dissipation of ultrarelativistic baryon-poor outflows[451, 452], while the associated supernova is strongly aspherical[268] and bright in X-ray line-emissions[17, 432, 613, 434, 454]. These observations further show the time of onset of the gamma-ray burst and the supernova to be the same within observational uncertainties.

This phenomenology reveals a baryon-poor active nucleus as the powerhouse of GRB supernovae in core collapse in massive stars. The only known baryon-free energy source is a rotating black hole. This presents an *energy paradox*: *the rotational energy of a rapidly rotating black hole is orders of magnitude larger than the energy requirements set by the observed radiation energies in GRB supernovae*. A rapidly rotating nucleus formed in core-collapse is relativistically compact and radiative primarily in "unseen" gravitational radiation and MeV-neutrino emissions. These channels provide a new opportunity for probing the inner engine of cosmological GRB supernovae.

The promise of a link between gravitational radiation and black holes in GRB supernovae provides a method for the gravitational wave-dectors LIGO and Virgo to provide first-principle evidence for Kerr black holes in association with a currently known observational phenomenon.

This book consists of three parts: gravitational radiation, waves in astrophysical fluids, and a theory of GRB supernovae from rotating black holes. Chapters 1–7 introduce general relativity and gravitational radiation. Chapters 8–10 discuss fluid dynamical waves in jets and tori around black holes. Gamma-ray burst supernovae are introduced in Chapter 11. A theory of gravitational waves created by GRB supernovae from rotating black holes is discussed in Chapters 12–15. Chapter 16 discusses GRB supernovae as observational opportunities for gravitational wave experiments LIGO and Virgo.

The author is greatly indebted to his collaborators and many colleagues for constructive discussions over many years, which made possible this venture into gravitational-wave astronomy: Amir Levinson, Eve C. Ostriker, Gerald E. Brown, Roy P. Kerr, Garry Tee, Gerard 't Hooft, H. Cheng, S.-T. Yau, Félix Mirabel, Dale A. Frail, Kevin Hurley, Douglas M. Eardley, John Heise, Stirling Colgate, Andy Fabian, Alain Brillet, Rainer Weiss, David Shoemaker, Barry Barish, Kip S. Thorne, Roger D. Blandford, Robert V. Wagoner, E. Sterl Phinney, Jacob Bekenstein, Gary Gibbons, Shrinivastas Kulkarni, Giora Shaviv, Tsvi Piran, Gennadii S. Bisnovatyi-Kogan, Ramesh Narayan, Bohdan Paczyński, Peter Mészáros, Saul Teukolsky, Stuart Shapiro, Edward E. Salpeter, Ira Wasserman, David Chernoff, Yvonne Choquet-Bruhat, Tim de Zeeuw, John F. Hawley, David Coward, Ron Burman, David Blair, Sungeun Kim, Hyun Kyu Lee, Tania Regimbau, Gregory M. Harry, Michele Punturo, Linqing Wen, Stephen Eikenberry, Mark Abramowicz, Michael L. Norman, Valeri Frolov, Donald

S. Cohen, Philip G. Saffman, Herbert B. Keller, Joel N. Franklin, Michele Zanolin, Masaaki Takahashi, Robert Preece, and Enrico Costa.

The author thanks Tamsin van Essen, Vince Higgs, Jayne Aldhouse, and Anthony John of Cambridge University Press for their enthusiastic editorial support.

The reader is referred to other texts for more general discussions on stellar structure, compact objects and general relativity, notably: *Gravitation* by C. W. Misner, K. S. Thorne and J. A. Wheeler[382], *Gravitation and Cosmology* by S. Weinberg[587], *The Membrane Paradigm* by K. S. Thorne, R. H. Price & A. MacDonald[534], *Stellar Structure and Evolution* by R. Kippenhahn and A. Weigert[295], *Introduction to General Relativity* by G. 't Hooft[527], *General Relativity* by R. M. Wald[577], *General Relativity* by H. Stephani[509], *Gravitation and Spacetime* by H. C. Ohanian and R. Ruffini[398], *A First Course in General Relativity* by Bernard F. Schutz[485], *Black Holes, White Dwarfs and Neutron Stars* by S. L. Shapiro and S. A. Teukolsky[490], *Black Hole Physics* by V. Frolov and I. D. Novikov[208], *Formation and Evolution of Black holes in the Galaxy* by H. A. Bethe, G. E. Brown and C.-H. Lee[53], and *Analysis, Manifolds and Physics* by Y. Choquet-Bruhat, C. DeWitt-Morette and M. Dillard-Bleick[120].

This book is based on research funded by NASA, the National Science Foundation, and awards from the Charles E. Reed Faculty Initiative Fund.

Notation

The metric signature $(-, +, +, +)$ is in conformance with Misner, Thorne and Wheeler 1974[382]. The Minkowski metric is given by $\eta_{ab} = \lceil -1, 1, 1, 1 \rfloor$.

Most of the expressions are in geometrical units, except where indicated. In the case of pair creation by black holes (Appendix D), we use mixed geometrical–natural units.

Tensors are written in the so-called abstract index notation in Latin script. Indices from the middle of the alphabet denote spatial coordinates. Four-vectors and p-forms are also indicated in small boldface. Three-vectors are indicated in capital boldface.

The epsilon tensor $\epsilon_{abcd} = \Delta_{abcd}\sqrt{-g}$ is defined in terms of the totally anti-symmetric symbol Δ_{abcd} and the determinant g of the metric, where $\Delta_{0123} = 1$ which changes sign under odd permutations.

Tetrad elements are indexed by $\{(e_\mu)^b\}_{\mu=1}^4$, where μ denotes the tetrad index and b denotes the coordinate index.

1

Superluminal motion in the quasar 3C273

General relativity endows spacetime with a causal structure described by observer-invariant *light cones*. This locally incorporates the theory of special relativity: the velocity of light is the same for all observers. Points *inside* a light cone are causally connected with its vertex, while points *outside* the same light cone are out-of-causal contact with its vertex. Light describes null-generators *on* the light cone. This simple structure suffices to capture the kinematic features of special relativity. We illustrate these ideas by looking at relativistic motion in the nearby quasar 3C273.

1.1 Lorentz transformations

Maxwell's equations describe the propagation of light in the form of electromagnetic waves. These equations are linear. The Michelson–Morley experiment[372] shows that the velocity of light is constant, independent of the state of the observer. Lorentz derived the commensurate linear transformation on the coordinates, which leaves Maxwell equations form-invariant. It will be appreciated that form invariance of Maxwell's equations implies invariance of the velocity of electromagnetic waves. This transformation was subsequently rederived by Einstein, based on the stipulation that the velocity of light is the same for any observer. It is non-Newtonian, in that it simultaneously transforms all four spacetime coordinates.

The results can be expressed geometrically, by introducing the notion of light cones. Suppose we have a beacon that produces a single flash of light in all directions. This flash creates an expanding shell. We can picture this in a spacetime

1

diagram by plotting the cross-section of this shell with the x-axis as a function of time – two diagonal and straight lines in an inertial setting (neglecting gravitational effects or accelerations). The interior of the light cone corresponds to points interior to the shell. These points can be associated with the centre of the shell by particles moving slower than the speed of light. The interior of the light cone is hereby causally connected to its vertex. The exterior of the shell is out-of-causal contact with the vertex of the light cone. This causal structure is local to the vertex of each light cone, illustrated in Figure (1.1).

Light-cones give a geometrical description of causal structure which is observer-invariant by invariance of the velocity of light, commonly referred to as "covariance". Covariance of a light cone gives rise to a linear transformation of the spacetime coordinates of two observers, one with a coordinate frame $K(t, x)$ and the other with a coordinate frame $K'(t', x')$. We may insist on coincidence of K and K' at $t = t' = 0$, and use geometrical units in which $c = 1$, whereby

$$\text{sign}(x^2 - t^2) = \text{sign}(x'^2 - t'^2). \tag{1.1}$$

The negative (positive) sign in (1.1) corresponds to the interior (exterior) of the light cone. The light cone itself satisfies

$$x^2 - t^2 = x'^2 - t'^2 = 0. \tag{1.2}$$

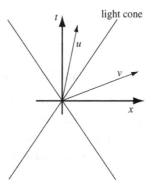

Figure 1.1 The local causal structure of spacetime is described by a light cone. Shown are the future and the past light cone about its vertex at the origin of a coordinate system (t, x). Vectors u within the light cone are timelike ($x^2 - t^2 < 0$); vectors v outside the light cone are spacelike ($x^2 - t^2 > 0$). By invariance of the velocity of light, this structure is the same for all observers. The linear transformation which leaves the signed distance $s^2 = x^2 - t^2$ invariant is the Lorentz transformation – a four-dimensional transformation of the coordinates of the frame of an observer.

A linear transformation between the coordinate frames of two observers which preserves the local causal structure obtains through Einstein's invariant distance

$$s^2 = -x^2 + t^2. \tag{1.3}$$

This generalizes Eqns (1.1) and (1.2). Remarkably, this simple ansatz recovers the Lorentz transformation, derived earlier by Lorentz on the basis of invariance of Maxwell's equations. The transformation in the invariant

$$x^2 - t^2 = x'^2 - t'^2 \tag{1.4}$$

can be inferred from rotations, describing the invariant $x^2 + y^2 = x'^2 + y'^2$ in the (x, y)-plane, as the hyperbolic variant

$$\begin{pmatrix} t' \\ x' \end{pmatrix} = \begin{pmatrix} \cosh\lambda & -\sinh\lambda \\ -\sinh\lambda & \cosh\lambda \end{pmatrix} \begin{pmatrix} t \\ x \end{pmatrix}. \tag{1.5}$$

The coordinates $(t, 0)$ in the observer's frame K correspond to the coordinates (t', x') in the frame K', such that

$$-\frac{x'}{t'} = \tanh\lambda. \tag{1.6}$$

This corresponds to a velocity $v = \tanh\lambda$ in terms of the "rapidity" λ of K' as seen in K. The matrix transformation (1.4) can now be expressed in terms of the relative velocity v,

$$t' = \Gamma(t - vx), \quad x' = \Gamma(x - vt), \tag{1.7}$$

where

$$\Gamma = \frac{1}{\sqrt{1 - v^2}} \tag{1.8}$$

denotes the Lorentz factor of the observer with three-velicity v.

The trajectory in spacetime traced out by an observer is called a world-line, e.g. that of K along the t-axis or the same observer as seen in K' following (1.8). The above shows that the tangents to world-lines – four-vectors – are connected by Lorentz transformations. The Lorentz transformation also shows that $v = 1$ is the limiting value for the relative velocity between observers, corresponding to a Lorentz factor Γ approaching infinity.

Minkowski introduced the world-line $x^b(\tau)$ of a particle and its tangent according to the velocity four-vector

$$u^b = \frac{dx^b}{d\tau}. \tag{1.9}$$

Here, we use a normalization in which τ denotes the eigentime,

$$u^2 = -1. \tag{1.10}$$

At this point, note the Einstein summation rule for repeated indices:

$$u^b u_b = \Sigma_{b=0}^3 u^b u_b = \eta_{ab} u^a u^b \tag{1.11}$$

in the Minkowski metric

$$\eta_{ab} = \begin{pmatrix} -1 & 0 & 0 & 0 \\ 0 & 1 & 0 & 0 \\ 0 & 0 & 1 & 0 \\ 0 & 0 & 0 & 1 \end{pmatrix}. \tag{1.12}$$

The Minkowski metric extends the Euclidian metric of a Cartesian coordinate system to four-dimensional spacetime. By (1.10) we insist

$$(u^x)^2 + (u^y)^2 + (u^z)^2 - (u^t)^2 = -1, \tag{1.13}$$

where $u^b = (u^t, u^x, u^y, u^z)$. In one-dimensional motion, it is often convenient to use the hyperbolic representation

$$u^b = (u^t, u^x, 0, 0) = (\cosh \lambda, \sinh \lambda, 0, 0) \tag{1.14}$$

in terms of λ, whereby the particle obtains a Lorentz factor $\Gamma = \cosh \lambda$ and a three-velocity

$$v = \frac{dx}{dt} = \frac{dx/d\tau}{dt/d\tau} = \frac{u^x}{u^t} = \tanh \lambda. \tag{1.15}$$

The Minkowski velocity four-vector u^b hereby transforms according to a Lorentz transformation ($d\tau$ is an invariant in (1.9)). We say that u^b is a covariant vector, and that the normalization $u^2 = -1$ is a Lorentz invariant, also known as a scalar.

To summarize, Einstein concluded on the basis of Maxwell's equations that spacetime exhibits an invariant causal structure in the form of an observer-invariant light cone at each point of spacetime. Points inside the light cone are causally connected to its vertex, and points outside are out-of-causal contact with its vertex. This structure is described by the Minkowski line-element

$$s^2 = x^2 + y^2 + z^2 - t^2, \tag{1.16}$$

which introduces a Lorentz-invariant signed distance in four-dimensional space-time (t, x, y, z) following (1.12). In attributing the causal structure as a property intrinsic to spacetime, Einstein proposed that *all* physical laws and physical observables are observer-independent, i.e. obey invariance under Lorentz trans-formations. This invariance is the principle of his theory of special relativity. Galileo's picture of spacetime corresponds to the limit of slow motion or, equiva-lently, the singular limit in which the velocity of light approaches infinity – back to Euclidean geometry and Newton's picture of spacetime.

1.2 Kinematic effects

In Minkowski spacetime, rapidly moving objects give rise to apparent kinematic effects, representing the intersections of their world-lines with surfaces Σ_t of constant time in the laboratory frame K. In a two-dimensional spacetime diagram (x, t), Σ_t corresponds to horizontal lines parallel to the x-axis.

Consider an object moving uniformly with Lorentz factor Γ as shown in Figure (1.2) such that its world-line – a straight line – intersects the origin. The *lapse* in eigentime τ in the motion of the object from Σ_0 to Σ_t is given by

$$\tau = \int_0^t \frac{ds}{dt}\, dt = \sqrt{-(t, vt)^2} = t\sqrt{1 - v^2}, \tag{1.17}$$

or

$$\frac{\tau}{t} = \frac{1}{\Gamma}. \tag{1.18}$$

Moving objects have a smaller lapse in eigentime between two surfaces of constant time, relative to the static observer in the laboratory frame. Rapidly moving elementary particles hereby appear with enhanced decay times. This effect is known as *time-dilation*.

The distance between two objects moving uniformly likewise depends on their common Lorentz factor Γ as seen in the laboratory frame K, as shown in

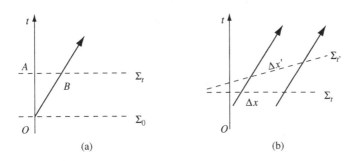

(a)   (b)

Figure 1.2 (*a*) Time dilation is described by the lapse in eigentime of a moving particle (arrow) between two surfaces of constant time Σ_0 and Σ_t in the laboratory frame K. The distance beteen these to surfaces in K is t, corresponding to O and A. The lapse in eigentime is t/Γ upon intersecting Σ_0 at O and Σ_t at B, where Γ is the Lorentz factor of the particle. Moving clocks hereby run slower. (*b*) The distance between two parallel world-lines (arrows) is the distance between their points of intersection with surfaces of constant time: Σ_t in K and $\Sigma_{t'}$ in the comoving frame K'. According to the Lorentz transformation, $\Delta x = \Delta x'/\Gamma$, showing that moving objects appear shortened and, in the ultrarelativistic case, become so-called "pancakes."

Figure (1.2). According to (1.7), the distance Δx between them as seen in K is related to the distance $\Delta x'$ as seen in the comoving frame K' by

$$\Delta x = \Delta x'/\Gamma. \tag{1.19}$$

Hence, the distance between two objects in uniform motion appears reduced as seen in the laboratory frame. This effect is known as the *Lorentz contraction*. Quite generally, an extended blob moving relativistically becomes a "pancake" as seen in the laboratory frame.

1.3 Quasar redshifts

Quasars are highly luminous and show powerful one-sided jets. They are now known to represent the luminous center of some of the active galaxies. These centers are believed to harbor supermassive black holes.

The archetype quasar is 3C273 at a redshift of $z = 0.158$. The redshift is defined as the relative increase in the wavelength of a photon coming from the source, as seen in the observer's frame: if λ_0 denotes the rest wavelength in the frame of the quasar, and λ denotes the wavelength in the observer's frame, we may write

$$1 + z = \frac{\lambda}{\lambda_0}. \tag{1.20}$$

The quasar 3C273 shows a relative increase in wavelength by about 16%. This feature is achromatic: it applies to any wavelength.

We can calculate z in terms of the three-velocity v with which the quasar is receding away from us. Consider a single period of the photon, as it travels a distance λ_0 in the rest frame. The null-displacement (λ_0, λ_0) *on* the light cone (in geometrical units) corresponds by a Lorentz transformation to

$$\Gamma(\lambda_0 + v\lambda_0), \quad \Gamma(\lambda_0 + v\lambda_0). \tag{1.21}$$

Note the plus sign in front of $v\lambda$ for a receding velocity of the quasar relative to the observer. The observer measures a wavelength

$$\lambda = \lambda_0 \Gamma(1 + v) = \lambda_0 \sqrt{\frac{1+v}{1-v}}. \tag{1.22}$$

It is instructive also to calculate the redshift factor z in terms of a redshift in energy. Let p_a denote the four-momentum of the photon, which satisfies $p^2 = 0$ as it moves along a null-trajectory on the light cone. Let also u^a and v^a denote the velocity four-vectors of of the quasar and that of the observer, respectively. The energies of the photon satisfy

$$\epsilon_0 = -p_a u^a, \quad \epsilon = -p_a v^a. \tag{1.23}$$

The velocity four-vectors u^a and v^a are related by a Lorentz transformation

$$v^a = \Lambda^b_a u^b, \quad \tanh \lambda = -v \tag{1.24}$$

in the notation of (1.5). It follows that

$$\epsilon = -p_a \Lambda^a_c u^c = -\eta_{ab} p^a \Lambda^a_c u^c. \tag{1.25}$$

This is a *scalar* expression, in view of complete contractions over all indices. We can evaluate it in any preferred frame. Doing so in the frame of the quasar, we have $p^a = \epsilon_0(1,1)$ and $u^b = (1,0)$. Hence, the energy in the observer's frame satisfies

$$\epsilon = \epsilon_0 (\cosh \lambda - \sinh \lambda) = \epsilon_0 \sqrt{\frac{1-v}{1+v}}. \tag{1.26}$$

Together, (1.22) and (1.26) obey the relationship $\epsilon = 2\pi/\lambda$, where $\epsilon_0 \lambda_0 = 2\pi$.

1.4 Superluminal motion in 3C273

The quasar 3C273 is a variable source. It ejected a powerful synchrotron emitting blob of plasma in 1977, shown in Figure (1.3)[412]. In subsequent years, the angular displacement of this blob was monitored. Given the distance to 3C273 (based on cosmological expansion, as described by the Hubble constant), the velocity projected on the sky was found to be

$$v_\perp = (9.6 \pm 0.8) \times c. \tag{1.27}$$

An elegant geometrical explanation is in terms of a relativistically moving blob, moving close to the line-of-sight towards the observer, given by R. D. Blandford, C. F. McKee and M. J. Rees[65]. Consider two photons emitted from the blob moving towards the observer at consecutive times. Because the second photon is emitted while the blob has moved closer to the observer, it requires less travel time to reach the observer compared with the preceding photon. This gives the blob the appearance of rapid motion. We can calculate this as follows, upon neglecting the relative motion between the observer and the quasar. (The relativistic motion of the ejecta is much faster than that of the quasar itself.)

Consider the time-interval Δt_e between the emission of the two photons. The associated time-interval Δt_r between the times of receiving these two photons is reduced by the distance $D_\parallel = v \cos \theta \Delta t_e$ along the line-of-sight traveled by the blob:

$$\Delta t_r = \Delta t_e - v \cos \theta \Delta t_e, \tag{1.28}$$

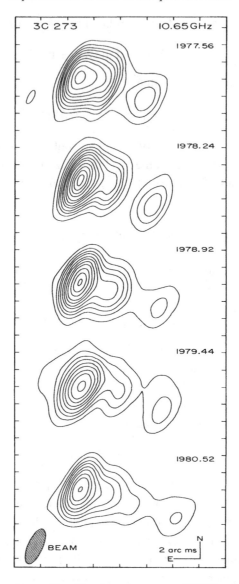

Figure 1.3 A Very Large Baseline Interferometry (VLBI) contour map of five epochs on an ejection event in the quasar 3C273 in the radio (10.65 GHz). (Reprinted by permission from the authors and *Nature*, Pearson, T. J. *et al.*, *Nature*, 280, 365. ©1981 Macmillan Publishers Ltd.)

where θ denotes the angle between the velocity of the blob and the line-of-sight. The projected distance on the celestial sphere is $D_\perp = \Delta t_e v \sin\theta$. The projected velocity on the sky is, therefore,

$$v_\perp = \frac{D_\perp}{\Delta t_r} = \frac{v\sin\theta}{1 - v\cos\theta}. \tag{1.29}$$

Several limits can be deduced. The maximal value of the apparent velocity $v_\perp$ is

$$v_\perp = v\Gamma. \tag{1.30}$$

Thus, an observed value for $v_\perp$ gives a minimal value of the three-velocity and Lorentz factor

$$v = \frac{v_\perp}{\sqrt{1 + v_\perp^2}}, \quad \Gamma = \sqrt{1 + v_\perp^2}. \tag{1.31}$$

Similarly, an observed value for $v_\perp$ gives rise to a maximal angle θ upon setting $v = 1$. With (1.27), we conclude that the blob has a Lorentz factor $\Gamma \geq 10$.

1.5 Doppler shift

The combined effects of redshift and projection are known as Doppler shift. Consider harmonic wave-motion described by $e^{i\phi}$. The phase ϕ is a scalar, i.e. it is a Lorentz invariant. For a plane wave we have $\phi = k_a x^a = \eta_{ab} k^a x^b$ in terms of the wave four-vector k^a. Thus, k^a is a four-vector and transforms accordingly. A photon moving towards an observer with angle θ to the line-of-sight has $k^x = \epsilon \cos \theta$ for an energy $k^0 = \epsilon$. By the Lorentz transformation, the energy in the source frame with velocity v is given by

$$k'^0 = \Gamma(k^0 - vk^1), \tag{1.32}$$

so that

$$\epsilon' = \Gamma\epsilon(1 - v\cos\theta). \tag{1.33}$$

The result can be seen also by considering the arrival times of pulses emitted at the beginning and the end of a period of the wave. If T' and T denote the period, in the source and in the laboratory frame, respectively, then $2\pi = \epsilon'T' = \epsilon'(T/\Gamma)$. The two pulses have a difference in arrival times $\Delta t = T(1 - v\cos\theta)$ and the energy in the observer's frame becomes

$$\epsilon = \frac{2\pi}{\Delta t} = \frac{\epsilon'}{\Gamma(1 - v\cos\theta)}. \tag{1.34}$$

This is the same as (1.33).

1.6 Relativistic equations of motion

Special relativity implies that all physical laws obey the same local causal structure defined by light cones. This imposes the condition that the world-line of any particle through a point remains inside the local light cone. This is a geometrical description of the condition that all physical particles move with velocities less than (if massive) or equal to (if massless) the velocity of light.

Newton's laws of motion for a particle of mass m are given by the three equations

$$F_i = m \frac{d^2 x_i(t)}{dt^2} \quad (i = 1, 2, 3). \tag{1.35}$$

We conventionally use Latin indices from the middle of the alphabet to denote spatial components i, corresponding to (x, y, z). The velocity $dx_i(t)/dt$ is unbounded in response to a constant forcing (m is a constant), and we note that (1.35) consists of merely three equations motion. It follows that (1.35) does not satisfy causality, and is not Lorentz-invariant.

Minkowski's world-line x^b of a particle is generated by a tangent given by the velocity four-vector (1.9). Here, we use a normalization in which τ denotes the eigentime, (1.10). We consider the Lorentz-invariant equations of motion

$$f^b = \frac{dp^b}{d\tau}, \tag{1.36}$$

where

$$p^b = mu^b = (E, P^i) \tag{1.37}$$

denotes the particle's four-momentum in terms of its energy, conjugate to the time-coordinate t, and three-momentum, conjugate to the spatial coordinates x^i. There is one Lorentz invariant:

$$p^2 = -m^2, \tag{1.38}$$

which is an integral of motion of (1.36). The forcing in (1.36) is subject to the orthogonality condition $f^b p_b = 0$, describing orthogonality to its world-line.

The non-relativistic limit corresponding to small three-velocities v in (1.38) gives

$$E = \sqrt{m^2 + P^2} \simeq m + \frac{1}{2} m v^2. \tag{1.39}$$

We conclude that E represents the sum of the Newtonian kinetic energy and the mass of the particle. This indicates that m (i.e. mc^2) represents rest mass-energy of a particle. As demonstrated by nuclear reactions, rest mass-energy can be released in other forms of energy, and notably so in radiation. In general, it is important to note that energy is the time-component of a four-vector, and that it transforms accordingly.

Exercises

1. Maxwell's equations in a vacuum are the first-order linear equations $\nabla \times \mathbf{H} = -\partial_t \epsilon_0 \mathbf{E}$ and $\nabla \times \mathbf{E} = \partial_t \mu_0 \mathbf{H}$ on the electric field $\mathbf{E}$ and the magnetic field $\mathbf{H}$, where ϵ_0 and μ_0 denote the electric permittivity and magnetic permeability of vacuum. Show that both $\mathbf{E}$ and $\mathbf{H}$ satisfy the wave-equation $\nabla^2 \{\mathbf{E}, \mathbf{H}\} = 0$ in terms of the d'Alembertian

$$\nabla^2 = -c^{-2}\partial_t^2 + \partial_x^2 + \partial_y^2 + \partial_z^2, \tag{1.40}$$

where $c = (\epsilon_0 \mu_0)^{-1/2}$ denotes the velocity of light.

2. *Simple wave* solutions $\mathbf{E} = \Phi(k_a x^a)$ to (1.40) are plane-wave solutions satisfing the characteristic equation

$$k^a k^b \eta_{ab} = 0, \tag{1.41}$$

where k^a denotes the wave-vector and η_{ab} denotes the Minkowski metric (1.12). Verify that the null-surface (1.41) describes a cone in spacetime with vertex at the origin. Coordinate transformations which leave the Minkowski metric explicitly in the form (1.12), and hence the d'Alembertian in the form (1.40), are the Lorentz transformations. The postulate that c is constant hereby introduces Lorentz transformations between different observers. Verify geometrically that the Lorentz transformations form a group.

3. Obtain explicitly the product of two Lorentz transformations representing boosts along the x-axis with velocities v and w.

4. Derive the general class of infinitesimal Lorentz transformations for (1.16), consisting of small boosts and rotations. What is their dimensionality and do they commute?

5. Consider two world-lines with velocity four-vectors u^b and v^b which intersect at $\mathcal{O}$. In the wedge product

$$(\mathbf{u} \wedge \mathbf{m})^{ab} = u^a m^b - u^b m^a \tag{1.42}$$

we may assume without loss of generality $m^c u_c = 0$. Show that

$$\mathbf{n} = (\mathbf{u} \wedge \mathbf{m}) \cdot \mathbf{v} \qquad (1.43)$$

produces a vector field n^b such that $n^c v_c = 0$. If u^b and v^b represent a boost between two observers, show that n^b are related by the same if (n^b, u^b, v^b) are coplanar. In this event, (1.43) represents a finite Lorentz transformation between two four-vectors m^b and n^b.

6. An experimentalist emits a photon of energy ϵ onto a mirror, which moves rapidly towards the observer with Lorentz factor Γ. What is the energy of the reflected photon received by the observer? (This is the mechanism of inverse Compton scattering, raising photon energies by moving charged particles below the Klein-Nishina limit[468].)

7. Generalize the results of Section 1.4 by including the redshift factor of the quasar.

8. Consider a radiation front moving towards the observer with Lorentz factor Γ. If the front is time variable on a timescale $\delta\tau$ in the comoving frame, what is the observed timescale of variability?

2

Curved spacetime and SgrA*

General relativity extends Newton's theory of gravitation, by taking into account a local causal structure described by coordinate-invariant light cones. This proposal predicts some novel features around stars. Ultimately, it predicts black holes as fundamental objects and gravitational radiation.

It was Einstein's great insight to consider Lorentz invariance of Maxwell's equations as a property of spacetime. All physical laws hereby are subject to one and the same causal structure. To incorporate gravitation, he posed a local equivalence between gravitation and acceleration. This introduces the concept of freely falling observers in the limit of zero acceleration and described by geodesic motion.

The accelerated motion of the proverbial Newton's apple freely falling in the gravitation field is fundamental to gravitation. The weight of the apple when hanging on the tree or in Newton's hand is exactly equal to the body force when accelerated by hand at the same acceleration as that imparted by the gravitational field in free-fall. The mass of the apple as measured by its "weight" is unique whether gravitational or inertial.

Rapidly moving objects show kinematic effects in accord with special relativity. These effects may be attributed to the associated kinetic energies. In the Newtonian limit, the gravitational field may be described in terms of a potential energy. Kinetic energy and potential energy are interchangeable subject to conservation of total energy. Kinematic effects can hereby be attributed equivalently to a particle's kinetic energy or drop in potential energy. When viewed from the tree, *the apple in Newton's hand looks more red and flat than those still hanging in the tree.*

General relativity incorporates the Newtonian potential energy *of* a gravitational field *in* four-dimensional spacetime. This is covariantly described by curvature. In case of the spherically symmetric spacetime around the Sun, curvature is manifest in, for example, orbital precession of Mercury. Around extremely compact objects, particles may assume zero total energy: these objects are "black," wherefrom no particles or light can escape. While Newton's theory gives rise to black objects surrounded by flat spacetime predicted by J. Michell (1783) and P. Laplace (1796), general relativity gives rise to black holes: compact null-surfaces surrounded by curved spacetime.

Particularly striking observational evidence for black holes is based on proper motion studies of individual stars at the center of our galaxy, SgrA*[483], indicating a supermassive black hole mass of about $3 \times 10^6 M_\odot$.

2.1 The accelerated letter "*L*"

Figure (2.1) shows a pair of curved trajectories of two objects subject to constant acceleration. These trajectories gradually separate from the *t*-axis, and their velocity four-vectors satisfy

$$\frac{dx^b}{d\tau} = u^b = (\cosh \lambda, \sinh \lambda), \quad \lambda = g\tau. \tag{2.1}$$

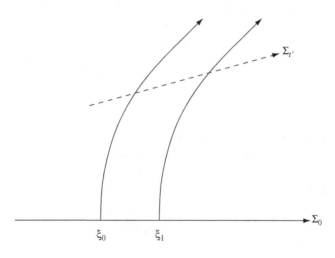

Figure 2.1 A Minkowski diagram of a pair of parallel world-lines subject to the same acceleration. Their initial positions are $\xi_{0,1}$ on Σ_0. This introduces a line-element $ds^2 = -dt'^2 + dx'^2 = -\Gamma^{-2}dt + \Gamma^2 d\xi^2$ expressing the instantaneous Lorentz transformation associated with the Lorentz factor $\Gamma = \cosh(gt')$, where t' denotes the time in the comoving frame and g denotes the constant of acceleration.

This descibes an acceleration

$$a^b = \frac{du^b}{d\tau} = g(\sinh \lambda, \cosh \lambda),$$ (2.2)

of constant strength $\sqrt{a^c a_c} = g$.

The trajectory of an accelerated observer initially at rest at $(0, \xi)$ results by integration of (2.1),

$$\begin{pmatrix} t \\ z \end{pmatrix} = \begin{pmatrix} g^{-1} \sinh(g\tau) \\ g^{-1} (\cosh(g\tau) - 1) \end{pmatrix} + \xi \begin{pmatrix} \cosh(g\tau) \\ \sinh(g\tau) \end{pmatrix}.$$ (2.3)

The accelerating observer carries along a frame of reference (t', x'). At any moment of time, these coordinates are related to the coordinates (t, x) of the laboratory through a Lorentz transformation plus a translation. As in Figure (1.2), kinematic effects are simultaneously time-dilation and Lorentz-contraction. This gives the following line-element relating observations in the laboratory frame and the comoving frame

$$ds^2 = -dt'^2 + dx'^2 = -\Gamma^{-2} dt^2 + \Gamma^2 d\xi^2.$$ (2.4)

Here, $d\xi$ refers to an *initial* spacelike separation and $\Gamma d\xi$ refers to the spacelike separation as seen in the comoving frame at time $t > 0$. With no motion along y- and z-axis, the line-element in (2.4) extends to three-dimensional spacelike coordinates (x, y, z) as

$$ds^2 = -dt'^2 + dx'^2 = -\Gamma^{-2} dt^2 + \Gamma^2 d\xi^2 + dy^2 + dz^2.$$ (2.5)

Suppose we accelerate a letter "L" along the x-axis. This is represented by a triple of world-lines in the Minkowski diagram (one for each vertex of the letter). The ratio of horizontal-to-vertical lengths of "L" (the aspect ratio) in the comoving frame equals Γ times the aspect ratio on the laboratory frame.

2.2 The length of timelike trajectories

If kinetic energy acumulated by acceleration affects the lapse of eigentime relative to a non-accelerating observer, then so does a change in potential energy due to gravitation by interchangeability of the two subject to conservation of energy. In the laboratory frame, we may, as in the previous section, describe the kinematic effects of curved trajectories of accelerating observers according to time-dependent Lorentz transformations. Similar results will hold due to variations in potential energy in an external gravitational field.

This equivalence is, in fact, familiar from the Coriolis effect. Here, inertial trajectories appear curved in the frame of rotating observers bound to the Earth's surface. Rotating observers attribute this curvature to a centrifugal Coriolis force or, since particles on these inertial trajectories do not experience any body-forces, to some gravitational field of same strength. The observed kinematic effects are the same. Einstein's treatment puts this in four-covariant form, and imposes Lorentz invariance.

According to the Lorentz transformation, the time-lapse over a finite trajectory obtains by integration of (2.1) and (2.2), i.e.:

$$\Delta t = \int_\gamma \frac{dt}{d\tau} d\tau = g^{-1} \sinh(g\tau). \tag{2.6}$$

For small Δt, we have

$$\tau = \Delta t - \frac{1}{6} g^2 (\Delta t)^3. \tag{2.7}$$

This shows that accelerated trajectories tend to be economical in bridging timelike distances between surfaces Σ_t of constant laboratory time t. The longest eigen-time lapse is reserved for non-accelerating, inertial observers. In the limit as g approaches infinity, the time-lapse of the accelerating observer vanishes.

2.3 Gravitational redshift

In the limit of $g\Delta t \ll 1$, the preceding result in (2.7) can be written in terms of a change in the mean kinetic energy E_k,

$$\frac{\Delta \tau}{\Delta t} = 1 - E_k, \quad E_k = gh. \tag{2.8}$$

Here, a is absorbed in the mean distance $h = g\Delta t^2/6$ between the accelerated and the inertial observer. Note the mean drop in potential energy $-gh$ in an external gravitational field providing the acceleration g, consistent with conservation of total energy $E_k + \Delta U = 0$. Thus, we also have

$$\frac{\tau}{t} = 1 + \overline{\Delta U}, \quad \overline{\Delta U} = -gh. \tag{2.9}$$

These results describe time-dilation between two observers in response to a potential energy drop. In what follows, we omit the overbar to ΔU.

Let us examine the above by summing the results N times over neighboring positions x_i at intervals $x_{i+1} - x_i = h$ in the external potential U. The ratio of time-lapses associated with observers at the endpoints x_N and x_1 over macroscopic separations $x_N - x_1 = Nh$ satisfies

$$\frac{d\tau_N}{d\tau_1} = \Pi_{i=1}^N \frac{d\tau_{i+1}}{d\tau_i} \simeq 1 + \Sigma_{i=1}^N \frac{\partial U}{\partial x} h. \tag{2.10}$$

Upon taking the continuum limit, we have

$$\frac{d\tau}{dt} = 1 + U. \tag{2.11}$$

Identifying the mean kinetic energy with the equivalent mean drop in potential energy and by (2.5) gives the line-element

$$ds^2 \simeq -(1+2U)dt^2 + (1+2U)^{-1}d\xi^2 + dy^2 + dz^2, \tag{2.12}$$

where we drop terms of order U^2. This illustrates one aspect of general relativity: the potential energy of the gravitational field is embedded *in* the metric of spacetime.

Photons travel on null-trajectories $ds = 0$, which satisfy

$$\frac{d\xi}{dt} = 1 + 2U. \tag{2.13}$$

It follows that $1 + 2U = 0$ forms a null-surface: the event horizon of a black hole.

Around a star with mass M and radius R and gravitational potential $U_S = M/R$ on its surface (using Newton's constant $G = 1$ in geometrical units), the gravitational redshift satisfies

$$\frac{d\tau}{dt} \simeq 1 - \frac{M}{R}, \tag{2.14}$$

where t denotes the time as measured at infinity. It follows that surface gravity slows down the time-rate of change of an observer on its surface, relative to an observer at infinity. Photons coming off the surface of a compact star with high surface gravity appear redshifted at infinity – it takes effort for these photons to escape the gravitational potential of the emitting star (a conversion between kinetic and potential energy). Conversely, the local clock speed of an observer is a measure for its potential energy. Both are relative concepts, and in similar ways.

Finally, (2.13) shows that when a particle reaches a null surface, it assumes a null-trajectory and becomes frozen. Such null-surfaces form at the Schwarzschild radius

$$R_S = 2M \tag{2.15}$$

in geometrical units. While our arguments leading to (2.13) were approximate in dropping higher-order terms U^2, the result (2.15) is nevertheless exact within the nonlinear equations of general relativity. It was predicted by Michel and Laplace within Newton's theory.

2.4 Spacetime around a star

The spherically symmetric gravitational field of a star has a Newtonian potential

$$U = U(r) \simeq -\frac{M}{r} \tag{2.16}$$

for large radius r in spherical coordinates (t, r, θ, ϕ). In Schwarzschild's line-element, r parametrizes the circumference of the equator of a shell concentric with the origin.

Consider two concentric shells of slightly different size. They may shrink, while their circumference reduces by a common factor. They then move deeper into the potential well of the central mass. By aforementioned equivalence to the kinematic effects in special relativity as illustrated by the letter "L" in Section 2.1, the separation Δs between two shells with circumference $2\pi r_0$ and $2\pi r_1$ ($\Delta r = r_2 - r_1$) satisfies $\Delta s = \Delta r/(1 + U)$ in the comoving frame. By (2.12) this gives the following spherically symmetric line-element

$$ds^2 \simeq -(1+2U)\, dt^2 + (1+2U)^{-1}\, dr^2 + r^2(d\theta^2 + \sin^2\theta d\phi^2), \tag{2.17}$$

neglecting higher order terms in U. At large distances, the line-element about a star of mass M satisfies

$$ds^2 = -\left(1 - \frac{2M}{r}\right) dt^2 + \left(1 - \frac{2M}{r}\right)^{-1} dr^2 + r^2(d\theta^2 + \sin^2\theta d\phi^2). \tag{2.18}$$

Remarkably, we shall later find that this represents the Schwarzschild line-element, the *exact* solution to the fully nonlinear Einstein equations of a point mass with zero angular momentum.

The Schwarzschild line-element (2.18) shows the existence of horizon surfaces. In dimensional units, we have

$$R_g = \frac{2GM}{c^2} = 3 \times 10^5 \, \text{cm} \left(\frac{M}{M_\odot}\right), \tag{2.19}$$

and a mass-density

$$\frac{3M}{4\pi R_g^3} = 2.2 \times 10^{16} \, \text{g cm}^{-3} \left(\frac{M}{M_\odot}\right)^{-2}. \tag{2.20}$$

Analogous to electromagnetism, consider the mass M as seen at infinity given by Gauss' integral over a sphere S of the Coulomb field $-g'_{rr}$:

$$M = \lim_{r \to \infty} \frac{1}{8\pi} \int -g'_{rr} dS. \tag{2.21}$$

The corresponding energy in the Coulomb field outside the star is

$$U_G = \lim_{r \to \infty} \frac{1}{16\pi} \int_R^\infty (g'_{rr})^2 \, 4\pi r^2 dr = \frac{M^2}{R}. \tag{2.22}$$

In general, energy in the gravitational field is hidden, not localized, and cannot so easily be identified. The limit of converting all mass-energy into a gravitational field characterizes the formation of a black hole. By direct extension of (2.22), we anticipate an energy-density in gravitational waves

$$T^{00} = \frac{1}{16\pi} \langle \Sigma_i \dot{A}_i^2 \rangle \tag{2.23}$$

over all polarizations i with amplitude A_i. This will be confirmed in a formal derivation of gravitational waves from the linearized Einstein equations, which further reveals the existence of two polarization modes.

2.4.1 Conserved quantities

The trajectory of a particle in a gravitational field satisfies an action principle. In Minkowski spacetime, a trajectory between two timelike separated points has maximal length if connected by a straight line. A trajectory is said to be geodesic if the particle moves in the absence of any body-forces (force-free motion). Equivalently, the particle moves on a geodesic if in free-fall in a gravitational field. The action principle for geodesic trajectories of a particle of mass m is therefore that of maximal distance between two points of spacetime,

$$\delta S = 0, \quad S = m \int_A^B ds, \tag{2.24}$$

where $ds = \sqrt{-dx^b dx_b}$. Following L. D. Landau & E. M. Lifschitz[318], evaluation gives

$$\delta S = m \int_A^B u_a \delta dx^a = -m u_a \delta x^a|_A^B + m \int_A^B \frac{du_a}{ds} \delta x^a ds, \tag{2.25}$$

Setting $\delta x^a = 0$ at the endpoints A and B, the equation of motion for geodesic trajectories is

$$\frac{du^b}{ds} = 0. \tag{2.26}$$

The energy-momentum four-vector $p_a = (E, P_i)$ is defined as a 1-form by

$$p_a = -\partial_a S = -\partial S / \partial x_B^a, \tag{2.27}$$

where S is evaluated along stationary trajectories. By (2.25), this recovers the familiar identity $p_a = m u_a$ and the invariant $p^2 = -m^2$ corresponding to the normalization $u^2 = -1$.

Geodesic motion derives from a Lagrangian $L = \sqrt{-u^c u_c}$ in (2.24) (or $L = 1/\Gamma$ in $S = -\int_{t_A}^{t_B} L\, dt$). This implies conserved quantities associated with cyclic coordinates. By the Euler-Lagrange equations of motion,[1]

$$\frac{d}{d\tau}\frac{\partial L}{\partial u^b} - \partial_a L = 0, \tag{2.28}$$

a momentum component

$$p_a = \frac{\partial L}{\partial u^a} \tag{2.29}$$

is conserved whenever $\partial_a L = 0$. In a spherically symmetric and time-independent spacetime around a star, there is conservation of the total energy

$$e = -p_t = (1+2U)p^t = m(1+2U)\frac{dt}{d\tau}. \tag{2.30}$$

Likewise, there is conservation of angular momentum

$$j = p_\phi = mr^2\frac{d\phi}{d\tau} \tag{2.31}$$

where the right-hand side represents the Newtonian limit. (Both e and j are energy and angular momentum per unit mass.) Furthermore, the normalization $u^2 = -1$ of the tangent four-vector u^b, e.g. in the equatorial plane $\theta = \pi/2$, gives

$$-1 = -\left(1-\frac{2M}{r}\right)\left(\frac{dt}{d\tau}\right)^2 + \left(1-\frac{2M}{r}\right)^{-1}\left(\frac{dr}{d\tau}\right)^2 + r^2\left(\frac{d\phi}{d\tau}\right)^2, \tag{2.32}$$

and hence three algebraic conditions, (2.30), (2.31) and (2.32) on the four velocity components u^b. This leaves one ordinary differential equation for $r(\phi)$.

2.5 Mercury's perihelion precession

Mercury shows a prograte precession in its elliptical orbit around the Sun. Mercury has a mean distance $a = 5.768 \times 10^{12}$ cm to the Sun. The Sun has a mass $M_\odot = 2 \times 10^{33}$ g, and hence a Schwarzschild radius

$$\frac{GM_\odot}{c^2} = 1.5 \times 10^5 \text{cm}, \tag{2.33}$$

where $G = 6.67 \times 10^{-8}$ cm^3 g^{-1} s^{-2} is Newton's constant. This introduces a small parameter

$$\frac{GM}{c^2 a} \sim 3 \times 10^{-7}, \tag{2.34}$$

---

[1] In this procedure, τ runs over a fixed interval $[\tau_A, \tau_B]$ for a family of trajectories about the geodesic, and does not represent the eigentime for each. We may normalize τ to correspond to the eigentime of the geodesic curve.

suggesting perturbation theory in analyzing the leading order corrections to Keplerian motion.

We set out to derive an equation in parametric form $r'(\phi) = (dr/d\tau)/(d\phi/d\tau)$. With (2.30–2.31), $r = 1/u$ and $dr/d\tau = -jdu/d\phi$ gives

$$\frac{d\phi}{d\tau} = Bu^2, \quad \frac{dt}{d\tau} = \frac{E}{1 - 2Mu}, \tag{2.35}$$

Substitution in (2.32) and differentiating once, we have

$$u'' + u = M/j^2 + 3Mu^2. \tag{2.36}$$

Rescaling $v = Mu/\epsilon$ with

$$\epsilon = \frac{M^2}{j^2} \simeq \left(\frac{M}{r}\right)\left(\frac{M/r}{v^2}\right) \simeq \frac{M}{r} \simeq 0.3 \times 10^{-7} \tag{2.37}$$

for approximately circular Keplerian motion, where v denotes the three-velocity, and (2.36) becomes the weakly nonlinear equation

$$v'' + v = 1 + 3\epsilon v^2. \tag{2.38}$$

To first approximation, we might try a regular perturbation expansion $v \simeq v_0 + \epsilon v_1 + \epsilon^2 v_2 + \cdots$. Then $v_0 = 1 + A\cos\phi$,

$$v_1'' + v_1 = 3\left[1 + 2A\cos\phi + \frac{1}{2}A^2 + \frac{1}{2}A^2\cos(2\phi)\right] \tag{2.39}$$

and

$$v_1 = 3\left(1 + \frac{1}{2}A^2\right) + \left(3A\phi\sin\phi - \frac{1}{6}A^2\cos\phi\right). \tag{2.40}$$

The second term $\phi\sin\phi$ on the right-hand side is secular and unbounded. However, the system is integrable,

$$H = \frac{1}{2}v'^2 + \frac{1}{2}v^2 - \frac{3}{2}\epsilon v^3 - v = \frac{1}{2}v'^2 + \frac{1}{2}(v - 1)^2 - \frac{3}{2}\epsilon v^3. \tag{2.41}$$

also since (2.38) was derived by differentiation of a constant of motion. This shows the existence of bounded solutions in an ϵ-neighborhood of $v = 1$. A "quick fix" to the secular term is

$$\cos\phi + 3\epsilon\phi\sin\phi \simeq \cos\phi + \sin\phi\sin(3\epsilon\phi) \simeq \cos(\phi - 3\epsilon\phi). \tag{2.42}$$

This gives the bounded solution

$$v = 1 + 3\left(1 + \frac{1}{2}A^2\right)\epsilon + A\cos(\phi - 3\epsilon\phi) \tag{2.43}$$

with a perihelion precession of

$$\Delta\phi = 6\pi\epsilon \tag{2.44}$$

per orbit.

Precessing orbits are two-timing problems. Rather than using an ad hoc approach, consider periodic solutions of the form $v = v(\phi, \psi)$ with a slowly varying angle $\psi = \delta\phi$ in the form

$$v_0 = 1 + \delta a + A(\psi)\sin\Phi, \quad \Phi = \phi + \Psi(\psi) \tag{2.45}$$

where $\delta = 3\epsilon$. This gives

$$v_0' = \delta A' \sin\Phi + A\cos\Phi + \delta\Psi'\cos\Phi$$
$$v_0'' = 2\delta A' \cos\Phi - A\sin\Phi - 2\delta\Psi'A\sin\Phi + O(\delta^2). \tag{2.46}$$

subject to

$$v_0'' + v_0 = 1 + \delta\left(1 + 2A\sin\Phi + \frac{1}{2}A^2 - \frac{1}{2}A^2\cos(2\Phi)\right). \tag{2.47}$$

It follows that $1 + \delta a = 1 + \delta\left(1 + \frac{1}{2}A^2\right)$, and, upon suppressing secular terms, $2A'\cos\Phi = 0$, $-2\Psi'\sin\Phi = 2A\sin\Phi$. The latter gives $A' = 0$, $\Psi' = -1$, and so

$$v_0 = 1 + 3\epsilon(1 + \frac{1}{2}A^2) + A\sin(\phi - 3\epsilon\phi), \tag{2.48}$$

giving rise to precession (2.44).

An exact solution to the precessional motion can be given in terms of elliptic functions[320].

The theoretical value of 43.1″ per century for the Mercury's precession is in perfect agreement with the observed value of 43.03″ per century[398]. While Mercury has an orbital period of 0.24 yr, the Hulse–Taylor binary neutron star system PSR 1913+16 has an orbital period $P = 7.75$ h, giving rise to a periastron precession of $4.2°$ yr^{-1}.

2.6 A supermassive black hole in SgrA*

Recently, R. Schödel *et al.*[483] reported on a discovery of a highly elliptical stellar orbit in Sagittarius A*, shown in Figures (2.2)–(2.4). In a decade of astrometric imaging, they discovered a star S2 with a period of 15.2 yr. It passes a central potential well at a velocity of about 5000 km s^{-1} at a pericenter radius of only 17 light hours (124 AU). The inferred central mass is hereby $3.7 \pm 1.5 \times 10^6 M_\odot$ Figure (2.4) – the sum of a point mass of $2.6 \pm 0.2 \times 10^6 M_\odot$ and a visible stellar cluster core of small radius 0.34 pc (Figure 2.3). Based on the central density of

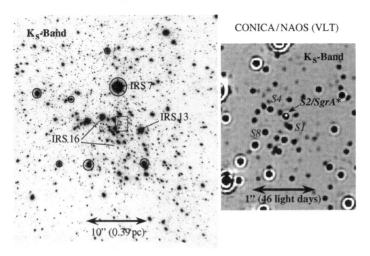

Figure 2.2 Optical observations of SgrA* and the identification of the star S2. (Reprinted with permission from [483]. ©2002 Macmillan Publishers Ltd.)

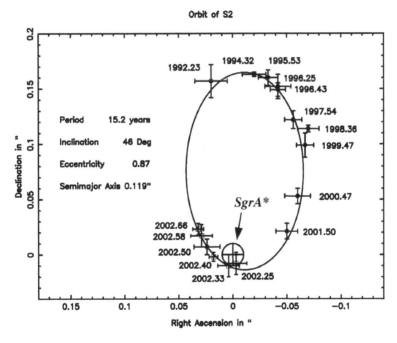

Figure 2.3 Kepler orbit of the star S2 around SgrA*. Note the different scales of length in the plane of projection for seconds of Declination and for seconds of Right Ascension in view of the inclination angle of 46°. The pericenter radius is 124 AU (17 light hours), the semimajor axis is 5.5 light days and the orbital period is 15.2 years. This implies a central mass of $(3.7 \pm 1.5) \times 10^6 M_\odot$. (Reprinted with permission from[483]. ©2002 Macmillan Publishers Ltd.)

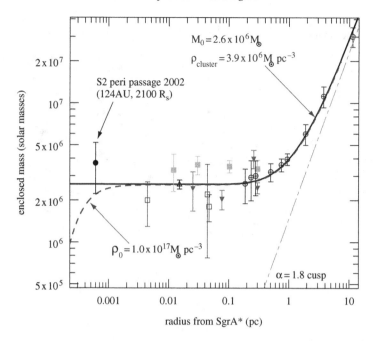

Figure 2.4 The mass–radius relationship around SgrA* is best described by a point mass of $(2.6 \pm 0.2) \times 10^6 M_\odot$ surrounded by a stellar mass cluster. (Reprinted with permission from[483]. ©2002 Macmillan Publishers Ltd.)

$10^{17} M_\odot \, pc^{-3}$, this leaves a supermassive black hole as the only viable alternative. This is perhaps the most blackening evidence to date of the existence of a supermassive black hole. M. Miyoshi *et al.*[383] present evidence for an extragalactic supermassive black hole in NGC4258, based on sub-parsec orbital motion in an accretion disk.

Exercises

1. Describe the evolution of the letter "L" dropped in the spherically symmetric spacetime around a star, with one leg along a radial direction.

2. Calculate the radius of a "black object" in Newton's theory of gravity, by considering the Hamiltonian of a particle and the condition of vanishing total energy.

3. By restoring dimensional units in (2.44), calculate the predicted precession rate in seconds of arc per century.

4. Compute the bending of light rays, whose trajectory passes a star. (*Hint:* Light rays are null-geodesics, satisfying $ds = 0$.)

5. Calculate the ratio of length scales in the plane of projection corresponding to seconds of declination and seconds of Right Ascension, given the inclination angle of 46° of SgrA*. Is the projection of S2 in Figure (2.3) reprinted to scale?

6. Calculate the precession of the star S2 around SgrA*. Compare your results with the precession of Mercury. Is this result measurable?

7. Compute the age difference between your feet and your head over a lifetime of 80 years, assuming a height of 170 cm.

3

Parallel transport and isometry of tangent bundles

We thus far considered transport of tangent vectors along their own integral curves – Minkowski's world-lines of particles. This naturally leads to transport of vectors along arbitrary curves in curved spacetime.

A Riemannian spacetime is endowed with a metric g_{ab} which introduces light cones at every point. These are known as "hyperbolic spacetimes". We may transport light cones $T_p(M)$ at p to $T_q(M)$ at q through transport of their null-generators. Parallel transport defines a mapping of $T_p(M)$ *onto* $T_q(M)$. This parallel transport introduces an isometry between tangent bundles at different points of spacetime. Specifically, this introduces invariance of inner products $\rho(u, v) = g_{ab} u^a v^b$ under such parallel transport. Light cones define the invariant local causal structure. Thus, vectors that are timelike (spacelike) remain timelike (spacelike) under parallel transport.

Parallel transport can be illustrated on the sphere. This is a surface of constant curvature, also of historical interest on which non-Euclidean geometry was first envisioned as recounted by S. Weinberg[587]. Moving a tangent vector along a triangle formed by three great circles, one on the equator and two through the north pole, the net result is a rotation over $\pi/2$. This example shows that parallel transport along closed curves generally returns a vector that is different from the initial vector. This is generic to the geometry of curved spacetime. It is at the root of energetic coupling between gravitation and angular momentum with some definite phenomenology around rotating black holes.

3.1 Covariant and contravariant tensors

A *manifold* M shown in Figure (3.1) allows its points to the labeled by coordinates x^b in any open subset of M. The latter assumes some point set topology on M, which we shall not elaborate on here. Coordinate functions are *maps* from M into $\mathcal{R}^n$, where $n = 4$ in case of four-dimensional spacetime. It is assumed that M allows four independent coordinates x^b in some open neighborhood at any of its points. We may therefore consider curves $p(\lambda)(\lambda\epsilon[0,1])$ on M in the coordinate form $x^b(\lambda) = x^b(p(\lambda))$ as maps from[0,1] into $\mathcal{R}^4$. Likewise, we introduce scalars: any function from M into $\mathcal{R}$ (or $\mathcal{C}$). A scalar field is hereby a function on M whose values are independent of the choice of coordinate system.

The tangent bundle $T_p(M)$ at $p\epsilon M$ is a linear vector space. Related to M, it defines the directional derivatives at p, i.e., $u^b\epsilon T_p(M)$ is associated with the derivative

$$\frac{d\phi}{ds}(p) = u^b\partial_b\phi \qquad (3.1)$$

of a scalar field ϕ. We can think of the various directions as tangent vectors to the family of curves passing through p. In particular,

$$\frac{d\phi}{ds}(p) = \partial_b\phi\frac{dx^b}{ds}, \qquad (3.2)$$

where $x^b(s)$ denotes a curve through p. The partial derivative $\partial_a\phi$ is a covariant tensor, satisfying the transformation rule (in the notation of[318])

$$\partial_{a'}\phi = A^a_{a'}\partial_a\phi \qquad (3.3)$$

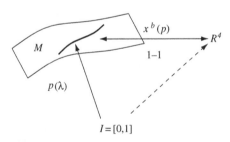

Figure 3.1 A four-dimensional spacetime manifold M can locally be parametrized by four coordinate functions $x^b(p) : M \rightarrow \mathcal{R}^4$. Curves in M are images $p = p(\lambda)$ of the unit interval I, and correspond to curves $x^b(\lambda) = x^b(p(\lambda))$ in coordinate space $\mathcal{R}^4$. Curves introduce tangents $dx^b/d\lambda$ and, collectively, introduce a tangent bundle $T_p(M)$ at points $p\epsilon M$ – linear vector spaces of dimension four. Coordinate derivatives of scalar fields $\phi : M \rightarrow \mathcal{R}$ (or $\mathcal{C}$) introduce 1-forms on M. M is hyperbolic if it has a Riemannian metric with signature $(-,+,+,+)$, corresponding to a Minkowski metric in the $T_p(M)$. The metric induces an inner product in $T_p(M)$ and, upon parallel transport, isometries between tangent bundles at different points.

for a coordinate transformation $x^a \rightarrow x^{a'}$. This follows from the identity

$$d\phi = \partial_a \phi dx^a = \partial_a \phi A^a_{a'} dx^{a'} = \partial_{a'} \phi dx^{a'}. \tag{3.4}$$

Generalizing, u_a is a covariant tensor if it satisfies the transformation rule (3.3). Similarly, the tangent vector $u^b = dx^b/ds$ of a curve satisfies the transformation rule

$$u^{b'} = A^{b'}_b u^b, \tag{3.5}$$

which is recognized as the inverse of (3.3). Generalizing, u^b is a contravariant tensor if it satisfies (3.5) – regardless of context.

Tensors can be combined, loosely speaking, by multiplication. In particular, we have a covariant–contravariant combination

$$T^b_a = \frac{dx^b}{ds}\partial_a \phi, \tag{3.6}$$

associated with a curve $x^b(s)$ and a scalar field ϕ. The upper index transforms contravariantly, and the lower index transforms covariantly. Again, this generalizes in the obvious manner, sometimes denoted as a tensor of type (1,1) – one covariant index and one contravariant index. Contravariant and covariant indices can be combined in the form of a contraction:

$$T = T^c_c = \Sigma^3_{c=0} T^c_c \tag{3.7}$$

which produces a scalar if no other indices are left. Indeed, in our example (3.6) gives a directional derivative

$$T = \frac{d\phi}{ds} \tag{3.8}$$

of ϕ along the curve $x^b(s)$. A further example of particular interest is the Kronecker δ–symbol

$$\delta^b_a = \begin{cases} 1 & (a = b) \\ 0 & (a \neq b) \end{cases} \tag{3.9}$$

for which contraction gives $\delta^c_c = 4$, i.e. a constant scalar field.

The Riemann tensor $R_{abc}{}^d$ is a tensor of type (3,1). The contraction between, for example, b and d, produces the Ricci tensor

$$R_{ac} = R_{abc}{}^b. \tag{3.10}$$

Defining $R_a{}^b = g^{bc} R_{ac}$, we further form the scalar curvature

$$R = R_c{}^c. \tag{3.11}$$

Because the Ricci tensor is symmetric $(R_{ac} = R_{ca})$, scalar curvature can be nonzero. We sometimes refer to the scalar curvature as the "trace of the

Ricci tensor." In this regard, note that the trace of the electromagnetic field tensor F_{ab} always vanishes by antisymmetry.

3.2 The metric g_{ab}

A manifold obtains additional structure in the presence of a symmetric covariant tensor field $g_{ab}(g_{ab} = g_{ba})$. If non-singular, it defines a metric through the line-element

$$ds^2 = g_{ab}dx^a dx^b. \tag{3.12}$$

A hyperbolic manifold – endowed with a light cone at each point – has a signed metric like Minkowski's η_{ab}. Such metrics are also referred to as pseudo-Riemannian. They are such that one of the eigenvalues is negative and the remaining are positive. We say the metric has signature $(-, +, +, +)$. There is no consensus on the choice of sign in the literature. Particle physicists often use the opposite sign convention with signature $(+, -, -, -)$.

At a given point $p \epsilon M$, we are at liberty to consider a smooth metric in the Taylor-series expansion

$$g_{ab}(q) = g_{ab}(p) + g_{ab,c}(p)x^c + \frac{1}{2}x^c x^d g_{ab,cd}(p), \tag{3.13}$$

where the comma denotes partial differentiation and $x^b = (q-p)^b$. Consider the Christoffel symbol

$$\Gamma^c_{ab} = \frac{1}{2}g^{ce}\left(g_{eb,c} + g_{ae,c} - g_{ab,e}\right). \tag{3.14}$$

In view of the identity

$$g_{ab,c} = g_{ae}\Gamma^e_{bc} + g_{be}\Gamma^e_{ac}, \tag{3.15}$$

there is a one-to-one correspondence between the components of $g_{ab,c}$ and those of Γ^c_{ab}. Note that the Christoffel symbol is symmetric in its lowest two indices. Consider now a coordinate transformation $x^b \leftrightarrow \bar{x}^b$,

$$x^a = \delta^a_b \bar{x}^b - \frac{1}{2}\Gamma^a_{bc}\bar{x}^b\bar{x}^c, \tag{3.16}$$

whereby

$$A^a_b = \delta^a_b - \Gamma^a_{bc}\bar{x}^c. \tag{3.17}$$

Applying (3.17) to (3.13) yields the metric in the new coordinates as

$$\bar{g}_{ab} = A^c_a A^d_b g_{cd} = g_{ab}(p) + \frac{1}{2}\bar{x}^c\bar{x}^d g_{ab,cd}(p) + O(\bar{x}^2). \tag{3.18}$$

We may continue with a subsequent coordinate transformation, which brings the metric in Minkowski form at p. In matrix notation, the coordinate transformation rule

$$\bar{\bar{g}}_{ab} = \bar{A}_a^a \bar{A}_b^b \bar{g}_{ab} \tag{3.19}$$

reads $\bar{\bar{G}} = \bar{A}\bar{G}\bar{A}^T$. In view of the symmetry of $G = g_{ab}$, there exists a symmetric factorization

$$\bar{\bar{G}} = LDL^T, \tag{3.20}$$

where L is a lower triangular matrix and $D = (\lambda_0, \cdots, \lambda_3)$ is a diagonal matrix that contains the eigenvalues of G. A coordinate transformation such that $[\bar{A}_a^a] = L$ at p, combined with an additional scaling of coordinates, hereby obtains the metric in the form

$$\bar{\bar{g}}_{ab}(p) = \eta_{ab} + O(\bar{\bar{x}}^2). \tag{3.21}$$

Without loss of generality, therefore, the metric is Minkowskian at a given point of interest up to second order. Such locally defined coordinate systems are referred to as "locally flat" or "geodesic."

3.3 The volume element

Integration of a scalar field ϕ over M with volume element $\sqrt{-g} = \det(g_{ab})$ is defined by

$$I(\phi) = \int_M \phi \sqrt{-g} d^4x = \int_M \phi \epsilon_{abcd} dx^a dx^b dx^c dx^d. \tag{3.22}$$

The ϵ-tensor ϵ_{abcd} is defined as the volume element on M in terms of the Levi-Civita symbol Δ_{abcd},

$$\epsilon_{abcd} = \sqrt{-g}\Delta_{abcd}. \tag{3.23}$$

This construction renders the integral (3.22) independent of the choice of coordinate system. In Minkowski space – a choice of local geodesic coordinates – $\sqrt{-g} = 1$. Hence, (3.23) defines generalization to metric tensors g_{ab}, wherein the Levi-Civita symbol acts as the Jacobian associated with a transformation of coordinates relative to a local geodesic coordinate system.

Performing a coordinate transformation, we have

$$\epsilon_{a'b'c'd'} = A_{a'}^a A_{b'}^b A_{c'}^c A_{d'}^d \epsilon_{abcd} = |A_{a'}^a| \epsilon_{abcd}, \tag{3.24}$$

where the factor on the right-hand side denotes the determinant $|\partial x / \partial x'|$ of the coordinate transformation. This determinant corresponds to the transformation of the Jacobian, i.e. $|\partial x / \partial x'| = \sqrt{-g'} / \sqrt{-g}$. It follows that

$$\epsilon_{a'b'c'd'} = \sqrt{-g'}\Delta_{a'b'c'd'}, \tag{3.25}$$

where we maintain the permutations $\Delta_{a'b'c'd'} = \pm 1$, depending on even or odd permutations of the indices (zero otherwise).

The coordinate derivative of the determinant g of the metric g_{ab} is closely related to the Christoffel symbol. Recall that for any a, we have the matrix identity $g = \Sigma_{b=1}^4 g_{ab} G_{ab}$ in terms of the cofactors G_{ab}, $g g^{ab} = G_{ab}$. For any (a, b), G_{ab} does not contain g_{ab}, whereby

$$\frac{\partial g}{\partial g_{ab}} = g g^{ab}. \tag{3.26}$$

The coordinate derivative gives $\partial g / \partial x^c = g g^{ab} g_{ab,c}$, so that

$$\frac{\partial \ln \sqrt{-g}}{\partial x^c} = \Gamma^a_{ac}. \tag{3.27}$$

3.4 Geodesic trajectories

Geodesic trajectories represent extrema of the action

$$S = m \int_A^B ds = m \int_A^B L(x^b, \dot{x}^b) d\lambda \tag{3.28}$$

with a Lagrangian given by the invariant length

$$L(x^b, \dot{x}^b) = \sqrt{-g_{ab}(x^c) \dot{x}^a \dot{x}^b}, \tag{3.29}$$

where $\dot{x}^b = dx^b / d\lambda$. The condition $\delta S = 0$ is *Fermat's principle* for geodesic trajectories.

We normalize the λ-parametrization such that extremal trajectories satisfy $d\lambda = ds$, where $ds^2 = g_{ab} dx^a dx^b$ denotes arclength. This gives $L \equiv 1$ on extremal trajectories, leaving a variation

$$\delta S = -\int_A^B \left[g_{ab} \dot{x}^a \delta \dot{x}^b + \frac{1}{2} \dot{x}^a \dot{x}^b \delta g_{ab} \right] d\lambda \tag{3.30}$$

about the extremum. Integration by parts factorizes out the variation δx^b ($\delta x^b = 0$ at the endpoints A and B)

$$\delta S = \int_A^B \left[\delta x^b (g_{ab} \ddot{x}^a + \dot{x}^a g_{ab,c} \dot{x}^c - \frac{1}{2} \dot{x}^a \dot{x}^b g_{ab,c} \delta x^c \right] d\lambda. \tag{3.31}$$

The extremal condition $\delta S = 0$ becomes

$$g_{ab} \ddot{x}^a + \frac{1}{2}(g_{ab,c} + g_{cb,a}) \dot{x}^a \dot{x}^c - \frac{1}{2} g_{ac,b} \dot{x}^a \dot{x}^c. \tag{3.32}$$

In terms of the aforementioned Christoffel symbol (3.14), the *geodesic equation* becomes

$$\ddot{x}^c + \Gamma^c_{ab} \dot{x}^a \dot{x}^b = 0. \tag{3.33}$$

3.5 The equation of parallel transport

Geodesic trajectories are integral curves of tangent vectors subject to parallel transport. Extending this notion, consider parallel transport of vectors along arbitrary curves which are not necessarily integral curves of the vectors at hand. In the Introduction, we observed that parallel transport of tangent vectors along closed curves introduces a map from $T_p(M) \to T_q(M)$. This map is nontrivial over finite distances when the surface at hand is not flat, e.g. the sphere S^2. In the presence of a metric, parallel transport obtains a unique definition, when we insist that for any two vectors their inner product is preserved (see also the discussion in R. M. Wald[577]). This implies that the length of a vector is preserved in parallel transport. In particular, a timelike vector remains timelike and a spacelike vector remains spacelike in parallel transport. Hence, parallel transport preserves the causal structure.

We derive a homogeneous linear first-order differential equation to describe parallel transport along a curve with tangent vector τ^b. Parallel transport of a scalar field is described by $\tau^a \partial_a \phi = 0$. Parallel transport of a vector field u^b is to be described by

$$\tau^a \nabla_a u^b = 0, \tag{3.34}$$

where $\tau^a \nabla_a u^b$ is a tensor for (3.34) – a covariant statement, provided that ∇_a is a suitable covariant operator. Note that coordinate derivative $\partial_a u^b$ is too rudimentary for this purpose, because $\partial_a u^b$ is not a tensor of type (1,1). This becomes explicit in case of parallel transport of tangent vectors on the sphere: $\partial_a u^b$ generally

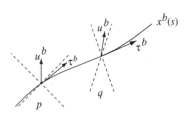

Figure 3.2 Parallel transport of vectors along a curve $x^b(s)$ with tangent $\tau^b = dx^b/ds$ is described by a homogeneous initial value problem: the vanishing covariant derivative $\tau^a \nabla_a = 0$ which takes initial values from $T_p(M)$ into $T_q(M)$. A geodesic is an integral curve obtained by parallel transport of a vector τ^b along itself. On a Riemannian manifold, the tangent bundle $T_p(M)$ is a Minkowskian spacetime with an associated light cone. We define parallel transport by the condition that the light cone at p maps onto the light cone at q, i.e. by invariance of the inner product $\rho(u, v) = g_{ab} u^a u^b$ under transport of the initial vectors $u^b, v^b \epsilon T_p(M)$ to corresponding vectors in $T_q(M)$. This isometry uniquely defines the covariant derivative operator ∇_a from the condition $\nabla_c g_{ab} = 0$.

contains components normal to S^2, which takes it *outside* the two-dimensional linear vector space $T(S^2)$.

The covariant derivative ∇_a can be defined by the proposed isometry between tangent bundles under parallel transport. Consider the tangent bundles at two different points p and q. Preserving the inner product $\rho = g_{ab} u^a v^b$ between two vectors u^b and v^b in the process of parallel transport implies a vanishing derivative

$$\frac{d\rho}{ds} = \lim_{p \to q} \frac{\rho(p) - \rho(q)}{\Delta s} = 0, \tag{3.35}$$

where u^b and v^b satisfy (3.34) and Δs denotes the distance between p and q. We insist that ∇_a satisfies the Leibniz rule, i.e.

$$0 = (\tau^c \nabla_c u^a) v^b g_{ab} + (\tau^c \nabla_c v^b) u^a g_{ab} + u^a v^b (\tau^c \nabla_c) g_{ab} = (u^a v^b \tau^c) \nabla_c g_{ab}. \tag{3.36}$$

Since u^a, v^b and τ^c are arbitrary, it follows that

$$\nabla_a g_{ab} = 0. \tag{3.37}$$

The derivative operator ∇_a will be linear upon taking it to be the sum of the coordinate derivative ∂_a plus a linear transformation acting on tensor indices. In case of contravariant vector fields, as in (3.34), we consider

$$\nabla_a u^b = \partial_a u^b + Q^b_{ac} u^c. \tag{3.38}$$

Operation on covariant tensors then derives from

$$u^c \partial_a w_c + w_c \partial_a u^c = \partial_a (u^c w_c) = \nabla_a (u^c w_c) = u^c \nabla_a w_c + w_c \nabla_a u^c, \tag{3.39}$$

i.e.

$$\nabla_a w_b = \partial_a w_b - \Gamma^c_{ab} w_c. \tag{3.40}$$

We conclude that (3.37) takes the form

$$\nabla_c g_{ab} = g_{ab,c} - Q^d_{ca} g_{db} - Q^d_{cb} g_{da} = 0. \tag{3.41}$$

Equations (3.37)–(3.41) imply (3.15), and by uniqueness recover the Christoffel symbols

$$Q^c_{ab} = \Gamma^c_{ab} \tag{3.42}$$

according to (3.14).

Using the covariant derivative ∇_a associated with a given metric g_{ab}, parallel transport (3.34) of a vector u^b now takes the form

$$\tau^a \nabla_a u^b = \tau^a \partial_a u^b + \Gamma^b_{cd} \tau^c u^d = 0. \tag{3.43}$$

Summarizing, parallel transport of vectors along a curve from P to Q with tangent vector τ^b gives:

1. A linear map of the light cone of $T_p(M)$ onto the light cone of $T_q(M)$.
2. A linear isometry between $T_p(M)$ and $T_q(M)$.
3. A vanishing covariant derivative of the metric: $\nabla_c g_{ab} = 0$.
4. Parallel transport of vectors: $(\tau^c \partial_c)\xi^b + \Gamma^b_{ac}\xi^a\tau^c = 0$.

3.6 Parallel transport on the sphere

Parallel transport along closed curves is perhaps best illustrated on the sphere S^2. Leaving it as an exercise to write out (3.43) in detail, we here follow a different route.

For transport along the boundary $\partial\Delta$ of a triangle Δ, consider the Gauss–Bonnet formula

$$\int_\Delta G + \int_{\partial\Delta} \kappa_g + \Sigma_{i=1}^3 (\pi - \alpha_i) = 2\pi, \tag{3.44}$$

where G denotes the Gaussian curvature of Δ, κ_g denotes the geodesic curvature on $\partial\Delta$, and the α_i denote the angles at the vertices. If the edges of $\partial\Delta$ are formed by great circles, two through the north pole and one on the equator, then $\alpha_Q = \alpha_R = \pi/2$ and $\Sigma\alpha_i = \pi + \alpha_P$ (see Figure 4.1); also, the geodesic curvature κ_g, defined as the projection of the the curvature κ onto S^2, vanishes on great circles. Hence, we have

$$\text{Area}(\Delta) = \alpha_P. \tag{3.45}$$

As illustrated in Figure (3.3), α_P corresponds with the change of angle between the initial and final state of a tangent vector at P, following parallel transport along $\partial\Delta$. The notion that the initial and final state of a vector upon parallel transport along closed curves differs in proportion to the enclosed surface area is generalized to surfaces in curved spacetime in terms of the Riemannian tensor in the next chapter.

3.7 Fermi–Walker transport

An observer may use four vectors at each point of its world-line as a basis for a local coordinate system. If the observer moves along a geodesic, then it is natural to employ a parallelly transported basis for every point along its world-line. If the observer does not move along a geodesic, what is the next best choice?

The observer may choose to transport an initial choice of vectors along with "free" rotation. Transport hereby reduces to pure boosts. Without change of lengths, an infinitesimal change in a vector e^b satisfies $e_b \delta e^b = 0$; linearity requires

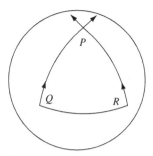

Figure 3.3 Parallel transport of a vector at P over a triangle on a sphere – formed by sections PQ, QR and RP of great circles in the figure – results in a change of angle upon return. The change of angle is proportional to the curvature of the surface and the area enclosed by the triangle, as follows from the Gauss–Bonnet formula.

$\delta e^b \propto e^b$. An infinitestimal boost is described by an antisymmetric tensor $H^{ab} = -H^{ba}$, and hence the type of transport of interest is given by

$$\delta e^b \propto H^{bc} e_c. \tag{3.46}$$

A non-geodesic trajectory with tangent $u^b = dx^b/ds$ deviates from a nearby geodesic trajectory in proportion to the acceleration $a^b = du^b/ds$. Insisting that the e^b maintain their cosines with respect to the observer's velocity four-vector u^b, δe^b lies within the two-dimensional surface spanned by u^b and a^b. This corresponds to the (only) antisymmetric tensor formed by the tensors u^b and a^b at hand, given by

$$H^{ab} = u^b a^a - u^a a^b. \tag{3.47}$$

Combining (3.46) and (3.47), Fermi–Walker transport of a vector satisfies

$$\frac{de^b}{ds} = \left(u^b a^a - u^a a^b\right) e_a. \tag{3.48}$$

It will be appreciated that (3.48) is norm- and cosine-preserving. For example, (3.48) can be used to drag along a tetrad $\{(e_\mu)^b\}_{\mu=0}^{\mu=3}$ of vectors along a world-line, satisfying

$$(e_\mu)^c (e_\nu)_c = \eta_{\mu\nu}, \quad \eta^{\mu\nu} (e_\mu)_a (e_\nu)^b = \delta_a^b, \quad \eta_{\mu\nu} = \lceil -1, 1, 1, 1 \rceil, \tag{3.49}$$

where δ_a^b denotes the Kronecker symbol.

3.8 Nongeodesic observers

The world-line of a non-geodesic observer is described by the equation of motion (3.43) with nonzero right-hand side. What does a local neighborhood look like?

An accelerating observer $x^b(\tau)$ can use a local geodesic coordinate system to map particle trajectories $y^b(\tau)$ in its neighborhood. In one-dimensional motion, a local timelike and spacelike geodesic are spanned by its velocity four-vector $u^b = (\cosh(\lambda\tau), \sinh(\lambda\tau))$ and acceleration $a^b = \lambda(\sinh(\lambda\tau), \cosh(\lambda\tau))$ with magnitude λ. A local "two-bein" is therefore $(u^b, a^b/\lambda)$. As illustrated in (3.8), the distance $p(\tau)$ to a neighboring particle as measured by the observer is defined by

$$x^b(\tau) + p(\tau)a^b(\tau)/\lambda = y^b(s(\tau)), \qquad (3.50)$$

where $s(\tau)$ is some function of τ. In the case of neighboring particles on geodesics, $dv^b/ds = 0$ where $v^b = \dot{y}^b$. Differentiation of (3.50) gives

$$(1 + \lambda p)u^b + \dot{p}a^b/\lambda = v^b \frac{ds}{d\tau}. \qquad (3.51)$$

Contraction of (3.51) with a^b gives the identity $\lambda\dot{p} = -v^c a_c/u^c v_c$, since $ds/d\tau$ equals the reciprocal of the Lorentz factor of the particle as seen by the observer. Differentiation once more of (3.51) gives

$$(1 + \lambda p)a^b + \ddot{p}a^b/\lambda + 2\dot{p}\lambda u^b = \ddot{s}v^b. \qquad (3.52)$$

Contracting this equation with u^b gives $-2\dot{p}\lambda = \ddot{s}v^c u_c$. Substitution into (3.52) and contraction with a^b gives

$$\ddot{p} - 2\dot{p}^2\lambda + \lambda^2 p = -\lambda \qquad (3.53)$$

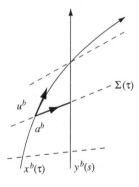

Figure 3.4 A non-geodesic observer may drag along a local "two-bein" (two orthonormal vectors) given by $(u^b, a^b/\lambda)$ at each point of its world-line x^b, where a^b denotes the accelaration of magnitude λ. Extension of a^b by parallel transport off its world-line creates an instantaneous geodesic which spans the surface $\Sigma(\tau)$ of constant eigentime τ. The observed distance to nearby particle with trajectory y^b is given by arclength of the heavy line-segment in $\Sigma(\tau)$.

Here, $v^i/(v^c u_c)$ denotes the three-velocity of the neighboring particle, as seen by the observer. About their point of intersection, $p = 0$, we are left with

$$\ddot{p} = -\lambda + 2\lambda \dot{p}^2. \tag{3.54}$$

The second term of the right-hand side in (3.54) is a relativistic correction due to "tilt" of $\Sigma(\tau)$ as seen in Minkowski spacetime.

More generally, an observer may use a local triad of spacelike vectors subject to Fermi–Walker transport for the purpose of setting up a local coordinate system. A rotating observer may do the same, while rotating the triad with its angular velocity. Counting shows a $3 + 3$ parametrization of rotations and boosts. This corresponds to the six degrees of freedom in Lorentz transformations. The following construction, adapted from H. Stephani[509], adds Coriolis effects to (3.54), but hides the second-order coupling λ^2 away from $p = 0$ in (3.53). The characteristic acceleration length $1/\lambda$ limits the applicability of the $p = 0$ approximation, see[365].

We consider a hypersurface Σ_t traced out by all geodesics that orthogonally intersect $x^b(t)$ at some point P. These are integral curves by parallel transport of tangent vectors v^b which satisfy $v_c u^c = 0$ at P as initial conditions. Here, $u^b = dx^b(t)/dt = 0$ denotes the velocity four-vector of the observer. Since parallel transport preserves causality, these vectors remain spacelike. Σ_t is hereby spacelike. At P, we introduce a tetrad consisting of $(e_0)^b = u^b$ and a triad $\{(e_\mu)^b\}^3_{\mu=1}$ of tangent vectors subject to (3.49). A point Q of Σ_t can be reached by a geodesic $\gamma(P, Q)$ from P, and uniquely so if Q is nearby P. Such geodesics correspond to a particular initial direction

$$v^b = \alpha^\mu (e_\mu)^b, \tag{3.55}$$

which corresponds to a rotation of the triad, i.e.

$$\alpha^1 = \cos\theta, \quad \alpha^2 = \sin\theta\cos\phi, \quad \alpha^3 = \sin\theta\sin\phi. \tag{3.56}$$

The geodesic distance $s = \int_{\gamma(P,Q)} ds$ provides a third coordinate in the scaling

$$(x^1, x^2, x^3) = s(\alpha^1, \alpha^2, \alpha^3) \tag{3.57}$$

as coordinates of Q. Using this set-up, the two conditions $d^2 x^i/ds^2 = 0$ and $x^0 = t$ are satisfied along any Σ_t-geodesic that emanates from P. This defines a locally flat coordinate system for Σ_t.

The four-dimensional metric obtained as an extension of the locally flat coordinate system for Σ_t becomes

$$g_{ab} = \begin{pmatrix} * & * \\ * & h_{ij} \end{pmatrix}, \quad g_{ab} = \eta_{ab} \text{ on } x^b(t), \tag{3.58}$$

where $h_{ij} = \eta_{ij} + O(x^i x^j)$ and, hence, $\Gamma^k_{ij} = 0$. It follows that $g_{ab,0} = 0$, $g_{ij,k} = 0$ on $x^b(t)$. A geodesic curve $\gamma(P, Q)$ from P to Q in Σ_t satisfies the equation for parallel transport

$$\frac{d^2 x^a}{ds^2} + \Gamma^a_{ij} \frac{dx^i}{ds} \frac{dx^j}{ds} = 0 \tag{3.59}$$

at P, showing that $\Gamma^a_{ij} = 0$, i.e. $g_{0i,j} = -g_{0j,i}$. The three degrees of freedom in these antisymmetric combinations can be expressed in terms of an angular velocity three-vector ω^i: $g_{0i,j} = -\epsilon_{ijk}\omega^k$. The remaining derivative $g_{00,i}$ is associated with the acceleration of the observer,

$$a^i = \frac{du^i}{dt} = \Gamma^i_{00} = -\frac{1}{2} g_{00,i}. \tag{3.60}$$

The line-element used by the observer becomes

$$ds^2 = -\alpha^2 dt^2 + h_{ij}(dx^i + \beta^i dt)(dx^j + \beta^j dt) + O(x^2) \tag{3.61}$$

with

$$\beta^i = -\epsilon^i_{jk} x^j \omega^k, \quad \alpha^2 = 1 + 2a_i x^i. \tag{3.62}$$

The associated nonzero Christoffel symbols at the origin are

$$\Gamma^j_{0i} = \beta^i_{,j}, \quad \Gamma^i_{00} = \Gamma^0_{0i} = a^i. \tag{3.63}$$

How does the observer using (3.61) describe the trajectory x^i of a particle under free-fall? In view of (3.63), we have

$$\frac{d^2 x^i}{ds^2} + \Gamma^i_{00} \frac{dt}{ds} \frac{dt}{ds} + 2\Gamma^i_{0i} \frac{dt}{ds} \frac{dx^i}{dt} = 0, \tag{3.64}$$

and

$$\frac{d^2 t}{ds^2} + 2\Gamma^0_{0i} \frac{dt}{ds} \frac{dx^i}{ds} = 0. \tag{3.65}$$

Using

$$\frac{dx^i}{ds} = \frac{dx^i}{dt} \frac{dt}{ds}, \quad \frac{d^2 x^i}{ds^2} = \left(\frac{dt}{ds}\right)^2 \frac{d^2 x^i}{dt^2} + \frac{dx^i}{dt} \frac{d^2 t}{ds^2}, \tag{3.66}$$

these equations become $\ddot{x}^i + a^i - 2\epsilon^i_{jk}\dot{x}^j \omega^k - 2a_j \dot{x}^j \dot{x}^i = 0$, or[509]

$$\ddot{\mathbf{x}} + \mathbf{a} + 2\boldsymbol{\omega} \times \dot{\mathbf{x}} - 2(\mathbf{a} \cdot \dot{\mathbf{x}})\dot{\mathbf{x}} = 0. \tag{3.67}$$

This is the acceleration seen by the observer, as a neighboring particle on a geodesic crosses its world-line. The terms following $\ddot{\mathbf{x}}$ are the guiding acceleration, the Coriolis acceleration and a relativistic correction of the order $\dot{\mathbf{x}}^2$ due to "tilt" of the Σ_t. The Coriolis acceleration – in response to rotation of the observer – defines a transformation in addition to that in (3.54). Notice a transformation with six degrees of freedom: three for a rotation and three for the acceleration, consistent with the six degrees of freedom of Lorentz transformations.

3.9 The Lie derivative

The directional derivative in terms of the covariant operator $u^a \nabla_a$ associated with a tangent vector u^b to a curve explicitly involves the Christoffel symbols, namely

$$u^a \nabla_a \xi^b = u^a \partial_a \xi^a + \Gamma^b_{ac} u^a \xi^c. \tag{3.68}$$

The permutation

$$\xi^a \nabla_a u^b = \xi^a \partial_a u^b + \Gamma^b_{ac} u^c \xi^a \tag{3.69}$$

gives rise to the difference

$$u^a \nabla_a \xi^b - \xi^a \nabla_a u^b = u^a \partial_a \xi^b - \xi^a \partial_a u^b \tag{3.70}$$

with no reference to parallel transport or an underlying metric. We define the Lie derivative as the commutator of two vector fields:

$$\mathcal{L}_u \xi^b = [u, \xi]^b = u^a \nabla_a \xi^b - \xi^a \nabla_a u^b. \tag{3.71}$$

The Lie derivative is a new type of directional derivative, which permits evaluation in terms of coordinate derivatives alone.

   The geometrical interpretation of the Lie derivative is as follows. A vector field u^b can be used to generate a coordinate translation

$$\bar{x}^b = x^b - \epsilon u^b. \tag{3.72}$$

This introduces a correspondence between constant $\bar{x}^b$ and $x^b + \epsilon u^b$. The associated coordinate transformation matrix is

$$\frac{\partial \bar{x}^b}{\partial x^a} = \delta^b_a - \epsilon \frac{\partial u^b}{\partial x^a}. \tag{3.73}$$

A vector field $\xi^b(x^b + \epsilon u^b)$ in the translated coordinate system corresponds to a tensor change

$$\Delta \xi^b = \frac{\partial \bar{x}^b}{\partial x^a} T^a(x^b + \epsilon u^b) - T^b(x^b). \tag{3.74}$$

Evaluation in the limit of $\epsilon \to 0$ gives

$$\lim_{\epsilon \to 0} \epsilon^{-1} \Delta \xi^b = u^a \partial_a T^b - T^a \partial_a u^b. \tag{3.75}$$

This forms a covariant expression (3.71).

Two settings of the Lie derivative are of particular interest.

1. Lie derivatives are important in studying symmetries in a metric. The covariant derivative can not be used for this purpose, since $\nabla_c g_{ab} \equiv 0$ by construction. We say ξ^b forms a Killing vector – a symmetry – if

$$\mathcal{L}_\xi g_{ab} = \nabla_a \xi_b + \nabla_b \xi_a = 0. \tag{3.76}$$

   For example, the previously discussed spherically symmetric and static spacetime around a star possesses the Killing vectors $(\partial_t)^b$ and $(\partial_\phi)^b$ – the flow field (of the coordinates) generated by the Killing vector forms an isometry of the metric. If p^b is the tangent vector to a geodesic (i.e. the four-momentum of a particle in free-fall) and ξ^b is a Killing vector, then $\xi^a p_a$ is conserved along the geodesic. In particular, the time- and azimuthal-Killing vectors of Schwarzschild spacetime introduce a conserved energy $E = -p_t$ and angular momentum p_ϕ.

2. If u^b and v^b are tangent vectors to two congruences of curves, then their Lie derivative vanishes. Indeed, points $x^b = x^b(s, t)$ on a pair of congruences may be coordinatized by two coordinates (s, t), with associated tangent vector fields $u^b = \partial x^b / \partial s$, and $v^b = \partial x^b / \partial t$. Then $u^a \partial_a \phi = \partial \phi / \partial s$ and $v^a \partial_a \phi = \partial \phi / \partial t$ are scalar fields generated by directional derivatives of $\phi(x^b) = \phi(x^b(s, t))$ along u^b and v^b, respectively. The Lie derivative $[u, v]^a \partial_a \phi$ of ϕ becomes

$$u^a \partial_a (v^b \partial_b) \phi - v^a \partial_a (u^b \partial_b) \phi = u^a \partial_a \phi_t - v^a \partial_a \phi_s = \phi_{ts} - \phi_{st} = 0 \tag{3.77}$$

   by commutativity of partial derivatives of scalar functions. We see that the Lie derivative $[u, v]^b$ of tangent vectors generated by coordinate functions is vanishing.

Exercises

1. The *Shapiro time delay*[489] represents the effect of curvature on the propagation of signals in curved spacetime. Calculate the time delay of a signal propagating between two planets around the Sun, due the the the Sun's gravitational potential well.

2. Consider an antisymmetric tensor field F^{ab}. Show that

$$\nabla_a F^{ab} = \frac{1}{\sqrt{-g}}(\sqrt{-g}F^{ab})_{,a}. \tag{3.78}$$

3. Show the following product rules

$$\epsilon_{abcd}\epsilon^{pqrs} = -24\delta^{[p}_a\delta^q_b\delta^r_c\delta^{s]}_d, \tag{3.79}$$

and, more explicitly,

$$\epsilon_{abcd}\epsilon^{apqr} = -\delta^p_b\delta^q_c\delta^r_d - \delta^q_b\delta^r_c\delta^p_d - \delta^r_b\delta^p_c\delta^q_d + \delta^p_b\delta^r_c\delta^q_d$$
$$+\delta^r_b\delta^q_c\delta^p_d + \delta^q_b\delta^p_c\delta^r_d, \tag{3.80}$$

$$\epsilon_{abcd}\epsilon^{abpq} = -2\left(\delta^p_c\delta^q_d - \delta^q_c\delta^p_d\right), \tag{3.81}$$

$$\epsilon_{abcd}\epsilon^{abcp} = -6\delta^p_d, \tag{3.82}$$

and

$$\epsilon_{abcd}\epsilon^{abcd} = -24. \tag{3.83}$$

4. Show that $\nabla_a\nabla_b\phi = \nabla_b\nabla_a\phi$ when ϕ is a scalar field.

5. Show that the Lie derivative agrees with the coordinate directional derivative in a locally flat coordinate sytem.

6. Derive (3.76) for a Killing vector $(\partial_p)^b$ in case of a metric which is independent of p.

7. If ξ^b is a Killing vector field and u^a a tangent vector to a geodesic, show that $\phi = \xi^a u_a$ is a conserved scalar along the geodesic.

8. Non-geodesic motion is described by (3.43) with a nonzero right-hand side. Show that the normalization $u^b u_b = -1$ implies that the forcing has only three independent components.

9. For large orbits derive the equation of *geodetic precession* of an orbiting gyroscope in the Schwarzschild metric.

4

Maxwell's equations

Electromagnism describes the dynamics of electromagnetic fields (E, B) in six degrees of freedom. These fields can be embodied covariantly in the six open "slots" of an antisymmetric tensor F_{ab}. As will be seen below, antisymmetry of F_{ab} embodies conservation of electric charge, and the existence of a vector potential $F_{ab} = \partial_a A_b - \partial_b A_a$ implies the absence of magnetic monopoles.

There are various representations of the electromagnetic field. We may choose to work with the anti-symmetric tensor field F_{ab} when describing magnetic and electric fields in classical interactions with charged particles; with A_a in describing wave-motion or quantum mechanical interactions with charged particles; or with the four-vectors (e_a, b_a) of the electromagnetic field in the comoving frame of perfectly conducting fluids. We shall discuss each of these in some detail.

4.1 *p*-forms and duality

A tensor $\omega_{a_1 \cdots a_p}$ is a totally antisymmetric contravariant tensor (all lower indices), if its sign is invariant (changes) under any even (odd) permutation of its indices, and vanishes if two or more of its indices are the same. For example, a scalar field ϕ is a 0-form, its derivative $\partial_a \phi$ is a 1-form and the induced 2-form $\partial_{[a}\partial_{b]}\phi = 0$ by commutativity of coordinate derivatives. In general, a 1-form is not the derivative of a scalar field. The electromagnetic field-tensor F_{ab} is a 2-form which, as indicated above, derives from a vector potential A_a in view of $\partial_{[a}F_{bc]} = 0$. Hodge duality introduces an algebraic relationship between p-forms and n-p-forms, where $n = 4$ denotes the dimensionality of spacetime. Thus, $*F_{ab}$ denotes the dual of F_{ab}, defined as

$$*F_{ab} = \frac{1}{2}\epsilon_{abcd}F^{cd} \tag{4.1}$$

in terms of the Levi-Civita tensor ϵ_{abcd},

$$\epsilon_{0123} = 1, \quad \epsilon_{1230} = 1, \quad \epsilon_{1023} = -1, \quad \epsilon_{0023} = 0, \quad \cdots; \tag{4.2}$$

ϵ changes sign for any odd permutation of 0123 and is zero when it contains repeated indices. Taking twice the dual of F_{ab} recovers F_{ab} with a minus sign. For example, we have

$$*^2 F_{01} = \epsilon_{0123}\epsilon^{2301}F_{01} = \epsilon_{0123}(-\epsilon_{2301})F_{01} = -F_{01}. \tag{4.3}$$

The dual of a 1-form j_a and a 3-form ω_{abc}, which is totally antisymmetric in its three indices, are

$$*j_{bcd} = j^a \epsilon_{abcd}, \quad *\omega_d = \frac{1}{3!}\omega^{abc}\epsilon_{abcd}. \tag{4.4}$$

In general, the square of the dual of a p-form satisfies

$$*^2 = -(-1)^{p(n-p)} \tag{4.5}$$

in an n-dimensional spacetime with hyperbolic metric (determinant of η_{ab} equal to -1).

4.2 Geometrical interpretation of F_{ab}

The electromagnetic tensor F_{ab} defines magnetic flux Φ through a two-dimensional surface S by

$$\Phi = \int_S F_{ab}\, dS^{ab}. \tag{4.6}$$

If p and q are two coordinate functions which cover the surface S with infinitesimal tangent vectors dp^b and dq^b, then

$$dS^{ab} = dp^a dq^b - dp^b dq^a \tag{4.7}$$

denotes the projection of a surface element onto the coordinate planes $x^a x^b$. A hypersurface Σ with boundary S has three linearly independent surface elements dS^{ab}. If Σ is spacelike – a volume in three-dimensional space – these degrees of

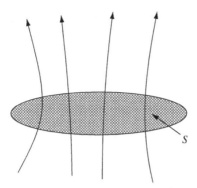

Figure 4.1 The electromagnetic field tensor defines a surface density of magnetic flux and electric flux, associated with the electromagnetic field $(\mathbf{E}, \mathbf{B})$ (arrows). Integration over a two-dimensional surface S obtains the magnetic flux $\Phi = \int_S F_{ab} dS^{ab}$ and electric flux $\Psi = \int_S *F_{ab} dS^{ab}$, where $*\mathbf{F}$ denotes the dual of $\mathbf{F}$.

freedom correspond to the three components of the magnetic field. Likewise, we define the electric flux Ψ through Σ according to

$$\Psi = \int_S *F_{ab} \, dS^{ab}, \tag{4.8}$$

similarly corresponding to three independent components of the electric field on a spacelike hypersurface Σ with boundary S.

As a result of the geometrical aspect (4.6) and (4.8), we can write F_{ab} in terms of coordinate 1-forms as derivatives of coordinate functions (0-forms), given by

$$
\begin{aligned}
(dt)_a &= \partial_a t = (1, 0, 0, 0),\\
(dx)_a &= \partial_a x = (0, 1, 0, 0),\\
(dy)_a &= \partial_a y = (0, 0, 1, 0),\\
(dz)_a &= \partial_a z = (0, 0, 0, 1).
\end{aligned}
\tag{4.9}
$$

According to (4.6), we shall have

$$\mathbf{F} = B_x \mathbf{dy} \wedge \mathbf{dz} + \cdots, \tag{4.10}$$

where the dots refer to cyclic permutations of (x, y, z) in the first term on the right-hand side, as well as remaining terms consisting of contributions of the electric field. Likewise, we have according to (4.8)

$$*\mathbf{F} = E_x \mathbf{dy} \wedge \mathbf{dz} + \cdots, \tag{4.11}$$

where the dots refer to cyclic permutations of (x, y, z) in the first term on the right-hand side, as well as remaining terms consisting of contributions of the magnetic field.

4.3 Two representations of F_{ab}

The electromagnetic field seen by a congruence of observers with velocity four-vector field u^b can be expressed in terms of the four-vectors (e_a, b_a) of electric and magnetic field, given by

$$e_a = u^b F_{ab}, \quad b_b = u^a * F_{ab}. \tag{4.12}$$

These four-vectors are subject to the algebraic constraints

$$e_a u^a = 0, \quad h_a u^a = 0 \tag{4.13}$$

resulting from the antisymmetric of F_{ab}. The constraints (4.13) ensure that the (e_a, b_a) have only six degrees of freedom. The electromagnetic field tensor is now the sum of two bivectors[343]

$$\mathbf{F} = \mathbf{u} \wedge \mathbf{e} + *\mathbf{u} \wedge \mathbf{b}. \tag{4.14}$$

This representation is convenient in applications to fluid dynamics. In the comoving frame, we have $u^b = (1, 0, 0, 0)$ and $e_a = (0, E_x, E_y, E_z)$, $b_a = (0, B_x, B_y, B_z)$. This gives the coordinate representation

$$F_{ab} = \begin{pmatrix} 0 & -E_x & -E_y & -E_z \\ E_x & 0 & B_z & -B_y \\ E_y & -B_z & 0 & B_x \\ E_z & B_y & -B_x & 0 \end{pmatrix}. \tag{4.15}$$

Maxwell's equations, in cgs units

$$\nabla \times \mathbf{B} = \frac{1}{c} \partial_t \mathbf{E} + 4\pi \mathbf{J}, \tag{4.16}$$

$$\nabla \times \mathbf{E} = -\frac{1}{c} \partial_t \mathbf{B}, \tag{4.17}$$

now become, in geometrical units with $c = 1$,

$$\partial_a F^{ab} = -4\pi j^b, \tag{4.18}$$

$$\partial_{[a} F_{cd]} = 0. \tag{4.19}$$

Here, $j^b = (\rho, J^i)$ denotes the electric four-current, in terms of the electric charge-density ρ and three-current J^i. Antisymmetry of F_{ab} implies conservation of electric charge, i.e.

$$0 \equiv \partial_b \partial_a F^{ab} = -4\pi \partial_b j^b = 0. \tag{4.20}$$

Here, we use $\equiv$ to denote algebraic identities. This can be written out more familiarly as

$$\partial_t \rho + \partial_i j^i = 0. \tag{4.21}$$

The second of Maxwell's equations implies that F_{ab} is generated by a vector potential A_a,

$$F_{ab} = \partial_a A_b - \partial_b A_a, \tag{4.22}$$

where A_a is defined up to a gradient $\partial_a \phi$ of a potential ϕ. Here, gauge invariance becomes explicit: F_{ab} is invariant under

$$A_a \rightarrow A_a + \partial_a \phi, \tag{4.23}$$

where ϕ denotes any smooth potential.

In summary, F_{ab} can be represented by electric and magnetic fields – in component form (4.15) or in bivector form (4.14) – or in terms of a vector potential A_a in (4.22). The first is convenient in applications to the electrodynamics of conducting fluids. The second is commonly used in radiation problems, and is essential in quantum mechanics and quantum field theory. It gives rise to the Lagrangian

$$\mathcal{L} = -\frac{1}{4} F^{ab} F_{ab} + j^a A_a. \tag{4.24}$$

4.4 Exterior derivatives

We create $(p+1)$–forms out of p-forms by exterior differentiation

$$\omega_{a_1 \cdots a_p} \rightarrow (\mathbf{d}\omega)_{ba_1 \cdots a_p} = (p+1)\partial_{[b}\omega_{a_1 \cdots a_p]}. \tag{4.25}$$

The derivative $\mathbf{d}$ acts on a scalar field ϕ by $(\mathbf{d}\phi)_a = \partial_a \phi$. It creates a 2-form out of a 1-form u_a by $(\mathbf{d}u)_{ab} = \partial_a u_b - \partial_b u_a = 2\partial_{[a}u_{b]}$, and, more generally, it creates a $p+1$-form out of a p-form ω by (4.25). Evidently, $\mathbf{d}^2 = 0$ by commutativity of coordinate derivatives

$$(\mathbf{d}^2\omega)_{bca_1 \cdots a_p} = (p+2)(p+1)\partial_{[b}\partial_c \omega_{a_1 \cdots a_p]} = 0. \tag{4.26}$$

For example, $\mathbf{d}^2\phi = 0$, as well as $\mathbf{dF} = 0$ for the electromagnetic field tensor $\mathbf{F} = \mathbf{dA}$.

Consider the three-form $\mathbf{d}*\mathbf{F}$ of the electromagnetic field F_{ab}. Its dual is the 1-form $*\mathbf{d}*\mathbf{F}$,

$$(*\mathbf{d}*\mathbf{F})_d = \frac{1}{2}\epsilon^{abc}{}_d 3\partial_{[a}*F_{bc]} = \frac{3}{4}\eta_{dg}\epsilon^{abcg}\epsilon_{ef[bc}\partial_{a]}F^{ef}. \tag{4.27}$$

We have

$$\epsilon_{abcd}\epsilon^{abef} = -2\left(\delta^e_c\delta^f_d - \delta^f_c\delta^e_d\right). \tag{4.28}$$

Hence,

$$(*d*F)_d = -\frac{1}{4}\eta_{dg}\epsilon^{bcag}\epsilon_{efbc}\partial_a F^{ef} = \frac{1}{2}\eta_{dg}\left(\partial_e F^{eg} - \partial_f F^{gf}\right) = \partial_a F^a_d. \quad (4.29)$$

The dual of a the current four-vector j^b is $*j_{bcd} = j^a \epsilon_{abcd}$. With (4.29) the dual of Ampère's law is

$$\mathbf{d}*\mathbf{F} = 4\pi*\mathbf{j}. \quad (4.30)$$

4.5 Stokes' theorem

The exterior derivative gives Stokes' theorem: in covariant form: if Σ is a compact oriented surface of dimension p with boundary S, then integration of a p-1-form ω satisfies

$$\int_\Sigma d\omega = \int_S \omega, \quad (4.31)$$

where the volume element is implicit. Here, Σ is oriented if it possesses a smooth vector field(s) which is (are) everywhere normal to Σ. Equivalently, Σ is defined by a smooth scalar field $\phi = 0$, whereby a normal is given by $\mathbf{d}\phi$.

Integration of a p-form over a p-dimensional A (either Σ or $\partial\Sigma$) is defined with respect to the p-dimensional volume element $dS_{a_1\cdots a_p}$: the one-dimensional "volume" element ds of arclength, the two-dimensional volume element (4.7), the three-dimensional volume element given by the determinants

$$dS_{abc} = \begin{vmatrix} dp^a & dq^a & dr^a \\ dp^b & dq^b & dr^b \\ dr^c & dq^c & dr^c \end{vmatrix} \quad (4.32)$$

of three independent 1-forms (dp, dq, dr) tangent to A, or the four-dimensional volume element $dxdydzdt$.

Stokes' theorem is equivalent to Gauss's law. This can be illustrated by (4.29) as follows. The three-dimensional volume Σ, i.e. a spacelike hypersurface Σ of $t = $const. with boundary $\partial\Sigma$ gives

$$\int_{\partial\Sigma} *F_{ab}dS^{abc} = \int_\Sigma d*F_{ab}dS^{abc} = 4\pi\int_\Sigma j^t d^3x = 4\pi Q, \quad (4.33)$$

which represents Gauss's law of electrostatics. In the same notation, Faraday's law $\mathbf{dF} = 0$ implies

$$\int_{\partial\Sigma} \mathbf{F} = \int_\Sigma \mathbf{dF} = 0. \quad (4.34)$$

There are no magnetic charges.

4.6 Some specific expressions

The stress-energy tensor of the electromagnetic field is defined by

$$T_{em}^{ab} = \frac{1}{4\pi} \left(F^{ac} F_c^b - \frac{1}{4} \eta^{ab} F_{cd} F^{cd} \right). \tag{4.35}$$

Note that T_{em}^{ab} is trace-free: $T_{em\ c}^c = 0$. Upon taking the 00-component,

$$4\pi T_{em}^{00} = F^{0i} F_i^0 + \frac{1}{4} F_{cd} F^{cd} = F^{0i} F^{0i} + \frac{1}{2} F_{0i} F^{0i} + \frac{1}{4} F_{ij} F^{ij}, \tag{4.36}$$

we have the exlicit expression for the energy density,

$$T_{em}^{00} = \frac{1}{8\pi} \left(\mathbf{E}^2 + \mathbf{B}^2 \right). \tag{4.37}$$

Likewise, the Poynting flux describing the flux of three-momentum is given by

$$T_{em}^{0j} = \frac{1}{4\pi} F^{0i} F_i^j = \frac{1}{4\pi} (\mathbf{E} \times \mathbf{B})^i, \tag{4.38}$$

and the stress-tensor

$$T_{em}^{ij} = \frac{1}{4\pi} \left(-(E^i E^j + B^i B^j) + \frac{1}{2} \eta^{ij} (\mathbf{E}^2 + \mathbf{B}^2) \right). \tag{4.39}$$

For a time-independent magnetic field, we recognize in (4.37) the expression for Maxwell stresses on a surface in the (x^i, x^j)-plane $(i \neq j)$,

$$T_{em}^{ij} = -\frac{B^i B^j}{4\pi}. \tag{4.40}$$

This spatial part of the stress-energy tensor is relevant to perfectly conducting fluids, wherein the electric field vanishes in the comoving frame. It describes the Lorentz force-density as currents cross surfaces of constant magnetic flux. For example, consider a bounday S to a region with a magnetic field, which conducts a surface density of electric current J_S. Let the magnetic field at the surface have a normal component B_x and a tangential component B_y, such that the magnetic field across the interface vanishes. For a surface current J_S along the z-direction, the tangential component satisfies the jump condition

$$4\pi J_S = [B_y] = (B_y)^+ - (B_y)^- = -(B_y)^- \tag{4.41}$$

and the Lorentz force per unit surface area is

$$J_S B_x = -\frac{B_x B_y}{4\pi}. \tag{4.42}$$

Invariant algebraic combinations of the electromagnetic field F_{ab} are

$$I_1 = F_{cd} F^{cd} = 2(\mathbf{B}^2 - \mathbf{E}^2), \quad I_2 = F_{cd} * F^{cd} = \mathbf{E} \cdot \mathbf{B}. \tag{4.43}$$

A null electromagnetic field generalizes plane waves, in having $I_1 = I_2 = 0$.

Magnetic flux is the integral counter part of the electromagnetic field tensor F_{ab}. Consider a sphere suspended in an axisymmetric magnetic field. Described in spherical coordinates, the electromagnetic field-tensor is

$$\mathbf{F} = B_r \mathbf{d}\theta \wedge \mathbf{d}\phi + \cdots \tag{4.44}$$

where the dots refer to contributions of the magnetic field components B_θ and B_ϕ. The magnetic flux through a polar cap with half-opening angle θ is

$$\Phi = \int_0^{2\pi}\int_0^\theta F_{\theta\phi}\, d\theta d\phi = 2\pi \int_0^\theta (\partial_\theta A_\phi - \partial_\phi A_\theta)d\theta = 2\pi A_\phi(\theta) \tag{4.45}$$

This shows that $A_\phi = $ const. labels surfaces of constant magnetic flux.

4.7 The limit of ideal MHD

A perfectly conducting fluid describes a medium for which the electric field vanishes in the comoving frame. The bivector representation of the electromagnetic field reduces to

$$\mathbf{F} = *\mathbf{u} \wedge \mathbf{b}, \tag{4.46}$$

since $e_a = 0$. The second of Maxwell's equations, Faraday's equation $d\mathbf{F} = 0$, becomes

$$K: \begin{cases} \partial_a \omega^{ab} = 0, \\ c = 0 \end{cases} \tag{4.47}$$

where $\omega = \mathbf{u} \wedge \mathbf{b}$ and $c = u^a b_a$.

4.7.1 The initial value problem for MHD

Time-dependent solutions to (4.47) can be computed numerically by solving an initial value problem. These solutions propagate physical initial data on a hypersurface Σ_t of constant time into a future domain of dependence $D^+(\Sigma)$. In general terms, K must be supplemented with other equations for the evolution of u^b and accompanying variables, e.g. conservation of energy-momentum and baryon number. Let us focus on the contribution of K to such a larger system of evolution equations, and count the number of independent equations it contributes. There are two issues: compatibility conditions for initial data to be consistent with the partial differential equation at hand, and the rank of the system. The first describes the problem of physical initial data, i.e. the magnetic field is divergence-free. The second refers to rank of the induced Jacobian. Let us describe these in turn.

Compatibility conditions. The divergence condition $\nabla \cdot \boldsymbol{B} = 0$ is a familiar constraint on the magnetic field in Maxwell's equations. The initial data must satisfy this condition in any initial value problem. To derive the covariant form of this constraint in the bivector representation K, we consider the linear decomposition

$$\partial_a = -\nu_a(\nu^c \partial_c) + (\partial_\Sigma)_a \qquad (4.48)$$

for the derivative operator on a spacelike initial hypersurface Σ with timelike normal ν_a, where $(\partial_\Sigma)_a$ denotes differentiation internal to Σ. Initial data compatible with K satisfy

$$-\nu_a(\nu^c \partial_c)\omega^{ab} + (\partial_\Sigma)_a \omega^{ab} = 0 \qquad (4.49)$$

on Σ. This implies that initial data on Σ must satisfy the two compatibility conditions

$$C: \begin{cases} \nu_b \nu_a (\partial_\sigma)_a \omega^{ab} = 0, \\ c = 0, \end{cases} \qquad (4.50)$$

where the first follows by antisymmetry of ω_{ab} in (4.49).

The rank of a system of equations. In the bivector representation, the initial value problem presents a mixed partial differential-algebraic system of equations

$$\partial_a H^{aA}(U^B) = 0, \quad c_i = 0, \qquad (4.51)$$

in terms of a system of N covariant expressions H^{aA} ($A = 1, \cdots N$) and constraints c_i ($i = 1, \cdots p$). Properly posed, the number $q \leq N$ of independent partial differential equations in (4.51) and the number of constraints (assumed to be independent) are consistent with the number of dependent variables U^B ($B = 1, \cdots, r$), i.e.

$$p + q = r. \qquad (4.52)$$

Upon expanding (4.51), we have

$$J_B^A \partial_t U^B + \partial_i H^{iA} = 0, \quad J_B^A = \frac{\partial H^{tA}}{\partial U^B}. \qquad (4.53)$$

Thus, we identify q from the rank of the Jacobian J_B^A. Here, $\boldsymbol{d}t$ refers to the normal to the initial hypersurface Σ, where the initial data are prescribed.

4.7.2 Rank of ideal magnetohydrodynamics

The partial differential equation in K is $\partial_a \omega^{ab} = 0$, where ω_{ab} is a 2-form. It satisfies the identity

$$\partial_b \partial_a \omega^{ab} \equiv 0, \qquad (4.54)$$

and thereby defines only *three* independent equations – its rank is three. (This identity carries over in curved spacetime.) This rank-deficiency of one is encountered similarly in the context of three-dimensional vector operations $\nabla \times \mathbf{A}$ and $\nabla \phi$, given the identities

$$\nabla \cdot (\nabla \times \mathbf{A}) \equiv 0, \quad \nabla \times \nabla \phi \equiv 0. \tag{4.55}$$

This shows that $\nabla \times \mathbf{A}$ and $\nabla \phi$ each have rank 2.

Following (4.53), consider the Jacobian $J = \partial f/\partial U$ of the density $f_{ab}v_a = 2u^{[a}b^{b]}v_a$ for a 1-form v_a, as a function of $U = (u^a, b^a)$. Here, v_a denotes the normal to the initial hypersurface. This gives a four by eight matrix

$$J = \left[v_a b^b - \delta_a^b (b^a v_a) - v_a b^b + \delta_a^b (u^a v_a) \right] \tag{4.56}$$

which satisfies

$$J \begin{pmatrix} v_b \\ v_b \end{pmatrix} = 0. \tag{4.57}$$

This is a direct consequence of the identity $u^{[a}b^{b]}v_a v_b = 0$.

With a rank-deficient Jacobian, equations for the magnetic field b^b do not define a unique propagation of initial data. Uniqueness is recovered by including the algebraic constraint $c = u^a b_a = 0$.

4.7.3 A hyperbolic formulation of MHD

Consider the new system

$$K' : \partial_a \left(\omega^{ab} + g^{ab} c \right) = 0. \tag{4.58}$$

Given the original physical initial data, K' forms an embedding of the initial value problem for K. Note that (4.58) is a system of partial differential equations without constraints. This follows from two observations.

The system K' has rank *four* through a rank-one update provided by the additional term $g^{ab}c$. This holds for all spacelike hypersurfaces ($v^2 \neq 0$). In (4.58), c generally satisfies the homogeneous wave equation

$$0 = \partial_b \partial_a \left(\omega^{ab} + g^{ab} c \right) = \partial^a \partial_a c. \tag{4.59}$$

In view of the compatibility conditions C, we have homogeneous initial conditions on the constraint c

$$c = 0, \quad v^a \partial_a c = 0 \quad \text{on} \quad \Sigma \tag{4.60}$$

in the initial value problem for K'.

In response (4.60), it follows that $c \equiv 0$ throughout the future domain of dependence $D^+(\Sigma)$ of Σ. Thus, K' provides an embedding of solutions to K in a

Table 4.1 *Table of symbols in electrodynamics.*

SYMBOL	ATTRIBUTE	EXAMPLE
ϕ	0-form	Coordinates (t, x, y, z)
A_a	1-form	$\partial_a \phi$
F_{ab}	2-form	$F_{ab} = \partial_a A_b - \partial_b A_a$
$\mathbf{d}$	$p-$ to $(p+1)$-form	$(\mathbf{dA})_{ab} = \partial_a A_b - \partial_b A_a$
ϵ_{abcd}	$p-$ to $(n-p)$-form	$*^2 = -(-1)^{p(n-p)}$
$dS^{ab} = dp^a dq^b - dp^b dq^a$	antisymmetric	$d\theta d\phi$
$\mathbf{F} = \mathbf{u} \wedge \mathbf{e} + *\mathbf{u} \wedge \mathbf{b}$	$u^c e_c = u^c b_c = 0$	$-*\mathbf{F} = \mathbf{u} \wedge \mathbf{b} + g\mathbf{c}$
$\mathbf{F} = B_x dy \wedge dz + \cdots$	$\Phi = \int_S F_{ab} dS^{ab}$	$\Phi = 2\pi A_\phi$
$\mathbf{F} = E_x dx \wedge dt + \cdots$	$\Psi = \int_S *F_{ab} dS^{ab}$	$\int_{\partial\Sigma} *\mathbf{F} = 4\pi Q$
$\mathbf{F} = \mathbf{dA}$	$\mathbf{dF} = 0$	$\int_{\partial\Sigma} \mathbf{F} = 0$
$I_1 = F_{ab}F^{ab}$	$2(\mathbf{B}^2 - \mathbf{E}^2)$	$I_1 = 0$ in equipartition
$I_2 = F_{ab} * F^{ab}$	$\mathbf{E} \cdot \mathbf{B}$	$I_2 = 0$ for plane wave
$T_{ab} = \frac{1}{4\pi}\left(F_a^c F_{cb} - \frac{1}{4}g_{ab}F_{cd}F^{cd}\right)$	$T_c^c = 0$	$T^{00} = \frac{1}{8\pi}\left(\mathbf{E}^2 + \mathbf{B}^2\right)$
		$T^{0i} = \frac{1}{4\pi}\left(\mathbf{E} \times \mathbf{B}\right)^i$
		$T^{ij} = -\frac{1}{4\pi}\left(B^i B^j - \frac{1}{2}\delta^{ij}B^2\right)$
$\mathcal{L} = -\frac{1}{4}F_{ab}F^{ab} + A_a j^a$		

system in divergence form without constraints. The system K' has full rank, and can be combined with supplementary equations describing the evolution of u^b and related variables. A complete system for ideal MHD includes conservation of energy-momentum, baryon number as well as an equation of state – the realm of ideal magnetohydrodynamics.

An overview of the symbols and expressions for expressing electrodynamics in covariant form is given in Table 4.1.

Exercises

1. Show by explicit evaluation that (4.19) describes Maxwell's equations $\nabla \times \mathbf{B} = \partial_t \mathbf{E} + 4\pi \mathbf{J}$ and $\nabla \mathbf{E} = -\partial_t \mathbf{B}$.

2. Show that the Lorentz gauge $\partial_a A^a = 0$ obtains a wave equation $\partial_c \partial^c A_a = -4\pi j_b$.

3. Verify that the variational principle applied to (4.24) recovers Ampère's law as given in the first equation of (4.18).

4. Derive the jump conditions for the electromagnetic field across a two-dimensional surface with surface current density J^i and surface charge σ. Interpret the Maxwell stresses (4.40) in terms of Lorentz forces.

5. Obtain an bivector expression for F_{ab} for an electromagnetic plane wave $\sim e^{ik_a x^a}$. Verify that both I_1 and I_2 vanish.

6. Show that $\mathbf{F} = E_x \mathbf{d} \times \wedge \mathbf{dt} + \cdots$, where the dots are as in (4.11).

7. Show that $\partial_b T^{ab}_{em} \equiv -F^{ab} j_b$, based on Maxwell's equations and interpret the right-hand side. Recall that Maxwell's equations combine the displacement current and the convective current of moving charged particles. Is this displacement current included in the four-covariant formulation of ideal MHD?

8. Show that the hyperbolic reformulation K' of K carries through in the presence of a current j^b by considering $\partial_a \omega^{ab} = j^b$.

5

Riemannian curvature

"Ubi materia, ibi geometria"

Johannes Kepler (1571–1630).

Gravitation is induced by the stress-energy tensor of matter and fields via curvature. This four-covariant description contains the Newtonian limit of weak gravity and slow motion. Subject to conservation of energy and momentum, this leads uniquely to the Einstein equations of motion, up to a cosmological constant. These equations admit a Lagrangian by the associated scalar curvature, as described by the Hilbert action.

Curvature of spacetime displays features similar to that of the sphere, as in the previous chapter. It generalizes to four-dimensional spacetime as in the discussion of the gravitational field of a star.

Spacetime curvature is described by the Riemann tensor. Given a metric, and so the light cones at every point of spacetime, the Riemann tensor is defined completely by the metric up to its second coordinate derivatives. Both the Riemann tensor and the metric, each in different ways, contain time-independent gravitational interactions, including the Newtonian limit of weak gravity, as well as gravitational radiation.

5.1 Derivations of the Riemann tensor

The Riemann tensor has various representations which bring about different aspects of spacetime.

Parallel transport over a closed loop. Continuing the discussion of parallel transport on the sphere, consider vectors carried along closed curves in spacetime. A vector is parallelly transported along a curve with tangent u^a if

$$u^a \nabla_a \xi^b = u^a \left(\partial_a \xi^b + \Gamma^b_{ac} \xi^c \right) = 0, \qquad (5.1)$$

55

where

$$\Gamma^c_{ab} = \frac{1}{2}g^{cd}\left(g_{eb,a} + g_{ae,b} - g_{ab,c}\right) \tag{5.2}$$

denotes the Christoffel connection in coordinate form. Parallel transport of ξ^b along a closed loop $\gamma[318, 527]$: $x^b(s)$ introduces a discrepancy between the initial and final state of a vector, given by

$$\delta\xi^b = \xi^b_f - \xi^b_i = \int_\gamma d\xi^b = \int_\gamma u^a \partial_a \xi^b ds = -\int_\gamma u^a \Gamma^b_{ac} \xi^c ds, \tag{5.3}$$

where $u^a = dx^a/ds$. The leading order contribution to the integral derives from the linear variations is the integrand. Upon taking a Taylor series expansion in case of small loops in the neighborhood of the origin, we write

$$\left(\Gamma^b_{ac} + \partial_e \Gamma^b_{ac} x^e\right) u^a \left(\xi^c + \partial_e \xi^c x^e\right). \tag{5.4}$$

Terms linear in x^e are

$$u^a x^e \left(\partial_e \Gamma^b_{ac} \xi^c + \Gamma^b_{ac} \partial_e \xi^c\right), \tag{5.5}$$

where the factor in parenthesis is constant, evaluated at the origin. By $\int_\gamma u^a x^e ds = -\int_\gamma u^e x^a ds$ and $u^e \partial_e \xi^c = -u^e \Gamma^c_{ef}\xi^f$, we have

$$\delta\xi^b = -\left(\int_\gamma u^a x^e ds\right)\left(\partial_e \Gamma^b_{af} - \Gamma^b_{ac}\Gamma^c_{ef}\right)\xi^f = \frac{1}{2}\left(\int_\gamma u^e x^a ds\right) R^b_{fea}\xi^f. \tag{5.6}$$

This linear transformation defines the Riemann tensor:

$$R^b_{fea} = \partial_e \Gamma^b_{af} - \partial_a \Gamma^b_{ef} + \Gamma^b_{ce}\Gamma^c_{af} - \Gamma^b_{ca}\Gamma^c_{ef}. \tag{5.7}$$

By construction, the Riemann tensor is antisymmetric in its last two indices.

Non-commutativity of covariant derivatives. Antisymmetric covariant differentiation reduces to a linear expression in the tensor at hand, similar but not identical to the Lie derivative, i.e.

$$\nabla_a \nabla_b \xi_c - \nabla_b \nabla_a \xi_c = R_{abc}{}^d \xi_d. \tag{5.8}$$

Indeed, by explicit calculation

$$\nabla_{[a}\nabla_{b]}\xi_c = \nabla_{[a}\left(\partial_{b]}\xi_c - \Gamma^e_{b]c}\xi_e\right). \tag{5.9}$$

The right-hand side expands into

$$\nabla_{[a}\nabla_{b]}\xi_c = \partial_{[a}\partial_{b]}\xi_c - \partial_{[a}\Gamma^e_{b]c}\xi_e - \Gamma^e_{c[b}\partial_{a]}\xi_e$$
$$- \Gamma^f_{[ab]}\left(\partial_f \xi_c - \Gamma^e_{fc}\xi_e\right) - \Gamma^f_{c[a}\left(\partial_{b]}\xi_f - \Gamma^e_{b]f}\xi_e\right) \tag{5.10}$$

i.e.

$$\nabla_{[a}\nabla_{b]}\xi_c = \left(\partial_{[b}\Gamma^e_{a]c} + \Gamma^f_{c[a}\Gamma^e_{b]f}\right)\xi_e. \tag{5.11}$$

This introduces (5.8).

5.2 Symmetries of the Riemann tensor

The Riemann tensor is highly degenerate due to a number of symmetries. These can be seen by inspection in a locally geodesic coordinate system. Consider $\Gamma^c_{ab} \equiv 0$ and $\partial_c g_{ab} = 0$ at a point. We have

$$R_{bfea} = g_{bc}R^c{}_{fea} = g_{bc}\left(\partial_e\Gamma^c_{af} - \partial_a\Gamma^c_{ef}\right). \tag{5.12}$$

Upon expansion, this gives

$$R_{bfea} = \partial_e\left(g_{bc}\Gamma^c_{af}\right) - \partial_a\left(g_{ac}\Gamma^c_{ef}\right) = \frac{1}{2}\left(g_{ab,fe} + g_{ef,ba} - g_{eb,fa} - g_{af,be}\right), \tag{5.13}$$

i.e.

$$R_{bfea} = \frac{1}{2}\left(g_{ba,fe} + g_{fe,ab} - g_{be,af} - g_{af,eb}\right). \tag{5.14}$$

By inspection, we draw two conclusions

$$R_{bfea} = -R_{fbea} = -R_{bfae} = R_{fbae} = R_{eabf} \tag{5.15}$$

$$R_{bfea} + R_{beaf} + R_{bafe} = 0. \tag{5.16}$$

The first (5.15) shows that R_{bfea} is represented by a symmetric 6×6 matrix, which has twenty-one independent components. The second (5.16) is independent of the first (5.15) only for $bfea = 0123$ (or any permutation thereof), so that combined, the Riemann tensor has twenty independent components (and $n^2(n^2 - 1)/12$ independent components in n-dimensional spaces.)

Working in the same locally flat coordinate system, consider the derivative

$$\nabla_d R_{bfea} = \partial_d\left(g_{bc}\left(\partial_e\Gamma^c_{ef} - \partial_a\Gamma^c_{ef}\right)\right), \tag{5.17}$$

i.e.

$$R_{bfea,d} = g_{bc}\left(\Gamma^c_{fa,de} - \Gamma^c_{fe,ad}\right). \tag{5.18}$$

This obtains the Bianchi identity

$$\nabla_{[e}R_{ab]cd} = 0, \tag{5.19}$$

which holds covariantly following general coordinate transformations.

The contractions

$$R_{ac} = R_{abc}{}^b, \quad R = R^c_c \tag{5.20}$$

define the Ricci and scalar curvature tensors. The Bianchi identity (5.19) defines the identity

$$\nabla^a G_{ab} \equiv 0 \tag{5.21}$$

for the Einstein tensor

$$G_{ab} = R_{ab} - \frac{1}{2} g_{ab} R. \tag{5.22}$$

This second form (5.21) of the Bianchi identity gives rise to the Einstein equations

$$G_{ab} = 8\pi T_{ab} \tag{5.23}$$

in the presence of a stress-energy tensor T_{ab} of matter and other fields, satisfying conservation of energy and momentum,

$$\nabla_a T^{ab} = 0. \tag{5.24}$$

Since (5.23) is a covariant expression, (5.21) implies that (5.23) does *not* impose conditions on the second time-derivatives on the four functions g_{0a}. The g_{0a} are not dynamical variables but represent freely specifiable functions: gauge functions which define slicing of spacetime in three-dimensional hypersurfaces.

5.3 Foliation in spacelike hypersurfaces

A time-coordinate t (with derivative vector $(\partial_t)^b$) and its hypersurfaces Σ_t of constant time come with two vectors:

$$\mathcal{N}_a = g_{ta}, \quad n_a = \partial_a t / \sqrt{-\partial_a t \partial^a t}, \tag{5.25}$$

where n_a denotes the unit normal ($n^2 = -1$) to Σ_t. (The vector $\mathcal{N}_a$ is commonly denoted by t_a, as in R. M. Wald[577].) Generally, the covariant vectors $\mathcal{N}_a$ and n_a are independent. Marching from one hypersurface to the next brings along a variation dt, along with the covariant displacement

$$ds_a = \mathcal{N}_a dt. \tag{5.26}$$

The displacement ds_a expresses $\mathcal{N}_a$ as a "flow of time." It can be expressed in terms of orthogonal projections along n_a onto Σ_t in terms of the lapse function N and shift functions N_a,

$$\mathcal{N}^a = Nn^a + N^a. \tag{5.27}$$

Here $N = -\mathcal{N}_a n^a$ and $N_a = h_a^b \mathcal{N}_b$, expressed in the metric

$$h_{ab} = g_{ab} + n_a n_b \tag{5.28}$$

as the orthogonal projection of g_{ab} onto Σ_t. Note that $ds^2 = \mathcal{N}^2 dt^2 = g_{tt} dt^2$ as the square of (5.26), so that $g_{tt} = -N^2 + N_c N^c$. With $n_a = (n_t, 0, 0, 0)$, it follows that

$$g_{ab} = \begin{pmatrix} N^c N_c - N^2 & N_j \\ N_i & h_{ij} \end{pmatrix}, \tag{5.29}$$

where i, j refer to the spatial coordinates x^i of (t, x^i). The lapse function satisfies $\sqrt{-g} = N\sqrt{h}$. The four degrees of freedom in the five functions (N, N_a) are algebraically equivalent to $\mathcal{N}_a$. An equivalent expression for the line-element, in so-called $3+1$ form[110, 534], is

$$ds^2 = -\alpha^2 dt^2 + \gamma_{ij}(dx^i - \omega^i dt)(dx^j - \omega^j dt), \tag{5.30}$$

where $\alpha = N$ is referred to as the redshift factor and $\gamma_{ij}\omega^j = -g_{it}$.

The line-element (5.30) is instructive. It contains the previous Schwarzschild line-element with $\omega^j = 0$, that of a non-geodesic observer with $\omega_i = \epsilon_{ijk} x^j \omega^k$ and, as will be seen later, the frame-dragging angular velocity ω^ϕ around a rotating black hole.

5.4 Curvature coupling to spin

Kepler discovered empirically that for each planet, its radius vector traces area increasing linearly with time. Newton realized that the projection of this rate of change on each of the coordinate surfaces $x^a x^b$ defines a vector, the *specific angular momentum.*

While test particles by definition move along geodesics, spinning objects bring along angular momentum and, by Kepler, a rate of change of surface area. They hereby couple to curvature on the basis of dimensionality. In geometric units, angular momentum per unit mass is described by an anti-symmetric two-tensor of dimension cm^2 which combines with curvature of dimension cm^{-2} to give a force – a dimensionless quantity in geometrical units.

To calculate these forces, we consider the time-rate of change in momentum of the center of mass of a particle in a bound, closed orbit. This could be a particle tied to a rod[534] or a continous mass-distribution in a solid ring.

The world-line x^a of a particle moving in a periodic orbit about the origin describes helical motion about the time-axis. Figure (5.1) shows the closed curve γ of a single orbit of period T as measured in a local restframe, consisting of an open curve plus closing line-segment

$$\gamma' : x^b(t)(0 < t < T), \quad \gamma'' : t(0 < t < T). \tag{5.31}$$

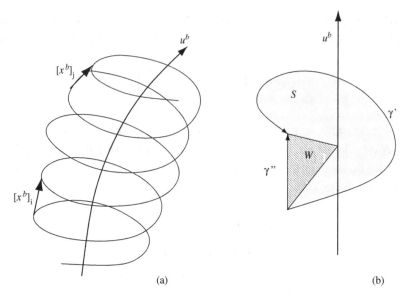

(a)   (b)

Figure 5.1 *Left*: Spacetime diagram of a particle in orbital motion. The orbital center and the orientation of the orbital plane are unconstraint. The separation vector $[x^b(t)] = x^b(t+T) - x^b(t)$ between successive orbits of period T is carried along by parallel transport. It defines the tangent $u^b = [x^b]/T$ to the world-line of the orbital center, and its evolution. *Right*: Curvature-spin coupling changes u^b proportional to the surface area of S and the wedge W in a single orbit γ', closed by γ''.

The surface enclosed by γ may be taken to be sum of the curved spiral surface S and a closing wedge W,

$$S^{ab} = \int_{\gamma'} x^{[a}v^{b]}ds, \quad W^{ab} = Tw^{ab} = Tx^{[a}u^{b]}, \tag{5.32}$$

where $v^b = dx^b/ds$ denotes the unit tangent to the particle world-line. This introduces the separation vector and four-velocity

$$[x^b(t)] = x^b(t+T) - x^b(t), \quad u^b = \frac{[x^b(t)]}{T} \tag{5.33}$$

of the particle between two consecutive orbits.

Consider parallel transport of a vector ξ^b along γ. According to (5.6), we have

$$\frac{\Delta\xi_c}{T} = \frac{1}{2T}R_{abcd}S^{ab}\xi^d + \frac{1}{2}R_{abcd}w_n^{ab}\xi^d. \tag{5.34}$$

By localizing to orbits of small radius x^a, the surface S^{ab} is orthogonal to u^b. Consider, therefore, the *spin-vector* s^a

$$\dot{s}^{ab} = \frac{1}{T}\int_{\gamma'} x^{[a}v^{b]}ds = \epsilon^{ab}{}_{cd}s^c u^d, \quad s_a = \frac{1}{2}\epsilon_{abcd}u^b S^{cd} \tag{5.35}$$

where the superscript dot indicates $d/d\tau$ expressed in terms of the specific angular momentum s^b: a spatial vector orthogonal to $\dot{S}^{ab}$ whose magnitude equals the orbital-averaged rate of change of surface area. The resulting variation satisfies

$$\dot{\xi}_c = \frac{1}{2}\epsilon_{abef}R^{ef}{}_{cd}s^a u^b \xi^d + \frac{1}{2}R_{abcd}w^{ab}\xi^d. \tag{5.36}$$

In case of a point symmetric mass-distribution about the orbital center, such as two particles attached to the end-points of a rod[534] or a continuous mass-distribution in a solid ring, we integrate (5.36) over the mass-distribution. Since w^{ab} is 2π-periodic in the angular position of the wedge, this leaves only the term coupled to s^b. Taking $\xi^b = u^b$, this gives the acceleration of A. Papapetrou[410] and F. A. E. Pirani[430]

$$(\dot{u}_c)_R = \frac{1}{2}\epsilon_{abef}R^{ef}{}_{cd}s^a u^b u^d. \tag{5.37}$$

due to curvature-spin coupling.

"Unfortunately, in practical situations (5.37) is so weak that nobody has ever found any significant application for it"[534]. Indeed, *spin–spin coupling* is typically weak, such as in the Earth's rotational interaction with the intrinsic spin $\hbar/2$ of electrons[430, 152, 382, 364]. *Spin–orbit coupling*, however, is arbitrarily strong, when s^a in (5.37) represents the specific angular momentum of charged particles in magnetic flux-tubes[558].

The left-hand side in the curvature spin-coupling (5.37) refers to the contribution by curvature. Spinning particles also feature a *drift velocity* in response to forces normal to their spin-vector. This is analogous to the electromagnetic drift velocity $\mathbf{v}_d/c = \mathbf{E} \times \mathbf{B}/B^2$ of particles with electric charge e gyrating in a magnetic field B[282, 156]. The particle angular momentum satisfies $J_e = \gamma m_e \omega R^2$, where $\omega = eB/\gamma m_e c$ denotes the Larmor frequency and R denotes the orbital radius. The specific angular momentum $j_e = J_e/\gamma mc^2$ and the acceleration $a = eE/\gamma m$ give $aj_e/c = (E/B)(1 - \gamma^{-2})$. In the ultrarelativistic limit, therefore, $\mathbf{v}_d = \mathbf{a} \times \mathbf{j}_e$. The drift velocity expresses conservation of total linear momentum, as an external potential U, $\mathbf{E} = -\nabla U$, introduces *high* momenta in the semi-orbit at low U and *low* momenta in the semi-orbit at high U. This symmetry breaking is compensated by a drift velocity of the center of mass of the particle orbits. Based on (5.35), we thus see that

$$u_c \dot{S}^{cd} = \epsilon_{ab}{}^{cd}s^a a^b u^c \tag{5.38}$$

represents a familar three-vector product $\mathbf{a} \times \mathbf{s}$. The complete left-hand side to (5.37) is given by the time-derivative of the total linear momentum vector (e.g.[509])

$$v^b = u^b + u_c \dot{S}^{cb}. \tag{5.39}$$

The particle trajectory becomes completely specified in the presence of a further prescription for the evolution of s^b. It is dragged along by Fermi–Walker transport (3.48), i.e. $\dot{s}^b = u^b(a_c s^c)$.

5.5 The Riemann tensor in connection form

Using the volume element $\epsilon_{abcd} = \Delta_{abcd}\sqrt{-g}$, where Δ_{abcd} denotes the totally antisymmetric symbol, we define the dual $*R_{abcd} = (1/2)\epsilon_{ab}^{ef}R_{efcd}$. The Bianchi identity becomes

$$\nabla^a * R_{abcd} = 0. \tag{5.40}$$

The Bianchi identity further gives $\nabla^d R_{abcd} = 2\nabla_{[b}R_{a]c}$. Combined with Einstein equations (5.23), we have

$$\nabla^a R_{abcd} = 16\pi\tau_{bcd}. \tag{5.41}$$

Here, we introduce

$$\tau_{bcd} = \left(\nabla_{[c}T_{d]b} - \frac{1}{2}g_{b[d}\nabla_{c]}T\right) \tag{5.42}$$

with T_c^c denoting the trace of the stress-energy tensor. This source term is divergence-free:

$$\nabla^b \tau_{bcd} \equiv 0. \tag{5.43}$$

The equations (5.40) and (5.41) are in many ways analogous to Maxwell's equations. This can be made more explicit as follows.

Introduce a tetrad $\{(e_\mu)^b\}$ as in (3.49). The tetrad elements have combined sixteen components. The metric g_{ab} has ten components, so that

$$g_{ab} = (e_\mu)_a(e^\mu)_b \tag{5.44}$$

is non-unique by six degrees of freedom. This internal gauge degree of freedom is associated with the improper Poincaré group SO(3,1), describing rotations and boosts of the tetrad elements. In writing equations in tetrad form, we are led to insist on such Poincaré gauge invariance, in addition to general coordinate invariance.

Tetrad elements bring along the connection one-forms

$$\omega_{a\mu\nu} = (e_\mu)^c \nabla_a(e_\nu)_c. \tag{5.45}$$

These Riemann–Cartan connections define a gauge covariant derivative

$$\hat{\nabla}_a = \nabla_a + [\omega_a, \cdot], \tag{5.46}$$

whereby in particular

$$\hat{\nabla}_a(e_\mu)^b = 0. \tag{5.47}$$

Here, the commutator is defined by its action on tensors $\phi_{a_1 \cdots a_k \alpha_1 \cdots \alpha_l}$ as

$$[\omega_a, \phi_{a_1 \cdots a_k}]_{\alpha_1 \cdots \alpha_l} = \Sigma_i \omega_{a\hat{\alpha}_i}^{\alpha_j} \omega_{a_1 \cdots a_k \alpha_1 \cdots \alpha_j \cdots \alpha_l}, \tag{5.48}$$

so that $[\omega_a, \omega_b]_{\mu\nu} = \omega_{a\mu}^\alpha \omega_{b\alpha\nu} - \omega_{a\nu}^\alpha \omega_{b\alpha\mu}$. The equations (5.40) and (5.41) become

$$\hat{\nabla}^a R_{ab\mu\nu} = 16\pi\tau_{b\mu\nu}, \tag{5.49}$$

wherein

$$R_{ab\mu\nu} = \nabla_a \omega_{b\mu\nu} - \nabla_b \omega_{a\mu\nu} + [\omega_a, \omega_b]_{\mu\nu}. \tag{5.50}$$

The tetrad elements satisfy the equations of structure

$$\partial_{[a}(e_\mu)_{b]} = (e^\nu)_{[b} \omega_{a]\nu\mu}. \tag{5.51}$$

It will be noted that $\partial_t(e_\mu)_t$ are undefined in (5.53). Let $\xi^b = (\partial_t)^b$. The four time-components introduce the tetrad lapse functions[566]

$$N_\mu = (e_\mu)_a \xi^a \tag{5.52}$$

as freely specifiable functions, whereby (5.53) becomes a system of ordinary differential equations

$$\partial_t(e_\mu)_b + \omega_{t\mu}{}^\nu(e_\nu)_b = \partial_b N_\mu + \omega_{b\mu}{}^\nu N_\nu. \tag{5.53}$$

The term $\omega_{b\mu\nu} N^\nu$ on the right hand-side of (5.53) shows that the tetrad lapse functions introduce different transformations on each of the legs; the term $\omega_{t\mu}{}^\nu(e_\nu)_b$ on the left-hand side introduces a transformation which applies to all four legs simultaneously. It is the infinitesimal Lorentz transformation $\omega_{t\mu\nu}$ which provides the internal gauge transformations. The tetrad lapse functions are algebraically related to the familiar lapse N and shift functions N_p in the Hamiltonian formulation[19, 577] through

$$g_{at} = N_\alpha(e^\alpha)_a = (N_q N^q - N^2, N_p). \tag{5.54}$$

Summarizing, the Riemann tensor has representations in Christoffel and Riemann–Cartan connections. The first gives rise to a representation in terms of second derivatives of the metric and leads to the Einstein equations for the metric. The second introduces a second-order equation of motion for the connections through (5.49), supplemented with the equations of structure (5.53) describing the evolution of the causal structure in the tangent bundle at each point.

5.6 The Weyl tensor

The Riemann tensor can be decomposed as the sum of a trace-free Weyl tensor C_{abcd} and remaining terms, involving the Ricci tensor and the scalar curvature tensor,

$$R_{abcd} = C_{abcd} + g_{a[c}R_{d]b} + g_{c[a}R_{b]d} - \frac{1}{3}g_{a[c}g_{d]b}R. \tag{5.55}$$

This applies to four-dimensional spacetime. In three dimensions, we have $C_{abcd} \equiv 0$; in two dimensions we are left with the last term on the right-hand side of (5.55). The Weyl tensor captures gravitational wave-motion in vacuum spacetimes.

5.7 The Hilbert action

The Lagrangian for the Einstein equations is given by the one scalar that can be constructed out of the metric: the scalar curvature. This gives the Hilbert action

$$S[g_{ab}] = \int_M R\sqrt{-g}d^4x. \tag{5.56}$$

Exercises

1. Calculate explicitly the surface integrals $S^{ab} = \int_\gamma x^a dx^b$ for a square and a circle in the Euclidean plane.
2. Express the force associated with w^{ab} with a time-rate of change of the moment of inertia relative to the orbital center.
3. Show that the wedge term in (5.36) contributes to collimation: a centriputal force on orbiting particles.
4. Show that the Fermi–Walker transport $\dot{s}^b = u^c \nabla_c s^b = u^b(a_c s^c)$ represents the precessional motion of a gyroscope, corresponding to $\mathbf{S} = \Omega_p \times \mathbf{S}$. Express Ω_p in terms of the frame-dragging angular velocity ω.
5. Show that the determinant of the metric satisfies $\sqrt{-g} = N\sqrt{h}$, by considering

$$
\begin{pmatrix} N^c N_c - N^2 & N_j \\ N_i & h_{ij} \end{pmatrix} = \begin{pmatrix} 1 & N^i \\ 0 & 1 \end{pmatrix} \begin{pmatrix} -N^2 & 0 \\ 0 & h_{ij} \end{pmatrix} \begin{pmatrix} 1 & 0 \\ N^j & 1 \end{pmatrix}. \tag{5.57}
$$

6. Verify the identity $(e_\mu)^c (e_\nu)^d \nabla^a R_{abcd} = \hat{\nabla}^a R_{ab\mu\nu}$.
7. Verify that the equations of structure obtain the system of ordinary differential equations

$$
\partial_t (e_\mu)_b + \omega_{t\mu\gamma}(e^\gamma)_b = \omega_{b\mu\gamma} N^\gamma + \partial_b N_\mu. \tag{5.58}
$$

Interpret the connection $\omega_{t\mu\nu}$.
8. Show that $\nabla^b T_{bcd} \equiv 0$ on account of the conservation law $\nabla^a T_{ab} = 0$ and consistent with divergence-free condition $\nabla^b \nabla^a R_{abcd} = 0$ on the left-hand side (5.41) (by anti-symmetry of the Riemann tensor in its first two indices).
9. Verify (5.56) by explicit calculation.
10. The *Palatini* action is given by[577]

$$
S[g_{ab}, \nabla_c] = \int_M R_{ab} g^{ab} \sqrt{-g} d^4 x. \tag{5.59}
$$

Show that extremizing S with respect to the metric g_{ab} and the operator ∇_a *independently* recovers the Einstein equations *and* the connection $\nabla_c g_{ab} = 0$.

11. Derive the following quadratic expression for the Hilbert action,

$$S[(e_\mu)^b] = -2 \int_M \nabla_a(e_\mu)^{[a}\nabla_b(e^\mu)^{b]} \tag{5.60}$$

Verify that $S[(e_\mu)^b] = S[g_{ab}]$ up to a boundary term.

12. The integrand in (5.60) is *not* SO(3,1) gauge-invariant, and is therefore not a proper Lagrangian. Consider the following extension based on (5.46)

$$S[(e_\mu)^b, \omega_{a\mu\nu}] = -2 \int_M \hat{\nabla}_a(e_\mu)^{[a}\hat{\nabla}_b(e^\mu)^{b]} \tag{5.61}$$

to obtain an SO(3,1) invariant Lagrangian density. Apply the variational principle to (5.61) with respect to both the tetrad elements and the connections *independently*, and derive the equations of motion.

13. The 4×4 Dirac matrices γ^μ satisfy

$$1_{4 \times 4}\eta_{\mu\nu} = \gamma_\mu\gamma_\nu + \gamma_\nu\gamma_\mu. \tag{5.62}$$

With $\gamma^a = (e^\mu)^a\gamma_\mu$, derive a quadratic action $S[\gamma_a]$ in terms of γ_a, analogous to the Hilbert action (5.61).

14. Discuss the introduction of an internal scale-factor λ in the tetrad elements, according to $(e_\mu)^b \to \lambda(e_\mu)^b$ and $(e^\mu)^b \to \lambda^{-1}(e^\mu)^b$, treated as an additional local symmetry in the tangent bundle.

6

Gravitational radiation

"To explain all nature is too difficult a task for any one man or even for any one age.' Tis much better to do a little with certainty, and leave the rest for others that come after you, than to explain all things."
Isaac Newton (1642–1727), in G. Simmons, *Calculus Gems*.

Hyperbolic spacetimes possess a local causal structure described by a light cone at every point. The metric obeys the second-order Einstein equations containing one parameter: the velocity of light. This suggests that infinitesimal perturbations of the metric itself propagate along the very same light cones. We have a separation theorem: gravitational radiation propagates in curved spacetime according to a four-covariant wave-equation, in response to which the metric evolves in the tangent bundle. The result is independent of the foliation of spacetime in spacelike hypersurfaces.

Recall that general relativity embodies the Newtonian gravitational potential energy embedded in the metric tensor. Gravitational radiation will be a novel feature which, for finite amplitudes, hereby carries off energy and momentum. As with waves in any field theory, the energy-momentum transport scales with the frequency and amplitude squared.

Gravitational radiation is a spin-2 wave, characterized by rotational symmetry over π in the plane orthogonal to the direction of propagation in the spin-classification of M. Fierz and W. Pauli[184]. The lowest-order mass-moment producing gravitational radiation, therefore, is the quadrupole moment. In this chapter, we derive the classical expressions for quadrupole emissions. Because coordinate invariance presents a unique gauge invariance to general relativity, some care is needed to identify the two physical degrees of freedom that carry the energy and momentum. This discussion is based on Wald[577], van Putten and Eardley[566, 556] and 't Hooft[527]. A special limit describes the emission of quadrupole gravitational radiation of lumps or blobs of matter swirling around black holes.

The quadrupole formula for gravitational waves has been observationally confirmed to within 1% in the Hulse–Taylor binary neutron star system PSR1916+13[271, 518, 519, 520], as shown by J. M. Weisberg and H. Taylor[589] in Figure 6.1, and in the recently discovered double pulsar system PSR0737-3039 reported by M. Burglay *et al.*[93, 350]. As of this writing, A. Lyne and M. Kramer report a time-rate-of-change $-1.1(3) \times 10^{-12}$s s^{-1} in the orbital period of 2.45 hours in PSR0737-3039, consistent with the expected

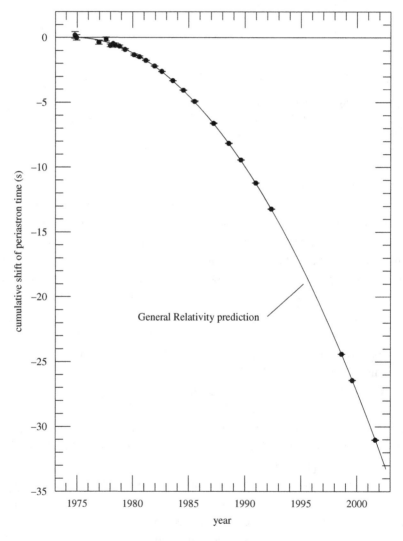

Figure 6.1 Comparison of measured orbital decay with theory in linearized general relativity for the pulsar binary system PSR1913+16. The agreement in cumulative shift in periastron time is within the thickness of the curve. (Reproduced with permission from[589]. ©2003 American Physical Society.)

value -1.24×10^{-12}s s^{-1}[351]. This double pulsar system adds to our sample of six known double neutron star systems, and promises an enhanced merger rate of double neutron star-systems as burst sources of gravitational radiation[93]. Hulse and Taylor were awarded the Nobel Prize in Physics in 1993 for their discovery and study of PSR1913+16.

6.1 Nonlinear wave equations

The Riemann tensor has been recognized for its importance in gravitational waves[431] and its connection to Yang–Mills formulations of general relativity[543, 544, 446]. The *interwoveness of wave motion and causal structure* distinguishes gravity from standard Yang–Mills theories, however. This becomes apparent in non-linear wave equations for the connections on the curved spacetime manifold side-by-side with equations of structure for the evolution of the metric in the tangent bundle.

Following F. A. E. Pirani[431], we take the view that gravitational wave-motion is contained in the Riemann tensor, R_{abcd}. As in the previous chapter, the Riemann tensor satisfies the Bianchi identity (5.19), which gives rise to the homogeneous divergence equation (5.40). In interaction with matter, the Ricci tensor satisfies $R_{ab} = 8\pi[T_{ab} - \frac{1}{2}g_{ab}T]$. The Bianchi identity hereby gives $\nabla^d R_{abcd} = 2\nabla_{[b}R_{a]c}$, and hence the inhomogeneous divergence equation (5.41).

In the Riemann–Cartan expression for the Riemann tensor of the previous chapter, we may impose a Lorentz gauge on the internal SO(3,1) symmetry of the tetrad elements[566]. This gauge choice is given by the six homogeneous conditions

$$c_{\mu\nu} = \nabla^a \omega_{a\mu\nu} = 0. \tag{6.1}$$

In a different context of compact gauge groups and a metric with Euclidean signature, a geometrical interpretation has been given by[339]. Through the linear combination

$$\hat{\nabla}^a \left(R_{ab\mu\nu} + g_{ab}c_{\mu\nu} \right) = 16\pi \tau_{b\mu\nu} \tag{6.2}$$

we arrive at

$$\hat{\nabla}^2 \omega_{a\mu\nu} - R_a^c \omega_{c\mu\nu} - [\omega^c, \nabla_a \omega_c]_{\mu\nu} = 16\pi \tau_{a\mu\nu}. \tag{6.3}$$

This is the separation theorem mentioned at the heading of this chapter: *gravitational waves propagate on a curved spacetime manifold by (6.3) in a Lorentz gauge on the Riemann–Cartan connections. In response, the causal structure of the manifold evolves in the tangent bundle by the equations of structure (5.53). The Hamiltonian lapse and shift functions find their algebraic counterparts in the tetrad lapse functions N_μ(5.54).*

In vacuo, the non-linear wave equations (6.3) reduce to

$$\hat{\nabla}^2 \omega_{a\mu\nu} - [\omega^c, \nabla_a \omega_c]_{\mu\nu} = 0, \tag{6.4}$$

which, in the linearized wave zone, become

$$\nabla^2 \omega_{a\mu\nu} = 0. \tag{6.5}$$

For a plane wave along the z-direction, only two of the $\omega_{a\mu\nu}$ are nonzero. Given the Lorentz condition (6.1), this leaves one connection with six degrees of freedom, e.g. $\omega_{z\mu\nu}$ for propagation along the z-direction. With four independent tetrad lapse functions which define the slicing of spacetime in (5.54), this leaves only *two* degrees of freedom in gravitational radiation. Linearizing small amplitude perturbations $h_{ij} = O(\epsilon)$ in the metric about the Minkowski metric η_{ab},

$$g_{ab} = \eta_{ab} + h_{ab}, \tag{6.6}$$

the Riemann tensor in connection form (5.50) reduces to

$$R_{0i0j} = \partial_t \omega_{i0j} = -\frac{1}{2} \partial_t^2 h_{ij}. \tag{6.7}$$

The tetrad approach[566, 173] bears some relation to but is different from A. Ashtekar's formulation[20, 21, 22, 23] of nonperturbative quantum gravity, and builds on Utigama's work[543, 544] on general relativity as a gauge theory[446, 270]. The original Ashtekar variables are SU(2,C) soldering forms and an associated complex connection in which the constraint equations become polynomial. The Riemann–Cartan variable is a real SO(3,1,R) connection. In Ashtekar's variables, a real spacetime is recovered from the complex one by reality constraints. See Barbero[28, 29] for a translation of Ashtekar's approach into SO(3,R) phase space with real connections. The main innovation in van Putten and Eardley[566] is the incorporation of the Lorentz gauge condition (6.1) which obtains new hyperbolic evolution equations above.

A number of very interesting and independent results on hyperbolic formulations in the Hamiltonian variables[19] have been considered, e.g., by Choquet-Bruhat and York[119, 3] and others (e.g., [71, 31, 74, 36, 83, 205, 206, 207, 8, 77, 133, 216, 458] and a review in[458]). These $3+1$ formulations are hyperbolic under restricted conditions on the Hamiltonian lapse function. In contrast, fully four-covariant formulations preserve hyperbolicity under arbitrary slicing of spacetime in spacelike hypersurfaces by separation of wave motion and evolution of causal structure[555, 173, 91].

6.1.1 Cosmological constant problem in SO(3,1)-gravity

The cosmological constant problem arose out of the observation that the Einstein equations allow an additional source term $-\Lambda g_{ab}$, where Λ is an arbitrary constant. Generally, a cosmological constant strongly affects the evolution of the universe, and therefore is observationally constrained. Independently, particle physics introduces a cosmological constant as a problem of ultraviolet divergence[588] in the summation of zero-point energies of matter fields. If energy couples to the metric tensor directly, as suggested by the Einstein equations, then this poses the challenge of finding suitable suppression or cancellation mechanisms. According to the previous separation theorem in the SO(3,1)-approach: *energy-momentum couples via $\tau_{a\mu\nu}$ to the Riemann–Cartan connections $\omega_{a\mu\nu}$ according to (6.3). The contribution of a cosmological constant $-\Lambda g_{ab}$ to $\tau_{a\mu\nu}$ vanishes identically. The Einstein equations represent integrals of motion, in which $-\Lambda g_{ab}$ represents a constant of integration.*

The aforementioned observation shows a hierarchy, in which ordering is important. In calculating the contribution of energy-momentum to gravitation, (1) calculate the energy-momentum tensor by summing the contributions from the various sources, (2) calculate $\tau_{a\mu\nu}$ and, if desired, (3) form the Einstein equations by integration. We do *not* skip (2). Thus, the cosmological constant problem is completely divorced from the problem of a divergent constant produced by zero-point fluctuations. (Mathematically, all this is equivalent to $\partial^{-1}\partial \neq 1$.)

Recent WMAP[45] observations have shown with remarkable precision that the universe has flat three-curvature and assumes the critical closure density

$$\Omega_{tot} = \Omega_b + \Omega_{CDM} + \Omega_\Lambda = 1.02^{+0.02}_{-0.02}, \tag{6.8}$$

where the closure densities of baryonic matter (b), Cold Dark Matter (CDM) and Dark Energy (the cosmological constant) are

$$\Omega_b = 0.044^{+0.004}_{-0.004}, \quad \Omega_{CDM} = 0.27^{+0.04}_{-0.04}, \quad \Omega_\Lambda = 0.73^{+0.04}_{-0.04} \tag{6.9}$$

and a Hubble constant of $71^{+0.04}_{-0.03}$ km s^{-1}/Mpc. Here, the Ω-values are defined by the respective energy densities relative to the closure density

$$\rho_c = \frac{3H^2}{8\pi G} \simeq 9.45 \times 10^{-30}\, \mathrm{g\,cm}^{-3}. \tag{6.10}$$

These data are consistent with BOEMERANG and MAXIMA[151, 249], distant Type Ia supernovae[418, 479], and previous estimates of the Hubble constant[201]. Thus, *most of the universe consists of CDM and Λ, wherein baryonic matter forms a mere small perturbation.*

These observations put the cosmological problem in a notably different and richer form: why do the three contributions of baryonic matter, CDM and Λ track cosmological evolution, and why at the ratios

$$\Omega_b : \Omega_{\text{CDM}} : \Omega_\Lambda \simeq 0.16 : 1 : 3? \tag{6.11}$$

This strong correlation is called the "coincidence problem." This would be a coincidence if CDM exists of exotic primordial particles (or radiation), in which case their decay $\propto t^{-4}(\propto t^{-3})$ is fundamentally different from $\rho_c \propto t^{-2}$. Perhaps Ω_Λ and Ω_{CDM} have a common origin in new microphysics, which gives rise to a combined stress-energy tensor approximately of the form

$$8\pi G T_{ab} \simeq 3H^2 \begin{pmatrix} 1 & 0 & 0 & 0 \\ 0 & -\frac{3}{4} & 0 & 0 \\ 0 & 0 & -\frac{3}{4} & 0 \\ 0 & 0 & 0 & -\frac{3}{4} \end{pmatrix}. \tag{6.12}$$

Alternatively, Ω_b and Ω_{CDM} may be related by new physics. At present, there is no consensus on how to approach this modern form of the cosmological constant problem.

6.2 Linear gravitational waves in h_{ij}

To leading order in the metric perturbation, the Riemann tensor satisfies

$$R_{abc}{}^d = \partial_b \Gamma^d_{ac} - \partial_a \Gamma^d_{bc} + O(\epsilon^2). \tag{6.13}$$

Using the expression

$$\Gamma^d_{dc} = \frac{1}{2} g^{de} g_{de,c} = \frac{1}{2} \frac{\partial_c g}{g}, \tag{6.14}$$

where g denotes the determinant of the covariant metric g_{ab}, the Ricci tensor becomes

$$R_{adc}{}^d = \partial_d \left(\frac{1}{2} g^{de} \left(g_{ec,a} + g_{ae,c} - g_{ac,e} \right) \right) - \frac{1}{2} \partial_a \partial_c h + O(\epsilon^2) \tag{6.15}$$

where $h = h^c_c$ denotes the trace of the metric perturbation. It follows that

$$R_{ac} = -\frac{1}{2} \nabla^2 h_{ac} + \eta^{de} \partial_d \partial_{(a} h_{c)e} - \frac{1}{2} \partial_a \partial_c h. \tag{6.16}$$

The second and third term in (6.16) can be rewritten as

$$\frac{1}{2} \partial^e \partial_a h_{ce} + \frac{1}{2} \partial^e \partial_c h_{ae} - \frac{1}{4} \partial_a \partial_c h - \frac{1}{4} \partial_c \partial_a h \tag{6.17}$$

and, to the same linear order,

$$\frac{1}{2}\partial_a\partial^e h_{ce} + \frac{1}{2}\partial_c\partial^e h_{ae} - \frac{1}{4}\partial_a\partial_c h - \frac{1}{4}\partial_c\partial_a h. \tag{6.18}$$

The Ricci tensor assumes the form

$$R_{ac} = -\frac{1}{2}\nabla^2 h_{ac} + \partial_{(c}\partial^e \bar{h}_{a)e}, \tag{6.19}$$

where

$$\bar{h}_{ab} = h_{ab} - \frac{1}{2}\eta_{ab}h. \tag{6.20}$$

Transverse traceless gauge. A reduction of (6.19) to a homogeneous wave equation obtains in a special gauge, by appropriate choice of coordinates. In the linearized approximation, consider the transformation rule of the metric perturbation by a coordinate transformation,

$$h_{ab} \rightarrow h_{ab} + \partial_a\xi_b + \partial_b\xi_a \tag{6.21}$$

subject to the four conditions

$$\nabla^2\xi_a = -\partial^c h_{ac}. \tag{6.22}$$

Note that (6.22) leaves ξ^b determined up to a linear combination with any vector μ^b, satisfying $\nabla^2\mu^b = 0$. Thus, (6.19) reduces to

$$R_{ac} = -\frac{1}{2}\nabla^2 h_{ac}. \tag{6.23}$$

By (6.23), the vacuum Einstein equations $R_{ac} = 0$ imply $\nabla^2 h = 0$. Consider, therefore, the transverse traceless (TT) gauge

$$\partial^a h_{ab} = 0, \quad h = 0. \tag{6.24}$$

The TT-gauge becomes more explicit in the plane wave approximation, whereby all quantities vary according to $e^{ik_a x^a}$, and hence $\partial_a = ik_a$. Here, k_a denotes the four-covariant wave vector that, for outgoing radiation along the z-direction, satisfies

$$k_a = (-1, 0, 0, 1)k. \tag{6.25}$$

Thus, $k^a h_{ab} = 0$ defines h_{ab} to be transverse. In the present linear theory in TT gauge, we are left with

$$R_{ac} = -\frac{1}{2}\nabla^2 h_{ac}, \quad R = 0. \tag{6.26}$$

Energy-momentum tensor of gravitational radiation. General relativity can be shown to derive from the Einstein–Hilbert action

$$S = \frac{1}{16\pi} \int R\sqrt{-g}\, d^4x. \tag{6.27}$$

According to the first variation of the integrand (Wald[577]), we have in transverse traceless gauge

$$S = \frac{1}{16\pi} \int G_{ab}h^{ab}\sqrt{-g}\, d^4x = \delta S = \frac{1}{16\pi} \int R_{ab}h^{ab}\sqrt{-g}\, d^4x \tag{6.28}$$

where $R_{ab} = R_{ab}(h_{cd}/2)$, and so

$$S = -\frac{1}{64\pi} \int \left(\nabla^2 h_{ab}\right) h^{ab}\sqrt{-g}\, d^4x = \frac{1}{64\pi} \int \left(\partial_c h_{ab}\right)^2 \sqrt{-g}\, d^4x. \tag{6.29}$$

Next, we use the additional freedom in (6.22) to impose the Coulomb or radiation gauge

$$h_{0a} = 0. \tag{6.30}$$

It follows that h_{ab} contains two physical degrees of freedom, representing the two independent polarization states of gravitational radiation

$$h_{ab} = \begin{pmatrix} 0 & 0 & 0 & 0 \\ 0 & h_+ & h_\times & 0 \\ 0 & h_\times & -h_+ & 0 \\ 0 & 0 & 0 & 0 \end{pmatrix} = h_+ e_{ab}^+ + h_\times e_{ab}^\times, \tag{6.31}$$

in terms of the polarization tensors e_{ab}^+ and $e_{ab}^\times$, here shown for propagation along the z-direction. This also makes explicit that gravitational radiation is a spin-2 wave, defined by the discrete rotational symmetry of rotation by π in the wavefront, i.e. the planes of constant phase orthogonal to the direction of propagation, for each of the two polarization states $+$ and $\times$[184]. These two polarization states describe the ellipsoidal strain deformations of geodesics which cross the wave-fronts, according to

$$\delta\xi_a = \frac{1}{2} h_{ab}\xi^b. \tag{6.32}$$

This derives by expressing a variation in length of ξ^b due to a change in metric in terms of a variation in ξ^b relative to the Minkowski metric, i.e. we define $\delta\xi^b$ according to

$$\xi^a\xi^b(\eta_{ab} + h_{ab}) = (\xi^a + \delta\xi^a)\eta_{ab}(\xi^b + \delta\xi^b), \tag{6.33}$$

which gives (6.32). The perturbed line-element becomes

$$ds^2 = \eta_{ab}dx^a dx^b + h_+(dx^2 - dy^2) + 2h_\times dxdy. \tag{6.34}$$

This shows that h_+ denotes the amplitude of ellipsoidal perturbations of a circular array of test particles in geodesic motion along the x- and y-axes, while $h_\times$ denotes the amplitude of such perturbations of the same along the diagonals $dx = \pm dy$.

Substitution of the representation (6.31) in (6.27) gives

$$S = \frac{1}{16\pi} \int \frac{1}{2} \left((\partial_a h_+)^2 + (\partial_a h_\times)^2 \right) \sqrt{-g} d^4 x. \qquad (6.35)$$

The integrand (6.35) hereby assumes a form similar to that of the Lagrangian of two scalar fields, except for the factor $1/16\pi$ – as anticipated in (6.36). By this identification, the stress-energy tensor of linearized gravitational radiation (in the TT radiation gauge) is inferred to be

$$t^{00} = t^{0z} = t^{zz} = \frac{1}{16\pi} \left\langle \dot{h}_+^2 + \dot{h}_\times^2 \right\rangle, \qquad (6.36)$$

where the dot refers to ∂_0 (or ∂_z) and the <> refers to a time average.

6.3 Quadrupole emissions

The lowest multipole moment of gravitational radiation takes the form of quadrupole emissions. To see this, we note the discussion in S. L. Shapiro and S. A. Teukolsky[490]:

No electric dipole radiation. The "electric" dipole moment of a mass distribution of particles of mass m_A at positions x^A is given by

$$d = \Sigma m_A x^A. \qquad (6.37)$$

The second time-derivative becomes the first time-derivative of the total momentum, which vanishes by momentum conservation:

$$\ddot{d} = \Sigma m_A \dot{p}_A = \dot{P} \equiv 0, \qquad (6.38)$$

where $p_A = \dot{x}_A$.
No magnetic dipole radiation. The "magnetic" moment of a similar distribution of particles satisfies

$$\mu = \Sigma x_A \times p_A = \Sigma j_A = J. \qquad (6.39)$$

The first time-derivative vanishes by conservation of total angular momentum J:

$$\dot{\mu} = \dot{J} \equiv 0. \qquad (6.40)$$

The next order of radiation is given by quadrupole emission. In the linearized wave-zone, the Einstein equations $G_{ab} = 8\pi T_{ab}$ reduce to $-(1/2)\nabla^2 h_{ab} = 0$.

Comparing with the analogous electromagnetic case in terms of the vector potential A_a, $\nabla^2 A_a = -4\pi j_a$, where j_a denotes the four- current, we find a corresponding Lienard–Wiechert potential[318]

$$h_{ij}^{TT}(r, t) = \frac{4}{r} \int_V T_{ij}^{TT}(t')d^3x, \tag{6.41}$$

where $t' = t - r$ denotes the retarded time in a source region with compact support V. The integral on the right-hand side of (6.41) can be seen to be equivalent to the second time-derivative of the integrated mass-density T^{00} in the source region. Indeed, we have[577, 527]

$$\int T^{ij}d^3x = \frac{1}{2}\int \left(T^{kj}\partial_k x^i + T^{ik}\partial_k x^j\right) = -\frac{1}{2}\int \left(x^i\partial_k T^{kj} + x^j\partial_k T^{ki}\right) \tag{6.42}$$

and, hence, by momentum conservation

$$\int T^{ij}d^3x = \frac{1}{2}\partial_0 \int \left(T^{0j}x^i + x^j T^{0i}\right) d^3x. \tag{6.43}$$

Proceeding in similar fashion, we find

$$\int T^{ij}d^3x = \frac{1}{2}\partial_0^2 \int T^{00}x^i x^j d^3x, \tag{6.44}$$

where the integral on the right-hand side refers to the integrated second moment of the mass distribution, I_{ij}. Upon considering the transverse traceless part, we have

$$h_{ij}^{TT} = \frac{2}{r}\ddot{I}_{ij}^{TT}(t'), \tag{6.45}$$

where the traceless part of I denotes the moment of inertia tensor, defined by

$$I_{ij}^T = I_{ij} - \frac{1}{3}\delta_{ij}I. \tag{6.46}$$

While both gravitational radiation and electromagnetism have two polarization modes, (6.45) reveals an additional factor 2 in the Lienard–Wiechert potential. The luminosity of gravitational waves is therefore four times the luminosity in the corresponding electromagnetic waves.

The energy flux in gravitational radiation becomes

$$T^{0z} = \frac{1}{32\pi} < \dot{h}_{jk}^{TT} \dot{h}_{jk}^{TT} > \tag{6.47}$$

(an additional factor of 1/2 arises because each of the polarization tensors has two nonzero components, which should not be double-counted towards the energy-flux), whereby

$$\frac{d^2E}{dtd\Omega} = \frac{1}{8\pi} < \partial_0^3 I_{jk}^{TT} \partial_0^3 I_{jk}^{TT} >. \tag{6.48}$$

Following 't Hooft[527], note that in each direction the two polarization modes in gravitational radiation represent a two-fifths fraction of the contribution by all five components in the (mere) traceless part I_{ij}^T. Thus, the calculations are facilitated by switching to the traceless gauge following

$$L_{gw} = \frac{2}{5} \times 4\pi \times \frac{1}{8\pi} < \partial_0^3 I_{jk}^T \partial_0^3 I_{jk}^T > = \frac{1}{5} < \partial_0^3 I_{jk}^T \partial_0^3 I_{jk}^T > . \tag{6.49}$$

Consider a binary system of two stars, consisting of two point masses of mass m_1 and m_2 and radii a_1 and a_2 to the center of mass ($m_1 a_1 = m_2 a_2 = \mu a$, $\mu = m_1 m_2/(m_1 + m_2)$). Their orbital separation is $a = a_1 + a_2$ with angular velocity Ω, $\Omega^2 = (m_1 + m_2)/a^3$ (Kepler's 3rd). For circular motion, we have

$$I_{xx} = (m_1 a_1^2 + m_2 a_2^2) \cos^2 \phi = \frac{1}{2} \mu a^2 \cos 2\phi + \text{const.}, \tag{6.50}$$

where $\phi = \Omega t$, and likewise

$$I_{yy} = -\frac{1}{2} \mu a^2 \cos 2\phi + \text{const.}, \quad I_{zz} = \text{const.} \quad I_{xy} = I_{yx} = \text{const.} \tag{6.51}$$

Because the trace I of I_{ij} reduces to a constant, we have $\partial_0^3 I_{ij}^T = \partial_0^3 I_{ij}$ and so

$$< \partial_0^3 I_{ij}^T \partial_0^3 I_{ij}^T > = (2\Omega)^6 \left(\frac{1}{2} \mu a^2 \right)^2 < 2\cos^2 2\phi + 2\cos^2 2\phi >$$

$$= 32\Omega^6 a^4 \mu^2. \tag{6.52}$$

We now write (6.49) as

$$L_{gw} = \frac{32}{5} \Omega^6 a^4 \mu^2 = \frac{32}{5} \frac{(m_1 + m_2)^3 \mu^2}{a^5} \tag{6.53}$$

in units of $c^5/G = 3.6 \times 10^{59} \, \text{erg}^{-1}$. Upon introducing the chirp mass $\mathcal{M} = m_1^{3/5} m_2^{3/5} (m_1 + m_2)^{-1/5}$, we may equivalently write

$$L_{gw} = \frac{32}{5} (\Omega \mathcal{M})^{10/3}. \tag{6.54}$$

This expression has been confirmed to within 1% by the observed orbital decay of the Hulse–Taylor binary neutron star system PSR1913+16[520]. The case of elliptical orbits has been worked out by P. C. Peters and J. Matthews[419]. Thus, the theory of linearized quadrupole gravitational radiation has been confirmed to within 0.1% by the observed orbital decay of the Hulse–Taylor binary neutron star system PSR1913+16[520, 589].

Table 6.1 *Summary of tensors.*

Symbol	Attribute	Comment
$\xi_f^b - \xi_i^b\big|_\gamma$	$\frac{1}{2}\left(\int_\gamma u^e x^a ds\right) R^b_{fea}\xi^f$	vector change over a loop γ
$\nabla_{[a}\nabla_{b]}\xi_c$	$R_{abc}{}^d\xi_d$	non-commutativity of ∇_a
$R_{abc}{}^d$	$\partial_b\Gamma^d_{ac} - \partial_a\Gamma^d_{bc} + \Gamma^f_{ca}\Gamma^d_{bf} - \Gamma^f_{cb}\Gamma^d_{af}$	Christoffel form
$R_{abc}d$	$\partial_a\omega_{b\mu\nu} - \partial_b\omega_{a\mu\nu} + [\omega_a,\omega_b]_{\mu\nu}$	Riemann–Cartan form
R_{abcd}	$C_{abcd} + g_{a[c}R_{d]b} + g_{c[a}R_{b]d} - \frac{1}{3}g_{a[c}g_{d]b}R$	Weyl tensor C_{abcd}
R_{abcd}	$R_{abcd} = R_{cdab}$, $R^d_{[abc]} = 0$, $\nabla_{[a}R_{bc]d}{}^e = 0$	symmetries
R_{0i0j}	$-\frac{1}{2}\partial_t^2 h_{ij}$	linearized limit
$c_{\mu\nu}$	$\nabla^a\omega_{a\mu\nu} = 0$	Lorentz gauge
	$\hat{\nabla}^2\omega_{a\mu\nu} - R^c_a\omega_{c\mu\nu} - [\omega^c,\nabla_a\omega_c]_{\mu\nu} = 16\pi\tau_{a\mu\nu}$	nonlinear wave equation
G_{ab}	$R_{ab} - \frac{1}{2}g_{ab}R = 0$	Einstein equations
R_{ac}	$-\frac{1}{2}h_{ac}$	transverse traceless gauge
	$-\frac{1}{2}\nabla^2 h_{ij} = 0$	linear wave equation
h_{ab}	$h_+ e^+_{ab} + h_\times e^\times_{ab}$	spin-2 polarizations
h_{0i}	0	Radiation gauge
t^{ab}	$t^{00} = t^{0z} = t^{zz} = \frac{1}{16\pi}\left(\dot{h}_+^2 + \dot{h}_-^2\right)$	stress-energy tensor of GWs
$\mathcal{M}$	$\frac{m_1^{3/5} m_2^{3/5}}{(m_1+m_2)^{1/5}}$	chirp mass
L_{gw}	$\frac{32}{5}(\Omega\mathcal{M})^{10/3}$	Luminosity from binary motion
L_{gw}	$\frac{32}{5}(M/a)^5(\delta m/M)^2$	Lumps δm in Newtonian orbits
c^5/G	3.6×10^{59} erg s^{-1}	unit of GW luminosity

As a special limit (6.54), consider gravitational radiation produced by mass-inhomogeneity of mass δm in orbit around a large mass of mass M. The gravitational wave-luminosity is described by the limit of (6.54) of $\delta m = m_1 \ll M = m_2$. With $\Omega \simeq M^{1/2}/a^{3/2}$, we find

$$L_{gw} = \frac{32}{5}\left(\frac{M}{a}\right)^5\left(\frac{\delta m}{M}\right)^2. \tag{6.55}$$

The mass-inhomogeneity δm may be envisioned as a lump or blob of matter as part of a nonaxisymmetric torus around a black hole.

6.4 Summary of equations

The Riemann tensor describes spacetime curvature, observed by variations in a vector following parallel transport over a closed loop, or viewed as the linear operator $[\nabla_a, \nabla_b]$. The relevant expressions, properties and small-amplitude limits are listed in the Table 6.1.

Linearized plane waves for the connections have two degrees of freedom. In the transverse traceless gauge, the linearized Ricci tensor reduces to $R_{ac} = -\frac{1}{2}h_{ac}$. In the radiation gauge ($h_{0i} = 0$), the metric perturbation can be written explicitly in terms of the $+$ and $\times$ polarization modes $h_{ab} = h_{+}e_{ab}^{+} + h_{\times}e_{ab}^{\times}$ with stress-energy tensor (6.36). The lowest multipole moment of mass which generates gravitational radiation is the quadrupole moment. For binary motion, we have the quadrupole luminosity function (6.54). In the limit of a mass-inhomogeneity of mass δm in orbit around a large mass M with orbital radius a, it reduces to (6.55).

Exercises

1. Calculate the luminosity of gravitational radiation from the motion of the Earth around the Sun, PSR1913 + 16 (binary period 7.75 h), and PSR0737-3039 (binary period 2.45 h), assuming neutron star masses of 1.41 $M_\odot$. Estimate the time-to-coalescence for each of these binaries.

2. Consider a mass-inhomogeneity of about 10% in a torus of mass $0.1M_\odot$ at a radius of $6M$ around a black hole of mass $M = 10M_\odot$. What is the gravitational radiation luminosity?

3. Show that one or two lumps swirling around a compact object radiate at twice the angular frequency.

4. Anisotropic emission in gravitational radiation arises in the precession of a torus around a compact object, when the torus is tilted with respect to the axis of rotation. By inspection of the projections of the torus on the celestial sphere, derive the frequencies of gravitational radiation emitted along the axis of rotation and into the oribital plane.

5. Calculate the secular change $\dot{\mathcal{M}}/\mathcal{M}$ of a binary with chirp mass $\mathcal{M}$ due to the emission of gravitational radiation when $m_1 = m_2$ and when $m_1 \ll m_2$.

6. The Kozai mechanism[307] describes the secular evolution of the ellipticity of a binary, itself in orbit with a distant third partner. Write the resulting evolution equations in dimensionless form and identify the relevant small quantities. Use perturbation theory to calculate the leading-order term describing the secular evolution of the ellipticity of the (small) binary in case all three objects are coplanar. What are the implications for the lifetime of binaries in globular clusters?[591].

7

Cosmological event rates

"Everything that is really great and inspiring is created by the individual
who can labor in freedom."
Albert Einstein (1879–1955),
in H. Eves, *Return to Mathematical Circles.*

Cosmology – the study of the evolution of the universe as a whole – is becoming an
ever more exact science with the recent precision observations by BOEMERANG,
MAXIMA and WMAP. Within a few percent uncertainty, we know that the
universe is open, flat and contains only a few percent of baryonic matter. The
universe is primarily filled with Cold Dark Matter (CDM) and dark energy
(a cosmological constant). If this is not a coincidence, the cosmological constant
is time-varying, and exchanges energy and momentum with CDM and, possibly,
baryonic matter. The imprint of the earliest epoch of the universe that at present
can be probed, is the Cosmic Microwave Background (CMB). The CMB is a
relic of the last surface of scattering at time 379 kyr[45]. Its extreme homogeneity
is well accounted for by a preceeding inflationary phase. A recent review of
cosmometry is compiled by L. M. Kraus[308].

The early universe may well have produced a stochastic background in gravitational waves. If so, these relic waves could provide the earliest signature of
the universe at an epoch much earlier than the CMB and the preceding phase
which produced the initial light element abundances[360]. At present, this relic
in gravitational waves is largely unknown, except that its spectrum should be
smooth. It may or may not have a thermal component.

In this chapter, we review some basic elements of cosmology in its application
to the calculation of the stochastic background radiation in gravitational waves
produced by astrosphysical sources. These calculations have been pursued for a
number of candidate sources[180, 181, 137, 425, 138, 269, 573]. We summarize
here the these calculations for sources that are locked to the star-formation rate.

Figure 7.1 The Microwave Sky Image from the WMAP mission, showing temperature fluctuations in the 2.73 K CMB produced 379 kyr after the Big Bang. Colors indicate temperature fluctuations (blue-red is cold-hot) with a resolution of about 1 μK. The results show that the universe is flat ($\Omega_{tot} = 1.02^{0.02}_{0.02}$) comprising a cosmological constant ($\Omega_{\Lambda} = 0.73^{0.04}_{0.04}$), CDM ($\Omega_{CDM} = 0.27^{0.04}_{0.04}$) and baryonic matter ($\Omega_{b} = 0.044^{0.004}_{0.004}$). (Courtesy of NASA and the WMAP Science Team.)

7.1 The Cosmological principle

"Nature is an infinite sphere of which the center is everywhere and the
circumference nowhere."
Cardinal Nicholas of Cusa 1400–64, in Giorgio de Santilla,
The Age of Adventure: The Renaissance Philosophers.

The observed large-scale uniformity of visible matter in the sky allows *homogeneous and isotropic* models for the large-scale properties of the universe. These models embody the cosmological principle, in there being no preferred point of reference or orientation, as contemplated by Cusa in the quote above.

The symmetry conditions in the cosmological principle give rise to the Robertson–Walker line elements[463, 578] (also referred to as the Friedman–Robertson–Walker line-element after A. Friedman[202]).

$$ds^2 = -c^2 dt^2 + K^2 d\sigma^2, \tag{7.1}$$

where we reinstate the velocity of light c and where $d\sigma$ denotes the three-volume element of spacelike directions with either positive, zero or negative curvature. This may be expressed in various coordinates: in isotropic coordinates

$$d\sigma^2 = \frac{dx^2 + dy^2 + dz^2}{1 + \epsilon r^2/4}, \tag{7.2}$$

in spherical coordinates,

$$d\sigma^2 = \frac{dr^2}{1 - \epsilon r^2} + r^2(d\theta^2 + \sin^2\theta d\phi^2), \tag{7.3}$$

and in Robertson–Walker coordinates

$$d\sigma^2 = d\chi^2 + f^2(\chi)(d\theta^2 + \sin^2\theta d\phi^2). \tag{7.4}$$

Here, $\epsilon = -1, 0, 1$ describes the case of negative curvature (open universe), vanishing curvature (open universe) and positive curvature (closed universe), respectively, whereby

$$f(\chi) = \sinh\chi, \ \chi, \ \sin\chi. \tag{7.5}$$

7.2 Our flat and open universe

BOEMERANG and MAXIMA[151, 249], based on the power spectra of the cosmic microwave background and by observations of distant Type Ia super-novas[418, 479], and WMAP show that universe is well-described by a flat Λ-dominated CDM cosmology with a subdominant contribution in matter, satisfying

$$\Omega_m + \Omega_\Lambda = 1. \tag{7.6}$$

For practical calculations on astrophysical source-populations, it suffices to consider $\Omega_\Lambda = 0.70$ and $\Omega_m = 0.30$, neglecting the contribution of matter $\Omega_b = 0.044^{0.004}_{0.004}$ to the evolution of the universe. The Hubble parameter H_0 will taken to be $73\,\mathrm{km}\ \mathrm{s}^{-1}/\mathrm{Mpc}$[201].

Our current understanding, therefore, is that we live in a flat Robertson–Walker universe described by a line-element

$$ds^2 = -dt^2 + a(t)^2 d\sigma^2 \tag{7.7}$$

with $d\sigma^2$ as in (7.2), (7.3) or (7.4) with $\epsilon = 0$. The *proper distance r* between two points corresponds to the surface area $4\pi r^2$ of the sphere, which has one at its center and the other on its north pole. In the flat Robertson–Walker cosmology (7.7), the massless photons and gravitons emitted by a source appear redshifted at the observer due to cosmological expansion. This also implies they appear at the source at a reduced rate. This gives rise to two redshift factors in the *luminosity distance $d_L(z)$*, which the local energy flux S to the luminosity L as measured in the comoving frame of the source,

$$S = \frac{L}{4\pi d_L(z)^2}, \quad d_L(z) = (1+z)r. \tag{7.8}$$

The *comoving volume* element is given by

$$V'(z, \Lambda_i) = \frac{4\pi r^2 c}{H_0 E(\Omega_i, z)}, \quad r(z) = \int_0^z \frac{c}{H_0 E(\Omega_i, z')} dz', \tag{7.9}$$

where we restored the constants c and H_0 for computational reference. Upon neglecting Ω_b,

$$E(z, \Omega_\Lambda) = \frac{H(z)}{H_0} = \left[\Omega_M(1+z)^3 + \Omega_\Lambda\right]^{1/2} \tag{7.10}$$

represents the evolution of the Hubble parameter ($\Omega_m + \Omega_\Lambda = 1$). The time-evolution satisfies

$$\frac{dt_e}{dz} = \frac{1}{(1+z)E(z)}, \quad \frac{dt}{dt_e} = 1 + z. \tag{7.11}$$

The expression (7.10) for the Hubble expansion derives from the Einstein equations in the line-element (7.7). The single metric parameter $a(t)$ reduces the Christoffel symbols greatly. Upon using isotropic coordinates, the nonzero Γ^c_{ab} form out of $\dot{a}/a$, $\dot{a}a$ when one of the indices is t and the remaining two are equal to one of the three spatial coordinates:

$$\Gamma^t_{ii} = \dot{a}a, \quad \Gamma^i_{it} = \frac{\dot{a}}{a} (i = x, y, z). \tag{7.12}$$

Evaluation of the Ricci tensor (5.20) gives the expressions

$$R_{00} = -3\frac{\ddot{a}}{a}, \quad R_{ij} = \left(\frac{\ddot{a}}{a} + 2\dot{a}^2\right) g_{ij}, \quad R = 6\left(\frac{\ddot{a}}{a} + \frac{\dot{a}^2}{a^2}\right). \tag{7.13}$$

The Einstein equations dictate $G_{ab} = 8\pi T_{ab}$, where $G_{00} = R_{00} + R/2$, $G_{ii} = R_{ii} - R/2$, and T_{ab} is the stress-energy tensor comprising matter and a cosmological constant $T_{ab} = (r+P)u_a u_b + Pg_{ab} - \Lambda g_{ab}/8\pi$. Combined with (7.13), we have

$$3\frac{\dot{a}^2}{a^2} = 8\pi r + \Lambda, \quad -2\frac{\ddot{a}}{a} - \frac{\dot{a}^2}{a^2} = 8\pi P - \Lambda. \tag{7.14}$$

Following P. J. E. Peebles[414], we define the fractions

$$\Omega = \frac{8\pi G r_0}{3H_0^2}, \quad \Omega_\Lambda = \frac{\Lambda}{3H_0^2}, \tag{7.15}$$

where the subscript 0 refers to the quantities at the present epoch ($z = 0$), i.e. the present matter density r and the present Hubble constant H_0. Since non-relativistic matter (baryonic and dark) evolves according to the comoving volume $a(t)^{-3}$, and $1 + z = a_0/a(t)$, we obtain

$$\frac{\dot{a}}{a} = H_0 E(z) = H_0 \left[\Omega(1+z)^3 + \Omega_\Lambda\right]^{1/2} \tag{7.16}$$

as the definition for $E(z)$ in (7.10).

7.3 The cosmological star-formation rate

The star-formation rate $R_{SF}(z)$ has been modeled on the basis of deep redshift surveys. According to (7.9), we have the transformation rule used in the observational study of C. Porciani and P. Madau[439]

$$\frac{R_{SF}(z, \Omega_\Lambda)}{E(z, \Omega_\Lambda)} = \frac{R_{SF}(z, 0)}{E(z, 0)} \tag{7.17}$$

for the redshift distribution as a function of cosmological parameters.

Madau and Pozzetti[359] and Porciani and Madau[439] provide three models of the cosmic star formation rate (SFR) up to redshifts $z \sim 5$. In what follows, we use their model SFR2. In a universe dominated by Dark Matter ($\Omega_m = 1$), they determine

$$R_{SF2}(z; 0) = \frac{0.16 h_{73} U(z) U(5-z)}{1 + 660 e^{-3.4(1+z)}} M_\odot \, \mathrm{yr}^{-1} \, \mathrm{Mpc}^{-3} \tag{7.18}$$

with Hubble constant $H_0 = h_{73} 73 \, \mathrm{km \, s}^{-1} \, \mathrm{Mpc}^{-1}$ and Heaviside function $U(\cdot)$. According to (7.17), therefore

$$R_{SF2}(z, \Omega_\Lambda) = R_{SF2}(z, 0) \frac{E(z, \Omega_\Lambda)}{(1+z)^{3/2}}. \tag{7.19}$$

7.4 Background radiation from transients

The universe is essentially transparent in gravitational waves, starting from a very early phase of the universe. Consequently, the energy in gravitational waves emitted by astrophysical sources is conserved. The total energy in gravitational waves at present cosmic time is therefore the accumulated energy released during all past events, back to the earliest stages of the universe.

In the approximation of a homogeneous matter distribution, cosmological evolution depends only on redshift. The gravitational wave-energy density seen today is therefore a simple summation of gravitational waves emitted in the past, convolved with the expansion of the universe. The relevant quantity, therefore, is the cumulative number of transients $N(z)dz$ that have occurred as a function of redshift, and filled the universe with their gravitational wave-emissions as discussed by E. S. Phinney[425].

The spectral energy-density dE_{gw}/df of a single point source is a redshift-independent distribution. This follows from Einstein's adiabatic relationship $E_{gw}/f = \mathrm{const}$ and conservation of the number of gravitons in a redshift-corrected frequency bandwidth. The spectral energy $E_{gw}(f, z)$ hereby has a redshift invariant derivative,

$$E'_{gw}(f, z) = E'_{gw}((1+z)f, 0) \tag{7.20}$$

where $' = d/df$. The total energy in a given unit of comoving volume is the accumulated energy radiated by all past events – in a homogeneous universe, the net loss of radiation leaking out of a unit of comoving volume is zero. By the redshift invariance (7.20), we may sum $E'(f, z)$ over individual sources within a unit of comoving volume. Let the unit of comoving volume be defined by the unit volume at $z = 0$. The accumulated spectral energy-density per unit of volume (erg Hz^{-1} cm^{-3} in dimensionful units) at present time hereby satisfies

$$\epsilon'_B(f) = \int_0^{z_{max}} E'_{gw}(f, z)N(z)dz. \tag{7.21}$$

Phinney[425] arrives at (7.21) in slightly different form.

For events with a given event rate $R(z)$ per unit of comoving volume per unit of comoving time, we have

$$N(z)dz = R(z)dt_e = R(z)\frac{dt_e}{dz}dz = \frac{R(z)dz}{(1+z)E(z)}, \tag{7.22}$$

where the dependence on cosmological parameters Ω_i in the individual factors on the right-hand side is suppressed. For a distribution locked to the star-formation rate (7.19), this gives

$$N(z) = \frac{N(0)R_{SF2}(z, 0)}{R_{SF2}(0, 0)(1+z)^{5/2}}, \tag{7.23}$$

where $N(0)$ denotes the local event rate per unit volume. By (7.21), the spectral energy-density becomes

$$\epsilon'_B(f) = N(0) \int_0^{z_{max}} \frac{R_{SF2}(z, 0)}{R_{SF2}(0, 0)} \frac{E'_{gw}dz}{(1+z)^{5/2}}. \tag{7.24}$$

7.5 Observed versus true event rates

The redshift probability density $p(z)$ of events as seen in the observer's frame can be written in terms of the true event rate $dR_*(z)/dz$ per unit redshift[570, 138]

$$p_*(z) = \frac{dR_*(z)/dz}{\int_0^5 dR_*(z)}. \tag{7.25}$$

Likewise, we define the probability-density function of detection as a function of redshift

$$p_{detect}(z) = \frac{dR_{detect}/dz}{\int_0^5 dR_{detect}}. \tag{7.26}$$

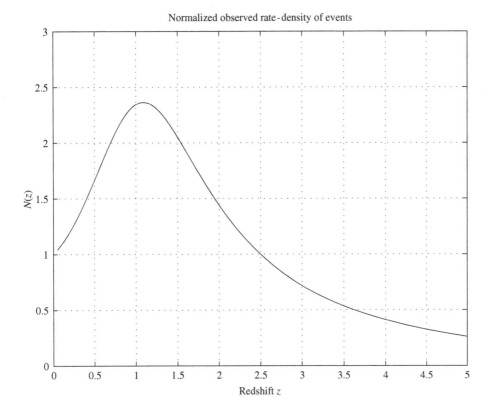

Figure 7.2 Normalized observed comoving rate-density $N(z)$ in the star-formation model of Porciani and Madau[439].

These two probabilities are related by dependence on the luminosity L of the sources. With $p(L)$ denoting the intrinsic luminosity distribution, J. S. Bromm and A. Loeb[84] introduce

$$dR_{detect} = dR_*(z) \int_{L_{lim}(z)} p(L)dL. \tag{7.27}$$

Here, $L_{lim}(z)$ denotes a luminosity threshold as a function of redshift, given by

$$L_{lim}(z) = 4\pi d_L^2(z)S_{lim} \tag{7.28}$$

where d_L is the luminosity distance to a source at redshift z and where S_{lim} denotes the sensitivity threshold of the instrument. For example, following Bromm and Loeb[84], the flux-density threshold of the Burst and Transient Source Experiment (BATSE) is 0.2 photon cm^{-2}/s.

Exercises

1. Plot the observed-to-true event rate as a function of redshift according to (7.27)▲ and (7.28).

8

Compressible fluid dynamics

"An expert is someone who knows some of the worst mistakes that can
be made in his subject, and how to avoid them."
Werner Heisenberg (1901–1976), *Physics and Beyond.*

Fluids in astrophysical systems show a variety of phenomena associated with
waves, shocks, magnetic fields, and instabilities. In what follows, we review
elements of non-relativistic fluid dynamics, before generalizing to relativistic
fluids.

Perhaps the most remarkable phenomenon in compressible fluid dynamics is
steepening. This is apparent in Burgers' equation, which models *dust*: a compressible fluid at zero temperature. In the absence of pressure, the equations of motion
are conservation of linear momentum

$$u_t + uu_x = 0 \tag{8.1}$$

for a Eulerian velocity field $u(t, x)$. Burgers' equation is commonly considered
in the context of an initial value problem: solve for $u(t, x)$ in response to initial
data $u = u_0(x)$ at $t = 0$. Burgers' equation has the simple *characteristic* solution

$$\frac{du}{ds} = 0 \quad \text{along} \quad \frac{dx}{dt} = u. \tag{8.2}$$

The surface area below the graph $u(t, \cdot)$ is a time-invariant[597]. This can be
seen by integration using horizontal slices, as in Lebesgue integration shown in
Figure (8.1).

Steepening is due to the convective derivative uu_x. Two characteristics – lines
of constant velocity in the (t, x)-plane – emanating from points ξ_0 and ξ_1 on the
x-axis meet at time $t_s\xi_0 + u(\xi_0)t_s = \xi_1 + u_0(\xi_1)t_s$, i.e. $t_s = -(\xi_1 - \xi_0)/(u_0(\xi_1) -
u_0(\xi_0))$. In the limit as $\xi_1, \xi_0 \to \xi$, this yields the time for shock formation

$$t_S(\xi) = -\frac{1}{u_0'(\xi)}. \tag{8.3}$$

89

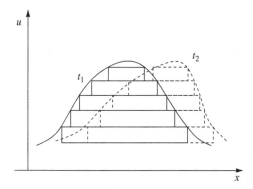

Figure 8.1 In Burgers' equation, the area under the curves $u(t, \cdot)$ remains invariant in view of a vanishing convective derivative of the velocity. This can be seen using horizontal partitioning (rectangles), as in Lebesgue integration of $u(t, \cdot)$. Time-evolution from $t = t_1$ to $t = t_2$ corresponds to a horizontal shift without change of size of this partitioning. This is like sliding the slices of shaped aluminum in Arthur Fiedler's sculpture on Storrow Drive, Boston. Sliding these slices sideways leaves the frontal area of the face invariant.

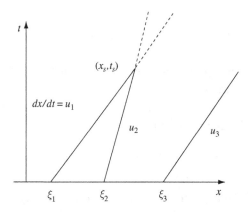

Figure 8.2 In Burgers' equation, a pair of convergent characteristics $dx/dt = u_i = u_0(\xi_i)$ emanating from ξ_i, $i = 1, 2$, meet at a finite time. The location of the resulting shock formation due to compression is their point of intersection (x_s, t_s). A pair of divergent characteristics $(i = 2, 3)$ never meets, and the associated expanding flow remains smooth.

Steepening creates shocks whenever $u'(\xi) < 0$. In traffic theory[597], this corresponds to faster vehicles taking over slower vehicles beyond. The opposite case of $u'(\xi) > 0$ corresponds to expansion, or faster vehicles moving ahead of slower vehicles. When a shock forms, the velocity field $u(t, \cdot)$ displays a discontinuity. In reality particles will collide, if the crest is sufficiently dense. This requires a description beyond Burgers' equation with supplementary input to the model: shock jump conditions, which are discussed later in this chapter, or through the addition of viscosity (below).

8.1 Shocks in 1D conservation laws

Burgers' equation is a special case of the more general one-dimensional conservation law

$$f_t + F'(f)f_x = 0, \tag{8.4}$$

where f denotes a density and $F(f)$ denotes a flux of a quantity of interest. Consider a surface of discontinuity S – a shock front – at location $x_S(t)$ and with velocity $U = x_S'(t)$. With f smooth to either side of S, we may consider integrating (8.4) on either side, according to

$$\int_{-\infty}^{x_S(t)} f_t dx + (F)^- - F_{-\infty} = 0, \quad \int_{x_S(t)}^{\infty} f_t dx + (F)_\infty - F^+ = 0 \tag{8.5}$$

where by Leibniz' rule

$$\int_{-\infty}^{x_S(t)} f_t dx = \partial_t \int_{-\infty}^{x_s(t)} f dx - (f)^- U,$$

$$\int_{x_S(t)}^{\infty} f_t dx = \partial_t \int_{x_S(t)}^{\infty} f dx + (f)^+ U. \tag{8.6}$$

As a conservation law, (8.4) satisfies

$$\partial_t \int_{-\infty}^{\infty} f dx = -[F]_{-\infty}^{\infty}. \tag{8.7}$$

Hence, by addition of these results we have

$$-[F]_{-\infty}^{\infty} + U[f]_S - [F]_S + [F]_{-\infty}^{\infty} = 0 \tag{8.8}$$

which obtains a relation for the shock velocity

$$U = \frac{[F]_S}{[f]_S}. \tag{8.9}$$

In case of Burgers' equation, $f = u$ and $F = u^2/2$ which obtains

$$U = \frac{1}{2} \frac{(u^+ - u^-)(u^+ + u^-)}{u^+ - u^-} = \frac{u^+ + u^-}{2}. \tag{8.10}$$

In the above, note that we referred explicitly to f as a conserved quantity in (8.7). This forms a supplementary condition to the differential equation (8.4).

The conservation law (8.4) can be given a weak formulation, which naturally incorporates discontinuous solutions. The adjective "weak" refers to weaker conditions on the smoothness of f. A weak formulation is open to a more general family of solutions. Integration of (8.4)

$$0 = \int \int_R (f_t + F'(f)f_x)\phi\, dxdt \qquad (8.11)$$

against functions of the form

$$\phi \epsilon C_0^1(R) = \{h\epsilon C^1(R) | h = 0 \text{ on } \partial R\}. \qquad (8.12)$$

These functions ϕ have smooth first derivatives and vanish on the boundary of the domain R. We shall take R to be the strip in the (x, t)-plane between the x-axis and a parallel of constant $t > 0$.

First, assume there is no shock front. Integration by parts on (8.11) gives

$$0 = -\int \int_R (f\phi_t + F\phi_x)\, dxdt + \int_{\partial R} (fn_t + Fn_x)\phi\, ds \qquad (8.13)$$

where $n = (n_x, n_t)$ denotes the outgoing unit normal to ∂R. For example, if ∂R is described by $\psi(x, t) = 0$, then $n_i = \partial_i \psi / \sqrt{\psi_x^2 + \psi_t^2}$. With the condition that $\phi = 0$ on ∂R, we conclude that

$$\int \int_R (f\phi_t + F\phi_x)\, dxdt = 0 \qquad (8.14)$$

for all $\phi \epsilon C_0^1(R)$. Within the assumption of no shock front, f is smooth. Hence, (8.14) implies that (8.4) holds pointwise everywhere in R.

The above shows that (8.14) contains the family of smooth solutions to (8.4). However, it is more general, in that it calls on f without derivatives. We now make the step to take (8.14) as our new formulation of (8.11), thereby extending the family of solutions to those that include discontinuities.

In the presence of a shock front S, consider (reverse) integration by parts on (8.14) following a partitioning $R = R_- \cup R_+$ of R into the the left-hand side R_- and the right-hand side R_+ of S. Thus,

$$0 = -\int \int_{R_-} (f_t + F'(f)f_x)\phi\, dxdt - \int \int_{R_+} (f_t + F'(f)f_x)\phi\, dxdt$$
$$+ \int_S ([f]_S n_t + [F]_S n_x)\phi\, ds. \qquad (8.15)$$

Here, the normal $n : n_x dx + n_t dt = 0$ is outgoing with respect to the sub-domains ∂R_-, which are separated by $S : x = x_s(t)$ from ∂R_+.

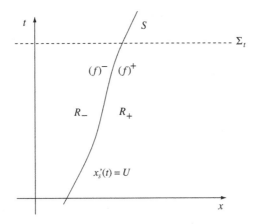

Figure 8.3 A shock surface S in the (x, t)-plane is described by a position $x_S(t)$ and velocity $U = x'_S(t)$. Jump conditions in a quantity f across the shock describe differences $[f]_S = (f)^+ - (f)^-$ between limiting values $(f)^+$ to the right and $(f)^-$ to the left of S at a given time t. Here, f is assumed to be smooth to either side of S, where it satisfies $f_t + F'(f)f_x = 0$. The weak formulation defines an integral formulation over the strip $R : 0 < t < \Sigma_t$ of global solutions in the presence of discontinuities, where R is the sum of the left side R_- and the right side R_+ of S.

The integral formulation (8.15) holds for all $\phi \epsilon C_0^1(R)$, i.e. smooth functions which vanish as $x \to \pm\infty$. First, consider functions $\psi \epsilon C_0^1(R_-)$: functions $\phi \epsilon C_0^1(R)$ which vanish on S and R_+. This leaves

$$\int \int_{R_-} (f_t + F'(f)f_x)\psi\, dx dt = 0 \qquad (8.16)$$

for all $\psi \epsilon C_0^1(R_-)$, whereby (8.11) holds in R_-. Similarly, we find that (8.11) holds in R_+. We are therefore left with

$$0 = + \int_S ([f]_S n_t + [F]_S n_x)\phi\, ds. \qquad (8.17)$$

This implies

$$U = \frac{dx}{dt} = -\frac{n_t}{n_x} = \frac{[F]_S}{[f]_S} \qquad (8.18)$$

as before.

It follows that the weak formulation (8.14) of the conservation law (8.11) comprises discontinuous solutions with the correct jump conditions. Conservation laws, therefore, are of particular interest as a starting point for shock capturing methods for numerical simulations.

8.2 Compressible gas dynamics

One-dimensional compressible gas dynamics at finite temperature is described by conservation of momentum and mass. A fluid with velocity u^i and density ρ hereby satisfies

$$u_t + uu_x = -P_x/\rho, \quad \rho_t + u\rho_x + \rho u_x = 0. \tag{8.19}$$

For a polytropic equation of state $P = K\rho^\gamma$, we define a velocity of sound

$$a = \sqrt{\frac{\gamma P}{\rho}}. \tag{8.20}$$

The polytropic index is defined formally by the ratio of specific heats c_P/c_V. To a good approximation, γ satisfies

$$\gamma = \frac{2+\alpha}{\alpha} = \begin{cases} 7/5 & \text{diatomic gas} \\ 5/3 & \text{monatomic gas} \end{cases} \tag{8.21}$$

corresponding to 5, respectively, 3 degrees of freedom. In equipartition, each degree of freedom shares the same fraction of total internal energy $e = \alpha kT/2$, where T denotes the temperature and k denotes Boltzmann's constant. For adiabatic changes, whereby the coefficient K remains constant in the presence of a constant entropy along streamlines, we can rewrite the equations of motion as follows.

The first equation of (8.19) can be written as

$$(\partial_t + u\partial_x)u + \frac{2a}{\gamma - 1}a_x = 0. \tag{8.22}$$

In the second equation of (8.20), we may use $d\rho/\rho = \gamma^{-1}dP/P$; substitution and multiplication by $\gamma P/\rho$ gives

$$\left(\frac{a^2}{\gamma - 1}\right)_t + u\left(\frac{a^2}{\gamma - 1}\right)_x + a^2 u_x = 0. \tag{8.23}$$

This reduces to

$$\left(\frac{2a}{\gamma - 1}\right)_t + u\left(\frac{2a}{\gamma - 1}\right)_x + au_x = 0. \tag{8.24}$$

Addition and subtraction of (8.22) and (8.24) gives

$$D_\pm\left(u \pm \frac{2a}{\gamma - 1}\right) = 0. \tag{8.25}$$

Here, $D_\pm = \partial_t + (u \pm a)\partial_x$ denotes differentiation in the directions

$$\left(\frac{dx}{dt}\right)_\pm = u \pm a \tag{8.26}$$

along which the Riemann invariants

$$J_{\pm} = u \pm \frac{2a}{\gamma - 1} \tag{8.27}$$

remain constant.

The Riemann invariants $J_{\pm}$ may be used in constructing solutions to initial value problems. In particular, we obtain *simple waves* when one of the Riemann invariants is constant throughout.

Simple waves have one Riemann-invariant constant throughout the fluid, e.g.

$$u + \frac{2a}{\gamma - 1} = c \tag{8.28}$$

This leaves a constant velocity

$$u = \frac{1}{2}(c + J_{-}) \tag{8.29}$$

along the characteristic

$$\left(\frac{dx}{dt}\right)_{-} = u - a = c\left(\frac{3 - \gamma}{4}\right) + J_{-}\left(\frac{\gamma + 1}{4}\right), \tag{8.30}$$

where J_{-} obtains from the initial data at the point of intersection of this characteristic with the x-axis. Alternatively, consider a vanishing Riemann-invariant $J_{-} = 0$. We then have $u = 2a/(\gamma - 1)$, whereby

$$\left(\partial_t + \frac{\gamma - 1}{2} u \partial_x\right) u = 0. \tag{8.31}$$

With $v = (\gamma + 1)u/2$, this corresponds to Burgers' equation.

8.3 Shock jump conditions

The jump conditions for compressible gas dynamics for the pressure and the density are the Rankine–Hugoniot condition for the pressure jump $P_2/P_1 - 1$ from upstream to downstream, given by

$$\frac{P_2}{P_1} - 1 = \frac{\gamma(1 - \rho_1/\rho_2)}{1 - \frac{\gamma+1}{2}(1 - \rho_1/\rho_2)} \tag{8.32}$$

in terms of the density jump $1 - \rho_1/\rho_2$. An immediate consequence is the maximal jump in density

$$\frac{\rho_2}{\rho_1} = \frac{\gamma + 1}{\gamma - 1} \tag{8.33}$$

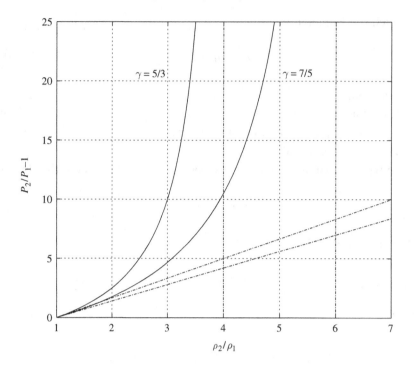

Figure 8.4 The pressure jump $P_2/P_1 - 1$ as a function of the density ratio ρ_2/ρ_1 of downstream values to upstream values, according to the Rankine–Hugoniot jump condition. Note the asymptotic value $\rho_2/\rho_1 = (\gamma + 1)/(\gamma - 1)$ in terms of the polytropic index γ (dashed lines) and the adiabatic tangents at the origin (dot–dashed lines).

across a strong shock. The Rankine–Hugoniot jump condition can be derived in the frame of the shock, where we have the jump conditions of conservation of mass, linear momentum and enthalpy given by

$$
\begin{cases}
\rho_1 u_1 = \rho_2 u_2 \\
\rho_1 u_1^2 + P_1 = \rho_2 u_2^2 + P_2 \\
\frac{\gamma}{\gamma - 1} \frac{P_1}{\rho_1} + \frac{1}{2} u_1^2 = \frac{\gamma}{\gamma - 1} \frac{P_2}{\rho_2} + \frac{1}{2} u_2^2.
\end{cases}
\tag{8.34}
$$

The combination $[\gamma/(\gamma - 1)](P/\rho) = (1/\rho)\left(P + (\gamma - 1)^{-1} P\right)$ denotes the sum of specific thermal and internal energy.

The first and second equation in (8.34) combine into the first form of Prandtl's relation

$$
\frac{P_2 - P_2}{\rho_2 - \rho_1} = u_1 u_2.
\tag{8.35}
$$

By the third equation of (8.34) – conservation of enthalpy – we obtain at the sound speed $u = a_* = \sqrt{\gamma P/\rho}$ ($P_1 = P_2 = P$, $\rho_1 = \rho_2 = 1$) the stagnation enthalpy

$$H = \frac{1}{\gamma - 1} a_*^2 + \frac{1}{2} a_*^2 = \frac{\gamma + 1}{2(\gamma - 1)} a_*^2.$$

Subtraction of the enthalpy conditions

$$\begin{cases} \dfrac{2\gamma}{\gamma - 1} P_1 + (\rho_2 u_2) u_1 = \rho_1 \dfrac{\gamma + 1}{\gamma - 1} a_*^2 \\[2ex] \dfrac{2\gamma}{\gamma - 1} P_2 + (\rho_1 u_1) u_2 = \rho_2 \dfrac{\gamma + 1}{\gamma - 1} a_*^2 \end{cases} \tag{8.36}$$

gives $[2\gamma/(\gamma - 1)](P_2 - P_1) + (\rho_1 - \rho_2) u_1 u_2 = (\rho_2 - \rho_1)[(\gamma + 1)/(\gamma - 1)] a_*^2$, and hence $[2\gamma/(\gamma - 1)] u_1 u_2 - u_1 u_2 = [(\gamma + 1)/(\gamma - 1)] a_*^2$. This yields the algebraic relation

$$u_1 u_2 = a_*^2 \tag{8.37}$$

between the up- and downstream values of the velocity. It forms an alternative statement to (8.35). A shock forms when

$$u_1 > a_*, \quad u_2 < a^*. \tag{8.38}$$

This anticipates that a shock forms when the shock propagates supersonically into the upstream fluid, and subsonically in the downstream fluid.

Using Prandtl's relation (8.35) and conservation of mass, we have for conservation of enthalpy (third equation in (8.34))

$$\frac{\gamma}{\gamma - 1} (P_1 \rho_2 - P_2 \rho_1) = \frac{1}{2} (\rho_1 + \rho_2)(P_1 - P_2). \tag{8.39}$$

Dividing the left- and right-hand side by $P_1 \rho_2$ gives

$$\frac{\gamma}{\gamma - 1} \left(1 - \frac{\rho_1}{\rho_2} \right) = \left(\frac{\gamma + 1}{2(\gamma - 1)} \frac{\rho_1}{\rho_2} - \frac{1}{2} \right) \left(\frac{P_2}{P_1} - 1 \right), \tag{8.40}$$

from which (8.32) readily follows. Let us make two observations.

1. In the limit of small $(\rho_2 - \rho_1)/\rho_2$, i.e. $\rho_1, \rho_2 \sim \rho$, (8.32) shows the asymptotic result

$$\frac{\Delta P}{P} \simeq \gamma \frac{\Delta \rho}{\rho} + \gamma \frac{\gamma + 1}{2} \left(\frac{\Delta \rho}{\rho} \right)^2 \tag{8.41}$$

as the sum of the adiabatic change plus a second-order correction. This may further be compared with the isothermal limit $\Delta P/P = \Delta \rho/\rho$.

2. The inverse of the Rankine–Hugoniot relation in terms of $z = P_2/P_1 - 1$ gives

$$\frac{\rho_1}{\rho_2} = \frac{1 + \frac{\gamma-1}{2\gamma}z}{1 + \frac{\gamma+1}{2\gamma}z}. \tag{8.42}$$

Evidently, the reciprocal ρ_2/ρ_1 ranges from 1 for $z \sim 0$ to the aforementioned limit $(\gamma+1)/(\gamma-1)$ as z becomes large.

8.4 Entropy creation in a shock

An ideal gas satisfies the polytropic equation of state

$$P = e^{S/C_v}\rho^\gamma \tag{8.43}$$

where S denotes the specific entropy and C_v denotes the specific heat at constant volume. The polytropic index satisfies $\gamma = C_p/C_v$, where $C_p = C_v + R$ denotes the specific heat at constant pressure. It corresponds to $\gamma = (\alpha+2)/\alpha$, where α denotes the number of degrees of freedom of each particle (atom or molecule). For example, for air we have the difference $R = C_p - C_v = 2.87 \times 10^{-6}\,\mathrm{cm^2\,s^{-2\circ}C^{-1}}$.

The entropy created in a shock is determined by the strength of the shock,

$$\frac{S_2 - S_1}{C_v} = \log\frac{P_2\rho_2^{-\gamma}}{P_1\rho_1^{-\gamma}} = \log\frac{P_2}{P_1}\left(\frac{\rho_1}{\rho_2}\right)^\gamma \tag{8.44}$$

$$\frac{\Delta S}{C_v} = \log(1+z) + \gamma\log\left(\frac{1 + \frac{\gamma-1}{2\gamma}z}{1 + \frac{\gamma+1}{2\gamma}z}\right). \tag{8.45}$$

The leading order expansion for small z satisfies

$$\frac{\Delta S}{C_v} \cong \frac{\gamma^2 - 1}{12\gamma^2}z^3 + O(z^4). \tag{8.46}$$

Generally, $dS/dz > 0\,(\gamma > 1)$, and hence $\Delta S > 0$ corresponds to $z > 0$. Therefore, (8.38) denotes the correct inequality for entropy creating shocks. We further note that the entropy increase correlates with γ as illustrated in Figure 8.5.

8.5 Relations for strong shocks

The shock jump conditions may also be expressed in the laboratory frame, where the upstream velocity is zero. Conservation of mass and momentum, the first and second equation of (8.34), show that

$$u_2 \cong \frac{2}{\gamma+1}u_s, \quad \frac{\rho_2}{\rho_1} \cong \frac{\gamma+1}{\gamma-1}, \quad P_2 \cong \frac{2}{\gamma+1}\rho_1 u_s^2 \tag{8.47}$$

where u_s denotes the shock velocity. Furthermore, the ratio of the sound velocities satisfies

$$\frac{a_2^2}{a_1^2} = \frac{P_2}{P_1}\frac{\rho_1}{\rho_2} = \frac{P_2}{P_1}\left(\frac{P_1}{P_2}\right)^{-\frac{1}{\gamma}} e^{\frac{S_2-S_1}{\gamma C_v}} = \left(\frac{P_2}{P_1}\right)^{\frac{\gamma-1}{\gamma}} e^{\Delta S/\gamma C_v} \geq 1. \qquad (8.48)$$

This shows that a positive entropy condition corresponds to a change in the thermodynamic state of the fluid, wherein the velocity of sound is larger in the shocked downstream fluid than in the initially unshocked fluid upstream. The entropy increase as a function of shock strength is shown in Figure (8.5) for two values of γ. It illustrates that entropy creation decreases with the number of degrees of freedom per particle.

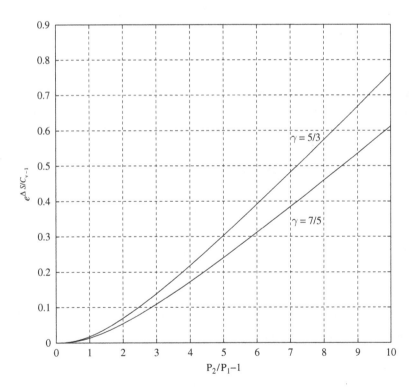

Figure 8.5 The entropy increase expressed as $e^{\Delta S/C_v} - 1$, where C_v denotes the specific heat at constant volume, as a function of shock strength $z = P_2/P_1 - 1$. Weak shocks are essentially adiabatic with $\Delta S = O(z^3)$. The entropy increase correlates with γ, and hence with the number of degrees of freedom per particle.

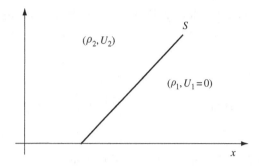

Figure 8.6 Shown is the one-dimensional shock problem viewed in the laboratory frame, in which the pre-shock fluid is at rest. The shock velocity U_s relative to the velocity of sound in the pre-shocked fluid can be expressed in terms of the Mach number $M = (U_s - U_1)/a_1$, which reduces to $-U_1/a_1$ in the laboratory frame.

8.6 The Mach number of a shock

Upon transforming back to an arbitrary frame of reference, (8.37) and (8.48), we find the general inequalities

$$u_1 + a_1 < u_s < u_2 + a_2. \tag{8.49}$$

Entropy creating shock fronts move faster than sound waves upstream and slower than sound waves downstream. Shock fronts result when positive characteristics downstream intersect positive characteristics downstream, as illustrated in Figure (8.7). The result (8.49) can be seen, by parametrizing shocks in an arbitrary frame of reference terms of the Mach number

$$M = \frac{U_s - u_1}{a_1} \tag{8.50}$$

relative to the flow velocity upstream. By momentum conservation, the second equation of (8.34), i.e. $P_2 - P_1 = \rho_1 u_1^2 - \rho_2 u_2^2 = \rho_1 u_1 (1 - \rho_1/\rho_2)$, gives the Rayleigh line

$$\frac{P_2}{P_1} - 1 = \gamma M^2 \left(1 - \frac{\rho_1}{\rho_2}\right). \tag{8.51}$$

The Rankine–Hugoniot relation (8.32) with (8.51) gives

$$M^2 = \frac{1}{1 - \frac{\gamma+1}{2}(1 - \rho_1/\rho_2)}. \tag{8.52}$$

Inverting (8.52) and using (8.32), we have

$$\frac{P_2}{P_1} - 1 = \frac{2\gamma(M^2 - 1)}{\gamma+1}, \quad 1 - \frac{\rho_1}{\rho_2} = \frac{2}{\gamma+1}\frac{M^2 - 1}{M^2}. \tag{8.53}$$

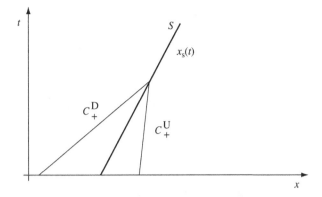

Figure 8.7 A shock front S creates positive entropy as it moves faster than sound waves upstream and slower than sound waves downstream. It therefore represents the intersection of positive characteristics upstream C_+^U and positive characteristics downstream C_+^D.

To finalize, introduce

$$M_2 = \frac{U_s - u_2}{a_2} = M\frac{\rho_1 a_1}{\rho_2 a_2} = M\left(\frac{\rho_1 P_1}{\rho_2 P_2}\right)^{1/2}.$$ (8.54)

This gives Euler's result

$$1 - M_2^2 = \frac{M^2 - 1}{1 + \frac{2\gamma}{\gamma+1}\left(M^2 - 1\right)}$$ (8.55)

showing that $M > 1$ – corresponding to the positive entropy condition – implies $M_2 < 1$. This shows (8.49) and, in the limit of weak shocks or transonic flow, the result may also be stated as $1 - M_2 \simeq M - 1$.

8.7 Polytropic equation of state

The first law of thermodynamics (conservation of energy) expresses the expulsion of heat dQ per particle in response to a change in specific internal energy de and specific work $pd(1/\rho)$: $dQ = de + pd\,(1/\rho)$. For adiabatic changes (i.e. changes in reversible processes) we have $de + pd\,(1/\rho) = 0$.

Entropy is "almost" energy: introducing the temperature T as an integrating factor, we have $dQ = TdS$ in terms of the specific entropy S (entropy per particle). This gives

$$de + pd\left(\frac{1}{\rho}\right) = T\,dS.$$ (8.56)

We next derive a partial differential equation for the temperature T.

Consider the total derivative

$$dS = \frac{1}{T}\left[de + p\,d\left(\frac{1}{\rho}\right)\right] = \frac{1}{T}\left(e_P dP + e_\rho d\rho - \frac{P}{\rho^2}d\rho\right) \tag{8.57}$$

whereby $S_P = T^{-1}e_P$ and $S_\rho = T^{-1}(e_\rho - P/\rho^2)$. The integrating factor T is such that $S_{P\rho} = S_{\rho P}$, or $\left(T^{-1}e_P\right)_\rho = \left(T^{-1}\left(e_\rho - \rho^{-2}P\right)\right)_P$. With $\tau = \ln T$, this produces a first-order partial differential equation for τ,

$$\tau_\rho e_P - \left(e_\rho - \rho^{-2}P\right)\tau_P = \rho^{-2}. \tag{8.58}$$

This can be written in characteristic form

$$\frac{d\tau}{ds} = 1 \text{ along } \frac{d\rho}{ds} = \rho^2 e_P, \quad \frac{dP}{ds} = P - \rho^2 e_\rho, \tag{8.59}$$

which can be solved once the constitutive relation $e = e(P, \rho)$ is specified.

For an ideal gas, we have $P = \rho RT$. Hence, $dS = T^{-1}de - R d\ln\rho$ and

$$T^{-1}de = dS + R\,d\ln\rho = d(S + R\ln\rho). \tag{8.60}$$

Because the right-hand side is a total derivative, we conclude that $e = e(T)$. We now define $C_v = e'(T)$ and for the change in specific enthalpy $dh/dT = d(e + P/\rho)/dT = C_v + R = C_P$. Experimentially, C_v and C_P are constant over a wide range of temperatures, so that $e = C_v T$, $h = C_P T$. We define $\gamma = C_P/C_v = 1 + R/C_v \geq 1$ to be the polytropic index, so that $R = (\gamma - 1)C_v$ and

$$e = \frac{1}{\gamma - 1}\frac{P}{\rho}, \quad h = \frac{\gamma}{\gamma - 1}\frac{P}{\rho}. \tag{8.61}$$

We now evaluate the change in entropy as

$$dS = \frac{de}{T} + \frac{P}{T}d\left(\frac{1}{\rho}\right) = C_v\frac{dT}{T} - R\frac{d\rho}{\rho}$$

$$= C_v\left(\frac{d(P/\rho)}{P/\rho}\right) - R\frac{d\rho}{\rho} = C_v\frac{dP}{P} - (C_v + R)\frac{d\rho}{\rho}$$

$$= C_v(dP/P - \gamma d\rho/\rho).$$

Upon integration, $S = S_0 + C_v\log(P/\rho^\gamma)$, so that

$$P = e^{\frac{S}{C_v}}\rho^\gamma \tag{8.62}$$

upon choosing $S_0 = 0$. This is the equation of state for a gas of constant specific heats, i.e. a polytropic gas.

8.8 Relativistic perfect fluids

The relativistic description of dust, a pressureless medium with zero viscosity and zero thermal conductivity, is described by a stress-energy tensor of the form

$$T^{ab} = ru^b u^a, \tag{8.63}$$

where r denotes the density of the fluid as seen in the frame comoving with the fluid with velocity four-vector u^b. For example, one-dimensional motion of a perfect fluid along the x-axis introduces an energy density, and convection of energy and momentum

$$\rho = T^{tt}, \quad \dot{E} = T^{tx}, \quad \dot{P} = T^{xx}. \tag{8.64}$$

At finite temperature and pressure, but still in the approximation of zero viscosity and zero thermal conductivity, we consider

$$T^{ab} = rfu^a u^b + Pg^{ab}, \tag{8.65}$$

where the specific entropy satisfies

$$f = 1 + \frac{\gamma}{\gamma - 1} \frac{P}{r} \tag{8.66}$$

for a polytropic equation of state with polytropic index γ,

$$P = Kr^\gamma. \tag{8.67}$$

Here, K is constant along the world-lines of the fluid-elements, in the absence of shocks. The specific enthalpy takes into account the mass-energy of both internal energy e and thermal pressure P according to

$$P = (\gamma - 1)e. \tag{8.68}$$

The single fluid description (8.65) is the result of leading-order moments of the underlying momentum distribution of the particles. For particles of mass m, we have

$$r = \int d\mu_p, \quad u^b = m^{-1} \int p^b d\mu_p, \quad T^{ab} = m^{-1} \int p^a p^b d\mu_p. \tag{8.69}$$

where $d\mu_p = f(p^b)dp^x dp^y dp^z/p^t$ denotes the invariant measure for integration over momentum space. In this covariant description, the polytropic index is formally defined through the definition of f in $rf = T^{ab} u_a u_b$.

In general, we have the first law of thermodynamics

$$dP = rdf - rTdS \tag{8.70}$$

in the presence of creation of entropy dS (per baryon) at a temperature T. The adiabatic law (8.67) is a special case with $dS = 0$ when K is constant. In the presence of shocks, entropy is created and K will vary along streamlines of the fluid.

A stress-energy tensor is subject to conservation of energy and momentum,

$$\partial_a T^{ab} = 0. \tag{8.71}$$

In the case of a perfect fluid, we further have conservation of baryon number

$$\partial_a (ru^a) = 0. \tag{8.72}$$

Together with the constraint $u^2 = -1$, (8.71) and (8.72) describe a partial differential-algebraic system of six equations in the six variables (u^b, r, P). There are five physical degrees of freedom; in the adiabatic limit, wherein K in (8.67) is constant, this reduces to four dynamical degrees of freedom.

In flows with shocks, entropy is created which changes K along streamlines. In the applications of some shock capturing schemes, it may be preferred to work with the full system of equations of six equations, writing $\partial_a(\xi^a u^2) = 0$ to incorporate $u^2 = -1$ with $\xi^b = (1, 0, 0, 0)$ in the laboratory frame. Leaving the system in covariant form (with no reductions) permits covariant generalizations to ideal magnetohydrodynamics, for example.

A finite temperature gives a finite sound speed. We can calculate the wave-structure of a one-dimensional perfect fluid somewhat analogously to the calculations on compressible gas dynamics in the Newtonian limit. The energy equation $u_b \partial_a T^{ab} = 0$ is automatically satisfied in the adiabatic limit (8.67). Consider, therefore, the momentum equation $v_b \partial_a T^{ab} = 0$, where $v^b = (\sinh \lambda, \cosh \lambda)$ is orthogonal to u^b: $v_b u^b = 0$. Together with adiabaticity $dP = rdf$, the momentum equation reduces to

$$\partial_a (fu^a) = 0. \tag{8.73}$$

With baryon conservation (8.72), we have a system of two equations

$$\partial_a v^a + a_s v^a \partial_a \phi = 0, \quad \partial_a u^a + a_s^{-1} u^a \partial_a \phi = 0, \tag{8.74}$$

where $\phi = \int a_s r^{-1} dr$ and

$$a_s^2 = \frac{rdf}{fdr} = \frac{dP}{fdr}. \tag{8.75}$$

Using $\partial_a v^a = u^a \partial_a \lambda$ and $\partial_a u^a = v^a \partial_a$, equations (8.74) can be combined by addition and subtraction, to arrive at the equations of motion in characteristic form

$$(u^a \pm a_s v^a) \nabla_a (\phi \pm \lambda) = 0, \tag{8.76}$$

The structure (8.76) is that of two first-order, quasi-linear partial differential equations of the form

$$(\partial_t + w\partial_x)\psi = 0. \tag{8.77}$$

The quantity $\psi(t, x) = \Psi(x - wt)$ is a Riemann-invariant along the directions $dx/dt = w$. In the case of (8.76), the Riemann invariants are the combinations $R_\pm = \phi \pm \lambda$ along the characteristic directions

$$\left(\frac{dx}{dt}\right)_\pm = \frac{u^x \pm a_s v^x}{u^t \pm a_s v^t} = \frac{v \pm a_s}{1 \pm v a_s}. \tag{8.78}$$

In the comoving frame, where $u^b = (1, 0, 0, 0)$ and $v^b = (0, 1, 0, 0)$, the characteristic directions become

$$\left(\frac{dx}{dt}\right)_\pm = \pm a_s, \tag{8.79}$$

which shows that a_s denotes the adiabatic sound speed of the fluid.

It is of interest to also look at the non-relativistic limit, consisting of non-relativistic temperatures ($f \simeq 1$) and velocities ($\tanh \lambda \simeq \lambda$) to recover the familiar equations of compressible gas dynamics

$$(\partial_t \pm a_s \partial_x)\left(\frac{2a_s}{\gamma - 1} + v\right) = 0. \tag{8.80}$$

The relativistic addition formula of parallel velocities (8.78) reduces to the Galilean transformation $(dx/dt)_\pm = v \pm a_s$.

Exercises

1. Write solutions to Burgers' equation in parametric form $u(x, t) = F(\xi) = u_0(\xi)$, $\xi = x - t u_0(\xi)$ subject to (8.2): $du/ds = 0$ on $dx/ds = u_0(\xi)$. Verify that Burgers' equation is satisfied.

2. Sketch the solution to a simple wave in case of zero pressure – the evolution of dust, described by Burgers' equation $u_t + u u_x = 0$ in response to an initial velocity $u(0, x) = \sin(x)$.

3. The shock jump conditions on Burgers' equation are consistent with a more detailed physical model that includes viscosity, described by the viscous Burgers' equation $u_t + u u_x = \varepsilon u_{xx}$. Stationary fronts can be analyzed as traveling waves $u = f(x - ct)$. Solve for f, and obtain an expression for c.

4. Simple waves can be used to calculate the solution to an expansion fan, describing the expansion wave in a pressurized tube with a moving piston. If the piston moves to the left (Figure 8.8), the fluid to the right is subject to a change of state only through the positive characteristics that intersect the piston. If initially the fluid is in a state of rest with constant sound speed a_0, we have a constant Riemann-invariant J_- throughout the fluid: $J_- = u - 2a/(\gamma - 1) = -2a_0/(\gamma - 1)$. Consider a point $C = (x, t)$ to the right of the piston, associated with a positive characteristic which intersects the piston at $B = (\xi(t'), t')$. The fluid state at C is coupled to the initial condition at $t = 0$ by Riemann invariants along two paths: directly along a negative characteristic with intersects $t = 0$ at A_1, as well as indirectly along a positive characteristic that reflects onto the surface of the piston at B into a negative characteristic which intersects $t = 0$ at A_2. (a) Derive the equations for (u, a) at P in terms of the initial conditions, the velocity $\dot{\xi}$ of the piston and the velocity of sound a_w on its surface. (b) Show that

$$u(x, t) = \dot{\xi}(t'), \quad a(x, t) = a_0 + \frac{\gamma - 1}{2} \dot{\xi}(t'). \tag{8.81}$$

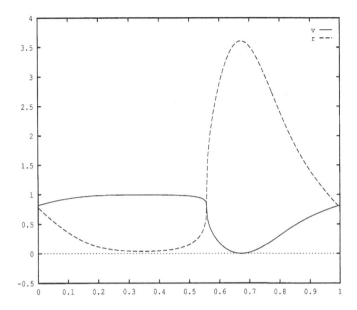

Figure 8.8 Shown is the velocity distribution v and rest mass density distribution r at the moment of breaking $t = t_B$. In this example $\lambda_0 = \lambda_1 = 7/5$, $J = 4.5$ and $t_B = 0.0963[548]$. (Reprinted from M. H. P. M. van Putten, 1991. ©1991 Springer-Verlag, Heidelberg.)

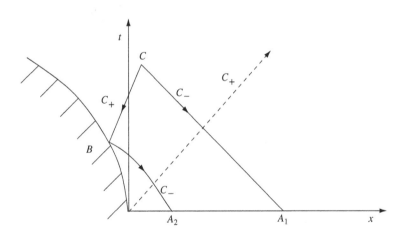

Figure 8.9 Construction in the (x, t)-plane an expansion fan in a pressurized tube with a moving piston by the method of characteristics. Tracing backwards in time over different characteristics, the fluid state at a point C is coupled to the initial data at $t = 0$ at two points A_1 and A_2. Tracing back over C_+ reaches A_1. When C is to the left of the C_+ emanating from the origin (dashed), tracing back over C_- reaches the surface of the piston at B and, upon reflection, over a C_+ reaches A_2. The two data thus propagated towards C define the local fluid velocity and sound-speed in terms of the initial data and the velocity of the piston at B.

Addition of the two equations (8.81) gives

$$C^+: \quad dx/dt = u + a = a_0 + \frac{\gamma+1}{2}\dot{\xi}(t'),\qquad (8.82)$$

showing that the C^+ are straight lines. (Note that this permits a simple prescription for $t'(t)$.) The C_- characteristics that reach B are generally curved; the expansion fan is a simple wave consisting of a divergent family of straight characteristics C_+.

5. In the limit of an instantaneous change of the piston to a constant velocity $\dot{\xi} = V < 0$ to the left, a Prandtl–Meyer expansion fan – a simple wave of diverging characteristics emanating from the origin – connects the fluid attached to the piston to the fluid at rest to the right, shown in Figure (8.8). (a) Show that the expansion fan has a constant Riemann-invariant J_-, whereby

$$\frac{dx}{dt} = a + u, \quad a = a_0 + \frac{\gamma-1}{2}u \qquad (8.83)$$

with $V < u < 0$. The expansion fan now consists of a multitude of straight characteristics C_+ all of which pass through the origin, i.e. $x/t = \text{const.}$; solving (8.83) gives

$$u = \frac{2}{\gamma+1}\left(\frac{x}{t} - a_0\right), \quad a = \frac{1}{\gamma+1}\left(\frac{x}{t} + 2a_0\right), \qquad (8.84)$$

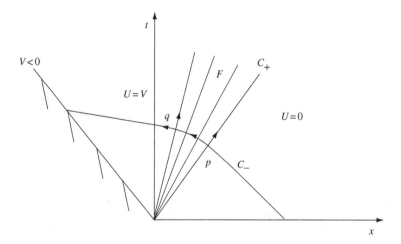

Figure 8.10 The one-dimensional problem of an expansion wave propagating into a fluid which is initially at rest, up instantaneous acceleration of a piston to a constant velocity $V < 0$ to the left. A region of uniform velocity $U = V$ attached to the piston is connected to the initial state of the fluid by an expansion wave – a simple wave for which the Riemann- invariant J_- is constant.

in the wedge $a_0 + V(\gamma+1)/2 \leq x/t \leq a_0$. (b) In the massless limit, a piston is suddenly released. What is its maximal velocity $V = -2a_0/(\gamma-1)$? This hydraulic analogue is the so-called dam-breaking problem with $\gamma = 2$.

6. Derive (8.74) from the first law of thermodynamics $TdS = de + Pdr^{-1}$.

7. The characteristic form (8.74) is due to Taub[516], originally using $\lambda = \ln\left(\frac{1+v}{1-v}\right)^{1/2}$. Verify this correspondence.

8. Use Schwarz's inequality on the definition $rf = T^{ab}u_a u_b$ to show that $\gamma \leq 5/3$ (Taub[516]). Note that $\gamma = 5/3$ is the Newtonian value of a monatomic gas.

9. Simple wave solutions are solutions in which one of the two Riemann invariants is constant throughout the fluid. Show that the special case of $\gamma = 3/2$ obtains $dx/dt = \tanh(5\lambda/4 - J/4)$ upon taking a constant Riemann-invariant $R_+ = \lambda + \phi$. Plot the solution in response to initial data $\lambda(x) = \lambda_0 + \lambda_1 \sin(2\pi x)$, using the method of characteristics, and describe the results.

10. Transverse magnetohydrodynamics describes a perfectly conducting fluid flowing along the x-direction with everywhere orthogonal magnetic field. It can be shown that the comoving specific magnetic field-strength $\kappa = h/r$ is a conserved quantity, in view of $\partial_a(hu^a) = 0$. This can be incorporated through a modified equation of state, given by $P = Kr^\gamma + \kappa^2 r^2$. Evaluate the magnetosonic sound speed.

11. The jump conditions of a gas about a shock front $\phi(t, x) = 0$ moving along the x-direction can be expressed covariantly in terms of $[F^b]v_b = 0$, where F^b is a covariant vector and $v_b = \partial_b \phi$ denotes the normal to the shock front. Apply this to T^{ab} and ru^b to derive the jump conditions. These are the relativistic Rankine–Hugoniot conditions. Show that the jump in the rest mass density across a shock is not bounded, and that the shock velocity approaches $c/\sqrt{3}$ in the ultrarelativistic limit[65, 317, 127].

12. Derive (8.46).

9

Waves in relativistic magnetohydrodynamics

"We have a habit in writing articles published in scientific journals to make the work as finished as possible, to cover up all the tracks, to not worry about the blind alleys or describe how you had the wrong idea first, and so on. So there isn't any place to publish, in a dignified manner, what you actually did in order to get to do the work."
Richard Philips Feynman (1918–88), Nobel Lecture, 1966.

Astrophysical outflows from stars, microquasars and active galactic nuclei (possible quasar remnants, D. Lynden–Bell[348, 349, 40, 615] show a prominent role of magnetic fields in rotation, radiation spectra, morphology, bright knotted structures, as well as long-term stability. Possibly, magnetic fields are relevant to the origin of these outflows (R. V. Lovelace[347] R. D. Blandford & R. L. Znajek[64], and E. S. Phinney[423]).

Extragalactic jets are observed over a broad range of wavelengths. They are luminous in radio emissions and typically display a remarkable correlation between morphology and radio luminosity, discovered by B. L. Fanaroff and J. M. Riley[178, 82, 94, 230, 114]. In their radio classification scheme, FR I sources are observed as relatively weak, two-sided, and edge-darkened with diffuse morphology, whereas FR II sources are observed as relatively strong, one-sided, edge-brightened with knotted structures terminating in a bright lobe or hot spot. Observed synchrotron emissions show preferred orientations of the magnetic field orthogonal to the jet (or a rapid transition from the source thereto) in FR I sources (e.g. 3C66B, $z = 0.0215$[251] with further polarization in the optical[200]), while parallel to the jet over an extended distance from the source in FR II sources (e.g. 3C273 at $z = 0.16$ discovered by M. Schmidt[480], reviewed in[132], QSO $0800 + 608$, $z = 0.689$[283], and 3C345, $z = 0.595$[89, 90]). Comprehensive reviews of FR I/II sources are given in[615, 11].

A few sources feature optical radio-jets: 3C66B [96, 199, 200, 284] and 3C31[96], 3C273 (T. J. Pearson $et\,al.$[412], R. C. Thomson, C. D. Mackay &

A. E. Wright[526], J. N. Bahcall *et al.*[26]), 3C346 (A. Dey & W. J. M. van Breugel [159]), M87 (J. A. Biretta, F. Zhou & F. N. Owen [57]) and PKS 1229-21 (V. Le Brun, J. Bergeron & P. Boisse[328, 329]). Radio features are typically more extended than optical emissions; in-situ particle acceleration mechanisms produce optical emitting electrons, the lifetime of emissions for which is shorter than the lower-energy radio emitting electrons. Multiwavelength observations also reveal a number of radio X-ray sources, notably Cygnus A[603]. In these jets or lobes, these X-ray emissions are probably Comptonized synchrotron emissions, (e. g. [38]).

Similar outflows on a smaller scale are seen in microquasars in our own galaxy such as GRS1915 + 105 (I. F. Mirabel and L. F. Rodríguez [378, 379, 381, 380, 464], R. M. Hjelming & M. P. Rupen[264]). These are also magnetized outflows, as studied by A. Levinson and R. D. Blandford[334]. Extragalactic jets and microquasars are both believed to be manifestations of active nuclei harboring black holes.

The most extreme ultrarelativistic sources are gamma-ray bursts. These gamma-rays are produced in the dissipation of ultrarelativistic baryon-poor outflows, probably in internal and external shocks due to time-variability and their interaction with the host environment as proposed by M. J. Rees and P. Mészáros[451, 452]. Their outflows also appear to be magnetized[126]. The observed association with supernovae notably in the observations by T. J. Galama *et al.*[224] of GRB980425/SN1998bw and by K. Z. Stanek[506] and J. Hjorth *et al.*[265] of GRB030329/SN2003dh created a new interest in the problem of understanding the relativistic hydrodynamics and magnetohydrodynamics of ultrarelativistic jets. In particular, it poses the problem of jets punching through a stellar envelope in the collapsar model of S. E. Woosley[608, 358, 87, 93].

Computer simulations of extragalactic jets and compact symmetric sources have been studied in higher dimensional simulations by parallel computing in various approximations. For relativistic hydrodynamical jets see[549, 165, 363, 217, 218, 219] and for relativistic magnetohydrodynamical jets, see[555, 389, 300, 301, 302, 370]. The reader is further referred to reviews[362, 220], and simulations of ultrarelativistic jets in gamma-ray bursts[592].

Large-scale computing of relativistic fluids in the presence of magnetic fields requires an accurate and stable numerical implementation of the equations of ideal magnetohydrodynamics. This includes the condition of maintaining a divergence-free magnetic field and allowing for the formation of shocks. The original covariant formulations of ideal magnetohydrodynamics are due to Y. Choquet-Bruhat[115, 116], and A. Lichnerowicz[343]. This formulation comprises a partial-differential algebraic system of equations.

A covariant hyperbolic formulation of magnetohydrodynamics consisting of conservation laws without algebraic constraints can be given by including the constraints as conserved quantities[548]. This belongs to a broader class of covariant hyperbolic formulations, including Yang–Mills magnetohydrodynamics in SU(N)[551, 553, 117, 118]. Hyperbolic formulations provided a suitable starting point for shock-capturing schemes by linear smoothing[555]. Linear smoothing preserves divergence-free magnetic fields to within machine round-off error[554]. This covariant hyperbolic formulation also serves as a starting point for characteristic-based shock-capturing schemes[303].

In this chapter, we study the infinitesimal wave-structure and well-posedness of the covariant hyperbolic formulation of relativistic magnetohydrodynamics. In the limit of weak magnetic fields, the slow magnetosonic and Alfvén waves are found to bifurcate from the contact discontinuity (entropy waves), while the fast magnetosonic wave is a regular perturbation of the hydrodynamical sound speed. The infinitesimal wave-structure of relativistic magnetohydrodynamicshas been considered previously by A. M. Anile[16], in particular that of Alfvén waves by S. S. Kommissarov[304]. The well-posedness proof presented here is new, based on an extension of the Friedrichs–Lax symmetrization procedure to Yang–Mills magnetohydrodynamics[551, 553].

We conclude with a simulation on a Stagnation-point Nozzle Mach disk morphology in a low-density, relativistic magnetized jet.

9.1 Ideal magnetohydrodynamics

A perfectly conducting fluid carries electric currents without dissipation, whereby magnetic diffusivity vanishes. Interactions between the fluid and the magnetic field energy are hereby *conservative*.

A conservative action on the magnetic field energy is instructive, as in the following example. Consider a magnetized perfectly conducting disk of fluid. Compression of the fluid in the radial direction shrinks the disk in surface area. Apart from work applied to the fluid against hydrostatic pressure, a change dA in surface area performs work against magnetic pressure $P_B = B^2/8\pi$ which alters the magnetic field-energy. When this process is conservative, the change in enclosed magnetic field-energy is related to the magnetic field energy density according to $d(AB^2/8\pi) = -(1/8\pi)B^2 dA$. Hence, we have

$$\frac{1}{8\pi}B^2 dA + \frac{1}{4\pi}AB \, dB = -\frac{1}{8\pi}B^2 \, dA. \tag{9.1}$$

It follows that $d\Phi = BdA + AdB = 0$, and hence the magnetic flux

$$\Phi = AB \tag{9.2}$$

is conserved. Here, conservation of magnetic flux has the same form as conservation of mass: $Ar = \text{const.}$ where r denotes the rest mass density of the fluid (measured in the comoving frame).

The above illustrates *transverse* magnetohydrodynamics, wherein the ratio of magnetic flux per unit mass $\kappa = B/r$ is constant along the world-lines of fluid elements. It follows that transverse magnetohydrodynamics is equivalent to hydrodynamics in the presence of a modified equation of state

$$P = Kr^\gamma + \frac{\kappa}{8\pi} r^2, \tag{9.3}$$

where $P = Kr^\gamma$ describes the polytropic equation of state of the unmagnetized fluid.

9.2 A covariant hyperbolic formulation

Ideal magnetohydrodynamics describes an inviscid, perfectly conductive plasma in a single fluid description with velocity four-vector, $u^b(u^c u_c = -1)$. It is given by the equations of energy-momentum conservation,

$$\nabla_a T^{ab} = 0, \tag{9.4}$$

where T^{ab} is the stress-energy tensor of both the fluid and the electromagnetic field, Faraday's equations, $\nabla_a(u^{[a}h^{b]}) = 0$ subject to $u^c h_c = 0$, and conservation, $\nabla_a(ru^a) = 0$, of baryon number, r. For a polytropic equation of state with polytropic index γ, we have

$$T^{ab} = \left(r + \frac{\gamma}{\gamma-1}\frac{P}{r} + h^2\right) u^a u^b + \left(P + h^2/2\right) g^{ab} - h^a h^b, \tag{9.5}$$

where P is the hydrostatic pressure and g^{ab} is the metric tensor.

As described in Chapter 4, a constraint $c = 0$ and a four-divergence $\nabla^a \omega_{ab} = 0$ representing Faraday's equations can be combined according to

$$\nabla^a(\omega_{ab} + \lambda g_{ab}c) = 0, \quad \lambda \neq 0. \tag{9.6}$$

In an initial value problem with physical initial data, (9.6) conserves $c = 0$ in the future domain of dependence of the initial hypersurface[548]. From an algebraic point of view, (9.6) allows any choice of $\lambda \neq 0$. Applied to ideal magnetohydrodynamics, the questions are that of deriving right nullvectors of the characteristic matrix and establishing well-posedness. Remarkably, *both* analyses agree in their preferred choice: $\lambda = 1$.

The linear combination (9.6) establishes a rank-one update to its Jacobian. Symmetry conditions of the Jacobian may enter a particular choice of λ. In case of

$$\lambda = 1 \tag{9.7}$$

it follows that

$$
\begin{cases}
\nabla_a T^{ab} = 0, \\
-\nabla_a (h^{[a} u^{b]} + g^{ab} u^c h_c) = 0, \\
\nabla_a (r u^a) = 0, \\
\nabla_a \left(\xi^a (u^2 + 1) \right) = 0,
\end{cases}
\tag{9.8}
$$

where ξ is any time-like vector field and $U = (u^b, h^b, r, P)$. The minus sign in front of the present linear combination is chosen also in regards to the structure of the Jacobian of (9.8). This will be made explicit below.

Expanding (9.8) gives the system

$$
A^a \partial_a U + \cdots = 0,
\tag{9.9}
$$

where the matrices $A_B^{aA} = A_B^{aA}(U) = \frac{\partial F^{aA}}{\partial U^B}$ are 10 by 10, and the dots refer coupling terms to the Christoffel symbols. The infinitesimal wave-structure is given by characteristic wave-fronts at given U (since the A^a are tensors). The simple wave ansatz $U = U(\phi)$ obtains

$$
A^a \partial_a \phi U' + \cdots = 0.
\tag{9.10}
$$

These wave-fronts are characteristic surfaces, whenever the matrix $A^a \partial_a \phi$ is singular. The directions $\nu_a = \partial_a \phi$ then are the normals to these surfaces. Small amplitude simple waves are described by the relative perturbations of the physical quantities, given by right nullvectors R of $A^a \nu_a$. Thus, simple waves moving along the x-direction satisfy

$$
\left((A^t)^{-1} A^x - v \right) R = 0,
\tag{9.11}
$$

where v is the velocity of propagation.

The covariant hyperbolic formulation provides an embedding of the theory of ideal magnetohydrodynamics in ten partial differential equations. The original algebraic constraints are embedded as conserved quantities. This system propagates physical initial data without exiting non-physical wave-modes. Physical waves (entropy waves, Alfvén and magnetohydrodynamic waves) all exist inside the light cone. This ensures causality under appropriate conditions on the equation of state.

The addition of $g^{ab} u^c h_c$ to Faraday's equations provides a rank-one update to the characteristic matrix $A^c \nu_c$. On the light cone, we have $\nu^2 = 0$, and this linear combination no longer regularizes the characteristic determinant. (This results from insisting on covariance in the divergence formulation.) Attempts to discuss the covariant hyperbolic system of magnetohydrodynamics outside the context of the initial value problem with physical initial data[304] erroneously infer the presence of nonphysical wave-modes.

9.3 Characteristic determinant

Small amplitude waves are described by linearized equations,

$$A^{aA}(U) = \frac{\partial F^{aA}}{\partial U^B} v_a = \frac{\partial F^{aA} v_a}{\partial U^B}. \tag{9.12}$$

With total energy-density $\rho = r + \frac{\gamma}{\gamma-1}P + h^2 = rf + h^2$, they are

$$F^{cA}v_c = \begin{cases} \rho(u^c v_c)u^a + (P + h^2/2)v^a - (h^c v_c)h^a, \\ -\{(h^c v_c)u^a - (u^c v_c)h^a + v^a u^c h_c\}, \\ r(u^c v_c), \\ (\xi^c v_c)(u^2 + 1). \end{cases} \tag{9.13}$$

This system of 10×10 equations for $U^B = (u^b, h^b, r, P)$ can be reduced to 8×8 in the variables $V^B = (v^s, h^b, r)$ by expressing u^b in terms of the spatial three-velocity $u^b = \Gamma(1, v^s)$, $\Gamma = 1/\sqrt{1 - v^2}(1, v^s)$, $s = 1, 2, 3$. Note that small-amplitude wave-motion conserves entropy, so that $dP = \gamma \frac{P}{r}dr$. In V^B, the equation of energy conservation, $\nabla_a T^{at} = 0$ and the last equation of (9.8) are automatically satisfied, whence they can be ignored. In what follows, A^a shall denote the resulting 8×8 matrix, obtained from the original 10×10 matrix by deletion of the first and last row, addition of the last column (multiplied by $\gamma P/r$) to the one-but-last column (associated with r), followed by deletion of the first and last columns.

The linearized wave-structure is defined by the characteristic problem

$$A^c v_c z = 0 \tag{9.14}$$

for the right null-vectors $z = U'$. Without loss of generality, (9.14) can be studied in a comoving frame, in which $u^b = (1, 0, 0, 0)$. In this event, $\Gamma = 1$ and $\partial \Gamma / \partial v^s = 0$. Furthermore, the x-axis of the local coordinate system can be aligned with the magnetic field, so that $h^b = (0, H, 0, 0)$. Given the two orientations u^s and h^b, the wave-structure is rotationally symmetric about the x-axis, and hence v_y and v_z act symmetrically as $\sqrt{v_y^2 + v_z^2}$; we will put $v_z = 0$. For $A^c v_c$, we have

$$\begin{bmatrix} \rho v_1 & 0 & 0 & -v_1 H & -Hv_2 & -Hv_3 & 0 & \frac{\gamma P v_2}{r} \\ 0 & \rho v_1 & 0 & 0 & Hv_3 & -Hv_2 & 0 & \frac{\gamma P v_3}{r} \\ 0 & 0 & \rho v_1 & 0 & 0 & 0 & -Hv_2 & 0 \\ v_1 H & 0 & 0 & -v_1 & -v_2 & -v_3 & 0 & 0 \\ -Hv_2 & Hv_3 & 0 & v_2 & v_1 & 0 & 0 & 0 \\ -Hv_3 & -Hv_2 & 0 & v_3 & 0 & v_1 & 0 & 0 \\ 0 & 0 & -Hv_2 & 0 & 0 & 0 & v_1 & 0 \\ rv_2 & rv_3 & 0 & 0 & 0 & 0 & 0 & v_1 \end{bmatrix}. \tag{9.15}$$

Note that the lower diagonal block is ν_1 times the 4×4 identity matrix – a result from the sign choice in the given combination of Faraday's equations and the constraint in (9.8) and (9.13).

The third and seventh rows and columns act independently to give rise to the Alfvén waves. The remaining waves are described by the reduced problem

$$(A^c \nu_c)' z' = 0, \tag{9.16}$$

where $(A^c \nu_c)'$ is obtained from $A^c \nu_c$ by deleting the third and seventh rows and columns, thereby obtaining a problem in the 6-dimensional variable z'. Introducing

$$z' = \begin{pmatrix} x \\ y \end{pmatrix}, \tag{9.17}$$

(9.14) takes the form of a coupled system of 3×3 equations

$$\nu_1 Z x + X y = 0, \quad Y x + \nu_1 y = 0, \tag{9.18}$$

in which

$$Z = \begin{bmatrix} \rho & 0 & -H \\ 0 & \rho & 0 \\ H & 0 & -1 \end{bmatrix}, \quad X = \begin{bmatrix} -H\nu_2 & -H\nu_3 & \dfrac{\gamma P \nu_2}{r} \\ H\nu_3 & -H\nu_2 & \dfrac{\gamma P \nu_3}{r} \\ -\nu_2 & -\nu_3 & 0 \end{bmatrix},$$

$$Y = \begin{bmatrix} -H\nu_2 & H\nu_3 & \nu_2 \\ -H\nu_3 & -H\nu_2 & \nu_3 \\ r\nu_2 & r\nu_3 & 0 \end{bmatrix}. \tag{9.19}$$

There remains a single 3×3 eigenvalue problem in x,

$$XYx = \nu_1^2 Z x \quad \Leftrightarrow \quad Z^{-1} XY x = \nu_1^2 x. \tag{9.20}$$

Here, $Z^{-1} XY - \nu_1^2$ is given by the matrix

$$\begin{bmatrix} W_{1,1} & W_{1,2} & 0 \\ W_{2,1} & W_{2,2} & 0 \\ \dfrac{H \left(\gamma P \nu_2{}^2 - rf\,\nu_2{}^2 - rf\,\nu_3{}^2 \right)}{rf} & \dfrac{H \gamma P \nu_2 \nu_3}{rf} & \nu_2{}^2 + \nu_3{}^2 - \nu_1{}^2 \end{bmatrix} \tag{9.21}$$

where the upper diagonal 2×2 matrix W is given by

$$W = \begin{bmatrix} \dfrac{\gamma P \nu_2{}^2}{rf} - \nu_1{}^2 & \dfrac{\gamma P \nu_2 \nu_3}{rf} \\ \dfrac{\gamma P \nu_2 \nu_3}{rf + H^2} & \dfrac{H^2 \nu_3{}^2 + H^2 \nu_2{}^2 + \gamma P \nu_3{}^2}{rf + H^2} - \nu_1{}^2 \end{bmatrix}. \tag{9.22}$$

Here, the two zeros in the third column of (9.21) result from $\lambda = 1$.

Upon substitution $v_3^2 = v^2 + v_1^2 - v_2^2$, the determinant assumes the covariant expression

$$\rho \det W = (rf - \gamma P)(u^c v_c)^4 - (h^2 + \gamma P)v^2(u^c v_c)^2 + \frac{\gamma P}{rf}(h^c v_c)^2 v^2. \qquad (9.23)$$

The fact that $\det W$ is *not* identically equal to zero is a consequence of the rank-one update by addition of the constraint $c = 0$ to Faraday's equations.

9.4 Small amplitude waves

The small amplitude waves are determined by the roots of the characteristic determinant (9.23).

Alfvén waves. The eigenvalues for the Alfvén waves are given by $v_1 = \pm|h^c v_c|/\sqrt{\rho}$ with nullvector $z = (0, 0, Hv_2, 0, 0, 0, \rho v_1, 0)^T$, associated with Alfvén waves. Covariantly, we have

$$U^A = (v^a, \pm\sqrt{\rho}v^a, 0, 0)^T, \qquad (9.24)$$

where v_a may be taken to be $H(0, 0, v_4, -v_3) = \epsilon_{abcd}u^b h^c v^d \equiv v_a$. Thus, the Alfvén wave is a transversal in which h^2 is conserved (δh^b is orthogonal to h^b) and $\delta r = 0$.

Magnetohydrodynamic waves. The eigenvalues for the magnetohydrodynamic waves are given by the roots of the characteristic determinant in (9.23). Writing $n^b = v^b + (u^c v_c)u^c$, we have $v^2 = -t^2 + n^2$, $t = u^c v_c$, $n^2 = n^c n_c$. Let $\alpha = \frac{rf}{\gamma P}$ and $\beta = \frac{h^2}{\gamma P}$. Then

$$\frac{(h^c v_c)^2}{rfn^2} = \frac{\beta}{\alpha}\frac{(h^c n_c)^2}{h^2 n^2} \equiv \frac{\beta}{\alpha}\cos^2\phi. \qquad (9.25)$$

Consequently, (9.23) becomes

$$(\alpha - 1)v^4 - (1 + \beta)v^2(1 - v^2) + \beta\alpha^{-1}\cos^2\phi(1 - v^2) = 0, \qquad (9.26)$$

where $v^2 = t^2/n^2$. (9.26) has real solutions v for any given n^b, whenever

$$(\alpha + \beta)v^4 - (1 + \beta + \beta\alpha^{-1})v^2 + \beta\alpha^{-1} = 0 \qquad (9.27)$$

has real solutions v. But (9.27) has discriminant

$$D = (\alpha + \beta - \alpha\beta)^2 \geq 0. \qquad (9.28)$$

Weak magnetic fields are described by small β expansions as follows. Fast magnetosonic waves are a regular perturbation of sound waves in pure hydrodynamics, while the Alfvén and slow magnetosonic waves bifurcate from entropy waves (contact discontinuities), whose propagation velocities satisfy

$$v_f^2/v_h^2 \sim 1 + \beta \frac{\alpha - 1}{\alpha} \sin^2 \phi + O(\beta^2),$$

$$v_A^2/v_h^2 \sim \beta \cos^2 \phi [1 - \beta \alpha^{-1} + O(\beta^2)], \qquad (9.29)$$

$$v_s^2/v_h^2 \sim \beta \cos^2 \phi [1 - \beta (1 - \frac{\alpha - 1}{\alpha} \cos^2 \phi) + O(\beta^2)],$$

where $v_h^2 = \alpha^{-1}$ is the square of the hydrodynamical velocity, and which obey the inequalities

$$v_s^2 \le v_A^2 \le v_f^2. \qquad (9.30)$$

Inequalities of (9.30) remain valid for general β, e.g. J. Bazer and W. B. Ericson[37]; A. Lichnerowicz[343]; A. M. Anile[16].

9.5 Right nullvectors

Inspection of (9.22), together with (9.18), shows the nullvector

$$z = \begin{pmatrix} \nu_1 \nu_2 \nu_3^2 \\ -\nu_1 \nu_3 (\nu_2^2 - \alpha \nu_1^2) \\ 0 \\ H\nu_1 \nu_2 \nu_3^2 \\ H\nu_3^2 (\nu_2^2 - \alpha \nu_1^2) \\ -H\nu_2 \nu_3 (\nu_2^2 - \alpha \nu_1^2) \\ 0 \\ -\alpha r \nu_3^2 \nu_1^2 \end{pmatrix}. \qquad (9.31)$$

Of course, (9.31) can be stated covariantly by noting that $H^2 = h^2$, $H\nu_2 = h^c \nu_c$, $\nu_1 = u^c \nu_c$,

$$H^2 (\nu_2^2 - \alpha \nu_1^2) = (h^c \nu_c)^2 - \alpha h^2 (u^c \nu_c)^2 \equiv h^2 k_1, \qquad (9.32)$$

and introducing

$$H(0, \nu_4^2 + \nu_3^2, -\nu_2 \nu_3, -\nu_2 \nu_4)^T = \epsilon_{abcd} u^b \nu^c \nu^d \equiv w_a. \qquad (9.33)$$

Since $-\alpha r \nu_3^2 \nu_1^2$ is a scalar, ν^3 is to be treated as

$$H^2 (\nu_3^2 + \nu_4^2) = h^2 n^2 - (h^c \nu_c)^2 \equiv h^2 k_2, \qquad (9.34)$$

were $n_a = \nu_a + (u^c \nu_c) u_a$. Note that

$$k_1 = n^2 (\cos^2 \phi - \alpha v^2), \quad k_2 = n^2 \sin^2 \phi, \qquad (9.35)$$

where $v = v_S, v_f$. Clearly, z is formed from

$$\delta u^b = -t(k_1 n^b - (k_2 + k_1)(\hat{h}^c n_c)\hat{h}^b)$$
$$\delta h^b = k_1 w^b + k_2 t(h^c n_c)u^b,$$
$$\delta r = -\alpha r k_2 t^2,$$
$$\delta P = -r f k_2 t^2,$$
(9.36)

where $\hat{h}^b = h^b/|h|$, and

$$v_a = \epsilon_{abcd} u^b h^c v^d, \quad w_a = \epsilon_{abcd} u^b v^c v^d.$$
(9.37)

We thus have the following. Given a unit vector n^b orthogonal to u^b, and a root $v^b = n^b + vu^b$, $v = u^c v_c$ of (9.27), the right nullvectors for the hydrodynamical waves of (9.14), $U^A = (\delta u^b, \delta h^b, \delta r, \delta P)$, are

$$\delta u^b = v\left[\sin^2 \phi n^b - (1 - \alpha v^2)(n^b - \cos \phi \hat{h}^b)\right],$$
$$\delta h^b = |h|\left[(\cos^2 \phi - \alpha v^2)\tilde{w}^b + v\sin^2 \phi \cos \phi u^b\right],$$
$$\delta r = -v^2 \alpha r \sin^2 \phi,$$
$$\delta P = -v^2 r f \sin^2 \phi.$$
(9.38)

where $\tilde{w}^b = w^b/|h|$.

A. M. Anile[16] gives a different form of otherwise the same right nullvectors. Our preceding weak magnetic field-limit shows that

$$\cos^2 \phi - \alpha v_f^2 < 0, \quad \cos^2 \phi - \alpha v_s^2 > 0$$
(9.39)

for fast, respectively slow magnetosonic waves.

Inspection of (9.33) shows that the tangential component of the magnetic field is strengthened in fast magnetosonic waves, while it is weakened in slow magnetosonic waves. This distinguishing aspect of fast and slow magnetosonic waves was first noted by J. Bazer and W. B. Ericson[37] in their analysis of shocks in non-relativistic magnetohydrodynamics.

The limit of small β is of particular interest to computation, as when a magnetized fluid streams into a nearly unmagnetized environment. A characteristics-based scheme is to treat a large dynamic range in β. A full set of right nullvectors, including those of contact discontinuities, obtains for nonzero β. The limiting behavior of these nullvectors is somewhat nontrivial as β becomes small. In what follows, we consider small β in the sense of small $|h|/\sqrt{\gamma P}$, while keeping the direction $\hat{h}^b$ constant. Thus,

$$1 - \alpha v^2 \sim -\beta \frac{\alpha - 1}{\alpha} \sin^2 \phi + O(\beta^2),$$

$$1 - \alpha v^2 \sim 1 + O(\beta) \tag{9.40}$$

for the fast and slow magnetosonic speeds, respectively. It follows that in the limit of low magnetic field-strength, the fast magnetosonic waves are described by the right nullvectors

$$\delta u^b = v_f n^b + \beta \frac{\alpha - 1}{\alpha} (n^b - \cos \phi \hat{h}^b) v_f + O(\beta^2),$$

$$\delta h^b = |h|(-\tilde{w}^b + v_f \cos \phi u^b) + \beta \frac{\alpha - 1}{\alpha} w^b + O(\beta^2),$$

$$\delta r = -v_f^2 \alpha r, \tag{9.41}$$

$$\delta P = -v_f^2 r f,$$

and the slow magnetosonic waves by

$$\delta u^b = \cos \phi (\hat{h}^b - \cos \phi n^b) + O(\beta),$$

$$\delta h^b = \sqrt{\gamma P}(\cos \phi \tilde{w}^b + v_s \sin^2 \phi u^b) + O(\beta),$$

$$\delta r = -v_s \alpha r \sin^2 \phi, \tag{9.42}$$

$$\delta P = -v_s \alpha r f \sin^2 \phi.$$

The small β limit of the nullvectors can now be normalized.

9.5.1 Bifurcations from entropy waves

The behavior of the nullvectors in the limit of weak magnetic fields can be derived from (9.24) and (9.41–9.42). To this end, note that

$$v^a = |h|\tilde{v}^a = \sin \phi |h| \hat{v}^a, \tag{9.43}$$

where $\hat{v}^c \hat{v}_c = 1$, and ϕ denotes the angle between n^c and h^c,

$$n^b = \cos \phi \hat{h}^b + \sin \phi y^b, \tag{9.44}$$

$y^c u_c = h^c y_c = 0$, $y^c y_c = 1$ (n^b is normalized to be unit, as in the assumptions of (9.38). It follows that the Alfvén nullvectors can be normalized to

$$\delta \hat{U}^A = (\hat{v}^a, \pm\sqrt{\rho}\hat{v}^a, 0, 0). \tag{9.45}$$

In the limit of vanishingly small β, the pair of slow magnetosonic waves collapses to the single normalized nullvector

$$\delta \hat{U}^A = (y^b, \sqrt{\gamma P} y^b, 0, 0). \tag{9.46}$$

Note that $y^c \hat{v}_c = 0$, so that (9.45) and (9.46) are independent. Division by $\sin \phi$ thus provides a normalization of the original expressions (9.24) and (9.42).

The right nullvector associated with entropy waves $(u^c v_c = 0)$ is

$$\delta U^A = (0, 0, \delta r, 0) \tag{9.47}$$

if $h^c v_c \neq 0$, and

$$(0, \delta h^c, \delta r, \delta P), \quad (\delta u^c, 0, 0, 0), \tag{9.48}$$

if $h^c v_c = 0$, subject to

$$\delta P + h_c \delta h^c = 0, \quad v_c \delta h^c = 0, \quad v_c \delta u^c = 0. \tag{9.49}$$

The second case refers to transverse magnetohydrodynamics for which there holds continuity of total pressure, zero orthogonal magnetic field and transverse velocity. Note that transverse magnetohydrodynamics has two right nullvectors, similar to the case of pure hydrodynamics. With the exception of transverse magnetohydrodynamics, therefore, the contact discontinuity provides one right nullvector.

Transverse magnetohydrodynamics and pure hydrodynamics allow for shear along contact discontinuities. This gives rise to the two independent right nullvectors. Whenever magnetic field-lines cross a contact discontinuity, however, persistent coupling to the magnetic field-lines in ideal magnetohydrodynamics prohibits shear. Ideal magnetohydrodynamics responds to suppression of the original two-dimensional degree of freedom in shear with two new wave-modes. These two new wave-modes are the Alfvén wave and the slow magnetosonic wave. These two modes are distinct, as shown by (9.45) and (9.46). The Alfvén and slow magnetosonic wave may be regarded as one pair, bifurcating from the contact discontinuity (see, for example, Figure 6 of[550]). Conversely, *the limit of vanishing β recovers the two shear modes from the independent Alfvén and slow magnetosonic waves*. The Alfvén wave is purely rotational, while the slow magnetosonic wave is slightly helical, including a longitudinal variation of $\pm v_s \sin^2 \phi = \pm \beta \sin^2 \phi \cos \phi$. The fast magnetosonic wave remains a regular perturbation of the ordinary sound wave.

The weak magnetic field-limit thus obtains two right nullvectors from the fast magnetosonic waves, two from the Alfvén waves, one from the slow magnetosonic waves and generally one from the contact discontinuity – a total of six. This leaves an apparent degeneracy of one.

The degeneracy stems from neighboring to $O(v_S)$ of the two nullvectors of the slow magnetosonic waves. This would suggest ill-posedness to this order in projections. However, characteristic-based methods consider the product of the projections on the nullvectors *and* the associated eigenvectors. In the present case, therefore, the order of the degeneracy is precisely cancelled by multiplication with the eigenvalue v_S, which is computationally stable. The limit of arbitrarily small β *in the application of characteristic-based methods* is computationally well-posed.

9.6 Well-posedness

The theory of ideal relativistic magnetohydrodynamics was first shown to be well-posed by K. O. Friedrichs[203]. This proof is based on the Friedrichs–Lax symmetrization procedure[204]. The problem of constraints was circumvented by reduction of variables. The symmetrization procedure of Friedrichs[203] and P. D. Lax[204] applies to hyperbolic systems of equations of the form

$$\nabla_a F^{aB} = f^B \tag{9.50}$$

in the presence of a certain convexity condition. Constraints can be treated also by an extension of the Friedrichs–Lax symmetrization procedure with no need for an additional reduction of variables, by extending the linear combination used in the covariant hyperbolic formulation of ideal magnetohydrodynamics to Yang–Mills magnetohydrodynamics in SU(N)[551, 553]. Once in symmetric hyperbolic form, well-posedness results from standard energy arguments, e.g. Fischer and Marsden[189]). The main arguments of symmetrization in the presence of constraints are briefly recalled here, to highlight the same linear combination of (9.8), now from the point of view of well-posedness.

9.6.1 Symmetrization with constraints

Variations δV^A of (u^b, h^b, r, P) can either be unconstrained with respect to all ten degrees of freedom, or constrained, i.e. those variations obeying the algebraic constraints. For example, $\delta c \neq 0$ results from a total variation, while $\delta c = 0$ represents a constraint variation.

Symmetrization in the presence of constraints follows if there exists a vector field W_A which produces a total derivative in the modified main dependency relation[551, 553]

$$\text{YI}: \quad W_A \delta F^{aA} \equiv \delta z^a, \tag{9.51}$$

and which obtains constrained positive-definiteness in

$$\text{YII}: \quad \delta W_A \delta F^{aA} \xi_a > 0 \tag{9.52}$$

for some time-like vector ξ^a. Of course, the source terms f^B must satisfy the consistency condition

$$W_A f^A = 0 \tag{9.53}$$

whenever the constraints are satisfied. Allowing a nonzero total derivative in YI defines an extension to the Friedrichs–Lax symmetrization procedure[204].

Differentiation by V^C of the unconstraint identity YI obtains

$$\frac{\partial W_A}{\partial V^C} \frac{\partial F^{aA}}{\partial V^D} \nabla_a V^D + \frac{W_A \partial^2 F^{aA}}{\partial V^C \partial V^D} \nabla_a V^D = \frac{\partial^2 z}{\partial V^C \partial V^D} \nabla_a V^D. \tag{9.54}$$

This establishes symmetry of the matrices

$$A^a_{CD} = \frac{\partial W_A}{\partial V^C} \frac{\partial F^{aA}}{\partial V^D}. \tag{9.55}$$

Also,

$$\delta V^C A^a_{CD} \xi_a \delta V^D = \left(\delta V^C \frac{\partial W_A}{\partial V^C} \right) \left(\frac{\partial F^{aA} \xi_a}{\partial V^D} \delta V^D \right) = \delta W_A \delta F^{aA} \xi_a > 0 \tag{9.56}$$

for all constraint variations δV^A. Of course, given V^A, the constraint variations δV^A define a linear subspace $\mathcal{V}$ of dimension $N - m$, where m is the number of constraints $c = 0$, each giving rise to

$$0 = \delta c = \frac{\partial c}{\partial V^A} \delta V^A. \tag{9.57}$$

We have the following construction[551, 553]: *Given a real-symmetric $A \in \mathcal{L}(\mathbf{R}^n, \mathbf{R}^n)$ which is positive definite on a linear subspace $\mathcal{V} \subset \mathbf{R}^n$, there exists a real-symmetric, positive definite $A^* \in \mathcal{L}(\mathbf{R}^n, \mathbf{R}^n)$ such that*

$$A^* y = Ay (y \epsilon \mathcal{V}). \tag{9.58}$$

This may be seen as follows. Consider $A^* = A + \mu x^T x$, where x is a unit element from $V^\perp$. Then A^* is symmetric positive definite on $V' = \{z = y + \lambda x | y \epsilon V, \lambda \epsilon R\}$: $z^T A^T z \geq c' ||z||^2 = c' (||y||^2 + \lambda^2 ||x||^2)$ with $c' > 0$ upon choosing $\mu > M$, where $M = ||A||$ denotes the norm of A. This construction may be repeated until $V^\perp$ is exhausted, leaving A^* symmetric-positive-definite on $\mathbf{R}^n$ as an embedding of A on V.

The real-symmetric matrix $A^a_{CD} \xi_a$ is positive definite on the subspace of constrained variations $\mathcal{V}$; let $(A^a_{CD} \xi_a)^*$ be the positive definite, symmetric matrix obtained from the above. It follows that solutions to (9.50) (and its constraints) satisfy the *symmetric positive definite* system of equations

$$-(A^{aAB})^* \xi_a(\xi^c \nabla_c) V_A + A^{aAB} (\nabla_\Sigma)_a V_A = f^B, \tag{9.59}$$

where

$$\nabla_a = -\xi_a(\xi^c \nabla_c) + (\nabla_\Sigma)_a. \tag{9.60}$$

It remains to show that ideal MHD satisfies properties YI and YII.

9.6.2 Symmetrization of hydrodynamics

Relativistic hydrodynamics is symmetrizable, according to K. O. Friedrichs[203], T. Ruggeri and A. Strumia[467], and A. M. Anile[16]. They use the equations in the form

$$\nabla_a F_f^{aA} = \begin{cases} \nabla_a(rf u^a u^b + P g^{ab}) = 0, \\ \nabla_a(r u^a) = 0, \\ \nabla_a(r S u^a) = 0 \end{cases} \tag{9.61}$$

away from entropy-generating shocks. Then $W_A^f = (u_a, f - TS, T)$ and $V_C^f = (v_\alpha, T, f)$ with a reduction of variables on the velocity four-vector by $u^b = \Gamma(1, v^\alpha)$, where Γ is the Lorentz factor. With F_f^{aA} denoting the fluid dynamical equations $\nabla_a T_f^{ab} = 0$, $T_f^{ab} = rf u^a u^b + P g^{ab}$ with f the specific enthalpy, and $\nabla_a(r u^a) = 0$, it has been shown that[467, 16]

$$W_A^f \delta F_f^{aA} \equiv 0, \quad Q_f = \delta W_A \delta F_f^{aA} \xi_a > 0 \tag{9.62}$$

provided that the free enthalpy

$$G(T, P) = f - TS - 1 \tag{9.63}$$

is concave, and the sound velocity is less than the speed of light. Under these conditions, the hydrodynamical equations by themselves satisfy YI and YII. In fact, they satisfy the original homogeneous Friedrichs–Lax conditions CI and CII of Friedrichs and Lax[204], and hence they satisfy a symmetric hyperbolic system of equations.

9.6.3 Symmetrization of ideal MHD

In what follows, we set

$$\omega_{ab} = h^a u^b - u^a h^b + g^{ab} u^c h_c, \quad T_m^{ab} = h^2 u^a u^b + \frac{1}{2} h^2 g^{ab} - h^a h^b. \tag{9.64}$$

We then have the expansions

$$u_b \delta T_m^{ab} = u_b(h^2 u^a \delta u^b + h^2 u^b \delta u^a + 2u^a u^b h_c \delta h^c$$
$$+ g^{ab} h_c \delta h^c - h^a \delta h^b - h^b \delta h^a)$$
$$= -h^2 \delta u^a - u^a(h_c \delta h^c) - h^a(u_c \delta h^c) - c\delta h^a, \quad (9.65)$$
$$h_b \delta \omega^{ab} = h_b(h^a \delta u^b + u^b \delta h^a - h^b \delta u^a - u^a \delta h^b + g^{ab} \delta c)$$
$$= h^a(h_c \delta u^c) + c\delta h^a - h^2 \delta u^a - u^a(h_c \delta h^c) + h^a \delta c.$$

The above gives the identity

$$u_b \delta T_m^{ab} - h_b \delta \omega^{ab} \equiv \delta z^a, \quad (9.66)$$

where $z^a = -2h^a c$. It follows that the total derivative in (9.66) results from the unique linear combination $\omega^{ab} = h^a u^b - h^b u^a + g^{ab} c$, as in (9.8).

With $W_A = (u_a, h_a, f - TS, S)$ and F^{aA} given by (9.8) (rewritten according to (9.61)), it further follows that

$$W_A \delta(F_f^{aA} + F_m^{aA}) \equiv \delta z^a. \quad (9.67)$$

A similar calculation[551, 553] shows that the quadratic of constrained variations Q_m given by

$$\delta u_b \delta T_m^{ab} \xi_a - \delta h_b \delta \omega^{ab} \xi_a = (u^c \xi_c)[h^2(\delta u)^2 + (\delta h)^2]$$
$$+ 2[(\xi_c \delta u^c)(h_c \delta h^c) - (h^c \xi_c)(\delta u_c \delta h^c)] \quad (9.68)$$

is positive-definite (for $\delta h^a \neq 0$). Therefore, the sum

$$Q = \delta W_A \delta F^{aA} \xi_a = Q_f + Q_m \quad (9.69)$$

is *constrained positive-definite*, whenever Q_f is such (with respect to the fluid dynamical variables). It follows that both YI and YII are satisfied (with $W_A = (u_a, h_a, f - TS, S)$ and $V_A = (v_\alpha, h_a, T, f)$), and hence physical solutions to (9.8) satisfy the symmetric hyperbolic system (9.59) with $f^B = 0$.

9.7 Shock capturing in relativistic MHD

The covariant hyperbolic formulation of the theory of ideal magnetohydrodynamics (9.8) is in divergence form

$$\nabla_a F^{aA}(U^B) = 0 \ (A, B = 1, 2, \cdots, N), \quad (9.70)$$

where $U_B = (u_a, h_b, r, P)$ denote the fluid variables and $N = 10$ the number of equations. The nonlinear nature of ideal MHD typically introduces solutions with shocks, i.e. timelike surfaces of discontinuity, S, with normal one-form ν_a. The differential system of equations (9.70) describes solutions away from shocks, while any physical solution is subject to specific jump conditions across S. In particular, shocks are entropy-increasing. Solutions of this type may be computed using shock capturing schemes, which approximate jump conditions according to the weak formulation of (9.1),

$$[F^{aA}(U^B)]\nu_a = 0. \tag{9.71}$$

Smoothing using linear filtering of higher spatial harmonics is an effective method of creating entropy, whenever a shock forms. Smoothing operators that commute with finite-differencing operators provide an efficient shock-captering method which preserves divergence-free magnetic fields within machine round-off error. Alternative methods based on characteristics are complicated in view of the large number of equations in (9.70). Methods based on artificial viscosity are known to be surprisingly difficult in ultrarelativistic flows, because of the contribution of the thermal energy to the inertia of the fluid[395]. In the following steps, we shall outline that smoothing methods are computationally consistent with the continuum limit.

Conditions (9.71) impose the condition

$$0 = [\omega^{ab} + g^{ab}c]\nu_a = [\omega^{ab}]\nu_a + [c]\nu_a, \tag{9.72}$$

where $\omega^{ab} = u^a h^b - u^b h^a$. By antisymmetry of ω^{ab}, $\nu_a\nu_b[\omega^{ab}] = 0$, so that $\nu^2[c] = 0$. Since the normal of a timelike shock surface is spacelike, $\nu^2 > 0$, whence $[c] = 0$. It follows that the jump conditions preserve the jump condition $\nu_a[\omega^{ab}] = 0$ and the constraint $c = 0$. A stronger result applies, in that the homogeneous Maxwell equations $\nabla^\omega{}_{ab} = 0$ are preserved across S. To see this, consider a solution in the open region to the left of S. The jump condition $[c] = 0$ shows that $(c)^+ = 0$. We may decompose the derivative operator ∇_a on S according to a normal and internal derivative,

$$\nabla_a = \nu_a(\nu^c\nabla_c) + (\nabla_S)_a. \tag{9.73}$$

The jump condition $[\omega^{ab}]\nu_a$ shows that

$$0 = (\nabla_s)^a\nabla^b[\omega_{ab}] = \nu^b(\nabla_s)^a[\omega_{ab}] - [\omega_{ab}]K^{ab}, \tag{9.74}$$

where $K_a^b = \nabla_a\nu^b$ denotes the extrinsic curvature tensor of S upon using $[\omega_{ab}]\nu^a$ once more. (Any smooth extension can be used for ν^a off S in the definition of K_a^b.)

The extrinsic curvature tensor is symmetric[120]. This leaves $v^b(\nabla_S)^a[\omega_{ab}] = 0$. Furthermore, the assumption of satisfying Maxwell's equations to the left of S implies $0 = v^b(\nabla^a \omega_{ab})^- = v^b(\nabla_S)^a(\omega_{ab})^-$, so that

$$0 = v^b(\nabla_S)^a[\omega_{ab}] = v^b(\nabla_S)^a(\omega_{ab})^+. \tag{9.75}$$

Applied to the combination $0 = v^b\{\nabla^a(\omega^{ab} + g^{ab}c)\}^+$ in accord with the jump condition in (9.70), it follows that

$$v^a(\nabla_a c)^+ = 0. \tag{9.76}$$

We conclude that the condition that the homogeneous Maxwell equations are satisfied to the left of S together with the shock jump conditions for (9.70) implies that both c and its normal derivative vanish to the right of S. Since by assumption c satisfies the homogeneous wave-equation to the right of S, Holmgren's Uniqueness Theorem[225] forces $c = 0$ everywhere to the right of S in its past domain of dependence. The result can be generalized to the complete set of Maxwell's equations[548].

The condition that the magnetic field is divergence-free on a spacelike hypersurface Σ_t of constant time t is contained in the homogeneous Maxwell equation $F_{[\alpha\beta,\gamma]} = 0$, where F_{ab} denotes the electromagnetic field tensor and where the Greek indices refer to the three spatial coordinates of the hypersurface. The magnetic field in Σ_t is $H_\alpha = \frac{1}{2}[\alpha\beta\gamma]F_{\beta\gamma}$, where $[\alpha\beta\gamma]$ denotes the Levi-Civita tensor in Σ_t. The magnetic field in Σ_t is divergence-free if $\partial_\alpha \mathcal{H}^\alpha = 0$ according to the compatibility condition

$$\partial_\alpha \omega^{t\alpha} = 0. \tag{9.77}$$

A numerical scheme for the hyperbolic form of relativistic magnetohydrodynamics (9.70) is to preserve these continuum results by appropriate choice of numerical operators. In what follows, we consider a smoothing method with leapfrog time-stepping given by

$$(F^{tA})^m = S_w^{2D}\{(F^{tA})^{m-1}\} - 2\Delta t \delta_x (F^{xA})^m - 2\Delta t \delta_y (F^{yA})^m. \tag{9.78}$$

Here, S_w^{2D} denotes a two-dimensional linear smoothing operator, $\delta_{w,x}$ and $\delta_{w,y}$ denote finite-differencing operators in the x- and y-coordinates, and Δt denotes a time-step from t_m to t_{m+1}.

The operators $\delta_{w,x}$, $\delta_{w,y}$ and S_w^{2D} will be chosen to be mutually commuting. Explicit representations are

$$S_w^{2D}\{F^{tA}\}(x, y) = \lambda S_w\{F^{tA}\}(\cdot, y)(x) + (1 - \lambda)S_w\{F^{tA}\}(x, \cdot)(y) \tag{9.79}$$

where · refers to the argument on which the one-dimensional smoothing operator S_w and λ refer to a weighted average of smoothing in the x- and y-directions. For example, $\lambda = 1/2$ in case istropic smoothing for rotationally symmetric problems in Cartesian coordinates (x, y). Anisotropic smoothing operators can be constructed that are effective in the computation of jets in cylindrical coordinates, corresponding to $\lambda \neq 1/2$ which may further be chosen individually for each equation. In general, different problems may require different choices of S_w^{2D} for an optimal result. Similarly, we define

$$\delta_x(f)(x, y) = \delta_w\{f(\cdot, y)\}(x), \quad \delta_y(f)(x, y) = \delta_w\{f(x, \cdot)\}(y). \tag{9.80}$$

Time-updates (9.78) become iterative by application of the Newton–Raphson method: fluxes $(F^{\alpha A})^{m+1}$ may be obtained from U^{m+1} after numerical inversion of the densities $(F^{tA})^{m+1} \equiv F^{tA}(U^{m+1})$.

In a particular coordinate system $\{x^a\}$, the hyperbolic form of the homogenous Maxwell equations subject to constraint $c = u^c h_c = 0$ are

$$K': \{\sqrt{-g}\omega^{ab}\}_{,a} + \{\sqrt{-g}g^{ab}c\}_{,a} = -k^b, \tag{9.81}$$

where $k^b = \sqrt{-g}\,\Gamma^b_{cd}g^{cd}c$. We may assume $g^{tt} \neq 0$. Let $\varpi^{ab} = \sqrt{-g}\omega^{ab}$. We then have

$$\varpi^{tb}_{,t} + \varpi^{xb}_{,x} + \varpi^{yb}_{,y} + (\sqrt{-g}g^{tb}c)_{,t} + (\sqrt{-g}g^{tx}c)_{,x} + (\sqrt{-g}g^{ty}c)_{,y} = -k^b. \tag{9.82}$$

We will make the induction hypothesis $c^n = 0$ for $n \leq m$. Time-stepping according to (9.78) gives

$$(\sqrt{-g}g^{tb}c)^{m+1} + (\varpi^{tb})^{m+1} = S_w^{2D}\{(\varpi^{tb})^{m-1}\} - 2\Delta t\delta_x(\varpi^{xb})^m$$
$$-2\Delta t\delta_y(\varpi^{yb})^m, \tag{9.83}$$

where $(k^b)^m = 0$ according to c^m. Letting b run over t, x, y and using c^m, these equations give

$$(\sqrt{-g}g^{tt}c)^{m+1} + 2\Delta t\{\delta_x(\varpi^{xt})^m + \delta_y(\varpi^{yt})^m\} = 0,$$
$$(\varpi^{tx})^m - S_w^{2D}\{(\varpi^{tx})^{m-2} + 2\Delta t\delta_y(\varpi^{yx})^{m-1}\} = 0, \tag{9.84}$$
$$(\varpi^{ty})^m - S_w^{2D}\{(\varpi^{ty})^{m-2} + 2\Delta t\delta_x(\varpi^{xy})^{m-1}\} = 0.$$

Now apply δ_x to the second and δ_y to the third equation above, and use commutativity of the $\delta_{x,y}$ and S_w^{2D}, whereby

$$\delta_x(\varpi^{tx})^m - S_w^{2D}\{(\delta_x\varpi^{tx})^{m-2} + 2\Delta t\delta_x\delta_y(\varpi^{yx})^{m-1}\} = 0, \tag{9.85}$$
$$\delta_y(\varpi^{ty})^m - S_w^{2D}\{(\delta_y\varpi^{ty})^{m-2} + 2\Delta t\delta_y\delta_x(\varpi^{xy})^{m-1}\} = 0. \tag{9.86}$$

By antisymmetry of ϖ^{xy} and commutativity of the δ_x and δ_y, we have

$$\delta_x \delta_y (\omega^{yx})^{m-1} + \delta_y \delta_x (\omega^{xy})^{m-1} = 0. \tag{9.87}$$

Adding (9.85) and (9.86) hereby gives

$$(\delta_x \varpi^{tx} + \delta_y \varpi^{ty})^m = S_w^{2D}$$
$$\{\delta_x (\varpi^{tx})^{m-2} + \delta_y (\varpi^{ty})^{m-2}\} = 0 \tag{9.88}$$

by our induction hypothesis, i.e. the magnetic field remains divergence-free.

The identity (9.84) shows that divergence-free magnetic fields are equivalent to preserving the constraint $c^{m+1} = 0$.

The discrete operators S_w and δ_w on discrete functions $f \epsilon V_N$ can be constructed on a fixed grid $0 = x_1, \cdots, x_{N+1} = 1$, $N = 2^M$, with uniform grid spacing h as follows. Consider smoothing of a function f_i on this grid defined by the transformation in the Fourier domain

$$f_k' = f_k \frac{\sin 2\pi k h}{2\pi k h}, \tag{9.89}$$

where f_k, $k = -N/2, \cdots, N/2$, denotes the discrete Fourier transform of $f \epsilon V_N$. This defines a smoothed function $f' = L_{N,h}(f)$ on V_N. Notice that $L_{N,1/N}$ is Lanczos-smoothing[420, 98]. We define the finite-difference operators δ_N^R with Richardson extrapolation by

$$\delta_N^R = \frac{1}{N} \left(\frac{4}{3} \frac{f_{i+1} - f_{i-1}}{2} - \frac{1}{3} \frac{f_{i+2} - f_{i-2}}{4} \right). \tag{9.90}$$

In order to construct a weak smoothing operator S_w, such that the highest spectral coefficients are reduced only by a small amount, we shall work with interpolation $\iota_w : V_N \to V_{2N}$ of the form

$$\iota_w(\bar{f})_{2i-1} = f_i,$$
$$\iota_w(\bar{f})_{2i} = \frac{1}{2(w-1)} \{w(f_{i+1} + f_i) - f_{i+2} - f_{i-1})\}, \tag{9.91}$$

accompanied by the projection operator $\pi : V_{2N} \to V_N$ given by $\pi(\bar{f})_i = f_{2i-1}$. Thus, $\pi L_{2N,h} \iota_w$ is a map from V_N into itself. Smoothing S_w on V_N is now defined as

$$S_w = \pi \left(\frac{4}{3} L_{2N,h/2} - \frac{1}{3} L_{2N,h} \right) \equiv \pi S_{2N} \iota_w, \tag{9.92}$$

where $h = 1/N$. Commensurate with S_w, we take $\delta_w : V_N \to V_N$ to be

$$\delta_w(f) = \pi 2 \delta_{2N}^R \iota_w = (1+\eta) \frac{f_{i+1} - f_{i-1}}{2} - \eta \frac{f_{i+2} - f_{i-2}}{4}, \tag{9.93}$$

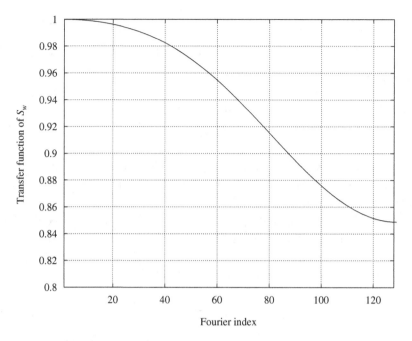

Figure 9.1 Shown is the transfer function of the smoothing operator S_ω versus Fourier index for $N = 256$ grid points and $\omega = 12$. Notice that the transfer function is remarkably flat and does not vanish at the high frequency end, where it is bounded below by $\frac{8}{3\pi}$. Combined with commuting finite-differencing operators, it gives a shock capturing method for the covariant hyperbolic equations of ideal magnetohydrodynamics (9.8), which preserves divergence free magnetic fields. (Adapted from M. H. P. M. van Putten, *SIAM J. Numer. Anal.* (1995), **32**, 1504. ©1995 Society for Industrial and Applied Mathematics.)

with $\eta = 8(w-1)/3$. In this fashion, S_w is a weak smoothing operator in the sense that its transfer function in the spectral domain is bounded between $8/3\pi \simeq 0.8488$ and 1. Because this transfer function is relatively flat and does not vanish at the high-frequency ends $k = \pm N/2$, $S_w\{f\}$ represents significantly weaker smoothing than Lanczos-smoothing. It will be appreciated that S_w is easily computed using integration of $\iota_w\{f\}$ using the discrete Fourier transform following by $\pi \delta^R_{2N}$. Figure (9.1) shows the spectral transfer function in case of $N = 256$ grid points.

The above can be adapted for cylindrical coordinates, i.e. using the line-element

$$ds^2 = -dt^2 + d\sigma^2 + \sigma^2 d\phi^2 + dz^2 \tag{9.94}$$

with uniform discretization

$$\{t, \sigma, \phi, z\} = \left\{ m\Delta t, \left(i + \frac{1}{2}\right)\Delta\sigma, (j + \frac{1}{2})\Delta z \right\}, \tag{9.95}$$

where $i = 0, \cdots, 2^N - 1$, $j = -2^N, \cdots, 2^N - 1$. In axisymmetric simulations, we may exploit symmetry by aligning the z-axis with the initial magnetic field through the center of a magnetized star[570] or along the direction of propagation of a magnetized jet[555]. Thus, the z-axis will be the axis of symmetry and $\partial_\phi = 0$.

The present cylindrical coordinate system introduces nonzero connection symbols in the updates

$$(F^{tA})^{m+1} = S^{2D}\{(F^{tA})^{m-1}\}$$

$$-2\Delta t\left[\frac{1}{\sqrt{-g}}\delta_\sigma(\sqrt{-g}F^{\sigma A})^m + \delta_z(F^{zA})^m\right] + \cdots \quad (9.96)$$

where the dots refer to $-2\Delta t \times$ further contributions from connection symbols associated with F^{tA}, given by

$$\begin{cases} -\sigma T^{\phi\phi} & \text{for } A = 2 \\ 0 & \text{otherwise.} \end{cases} \quad (9.97)$$

In the application of S_w^{2D} to the homogeneous Maxwell's equations, we take

$$S^{2D} = \begin{cases} S_w^{2D}\{F^{tA}\} & \text{for } A = 1, \cdots, 4, 9, 10 \\ \frac{1}{\sqrt{-g}}S_{\epsilon,w}\{\sqrt{-g}F^{tA}\} & \text{for } A = 5, \cdots, 8, \end{cases} \quad (9.98)$$

where $\sqrt{-g} = \sigma$.

The S_W^{2D} is as defined in [9.79], and $S_{\epsilon,w}^{2D}$ is a modification thereof to treat the coordinate singularity $\sigma = 0$. Notice that [9.98] applies smoothing to $\sqrt{-g}\omega^{ab}$ rather than ω^{ab}, in order to preserve divergence-free magnetic fields.

For numerical stability, we apply smoothing to functions extended to $\sigma < 0$ according to even or odd symmetry. We define regular extensions

$$\sigma F^{tA}(\sigma, z) = \begin{cases} -\sigma F^{tA}(-\sigma, z) & \text{if } F^{tA}(0, z) = 0, \\ \sigma F^{tA}(-\sigma, z) & \text{if } F^{tA}(0, z) \neq 0. \end{cases} \quad (9.99)$$

The first case concerns the radial magnetic field with $H^\sigma(0, z) = 0$, while the second concerns the z-magnetic field for which $h^z(0, z) \neq 0$ is allowed. These extensions preserve analyticity in σ, which is of advantage to numerical accuracy. Furthermore, for open boundary problems in the z-direction the extension in the z-coordinate may be obtained by simply taking z-cyclic boundary conditions.

In the course of extending the radial magnetic field, h^σ, to $\sigma < 0$, the function $\sigma H^\sigma(\sigma, 0)$ becomes even in σ. That is, if $H^\sigma = a_1\sigma + a_2\sigma^2 + \cdots$ is a Taylor series of H^σ about $\sigma = 0$, then $\sigma H^\sigma = a_1\sigma^2 + \cdots$ is convex about $\sigma = 0$. Application of S_w preserves the mean value, since attenuation of the zeroth spectral component equals 1, so that $S_w\{\sigma^2\}(0) > 0$. In case of the σ-component of the magnetic field, therefore, $(1/\sigma)S_w\{\sigma H^\sigma\}(\sigma)$ is no longer zero at $\sigma = 0$. This effect is compensated using the regularized smoothing operator

$$\sigma^{-1}S^{2D}_{\epsilon,w}\{\sigma f\}(\sigma, z) = \begin{cases} \sigma^{-1}S^{2D}_w\{\sigma f\}(\sigma, z) - \epsilon\sigma^{-1} & \text{if } f(0, z) = 0, \\ \sigma^{-1}S^{2D}_w\{\sigma f\}(\sigma, z) & \text{if } f(0, z) \neq 0. \end{cases} \qquad (9.100)$$

Here, $\epsilon\sigma^{-1}$ constitutes a correction for functions $f(0, z) = 0$ as to approximate $\sigma^{-1}S^{2D}_{\epsilon,w}\{\sigma^2\} \sim \sigma(\sigma \ll 1)$. With $\sigma_i = (1/2)\Delta\sigma, (3/2)\Delta\sigma, \cdots$ as specified above, $\epsilon = \epsilon_m$ is choosen numerically so as to satisfy, at each time-step,

$$\sigma_0^{-1}S^{2D}_{\epsilon,w}\{\sigma_0 f\} = \frac{1}{3}f(\sigma_1, z). \qquad (9.101)$$

This regularization preserves divergence-free magnetic fields as $\delta_\sigma S^{2D}_{\epsilon,w}\{\varpi^{\sigma t}\} = \delta_\sigma S^{2D}_w\{\varpi^{\sigma t}\}$ and $\delta_z S^{2D}_{\epsilon,w}\{\varpi^{\sigma t}\} = \delta_z S^{2D}_w\{\varpi^{z t}\}$, so that the previous discussion on S^{2D}_w applies with δ_σ and δ_z corresponding to δ_x and δ_y.

9.8 Morphology of a relativistic magnetized jet

Bright features ("knots") in extragalactic jets indicate regions of in-situ processes energizing charged particles. Notable processes are shocks and compression. Compression may be longitudinal through shocks as in M87[450, 604, 396, 183], transverse trough radial pinch by magnetic fields (in nonrelativistic fluid dynamics, e.g.[108, 124] and in time-independent solutions of relativistic magnetohydrodynamics[163]) or hydrodynamical instabilities. Magnetically driven pinches can be induced by toroidal magnetic fields, which generally tend to be destabilizing. In contrast, longitudinal magnetic fields contribute to stability, and suppress radial structures

Figure (9.2) reveals the formation of knotted structures induced by the toroidal magnetic pinch in a jet with Lorentz factor 2.46, extending simulations on the formation of jets in relativistic hydrodynamics[549]. Shown is the formation of a Stagnation-point Nozzle-Mach disk morphology (SNM). The stagnation point defines the root of an extended nose-cone. The Mach disk in the nose-cone oscillates periodically, leaving behind "knots" in the form of nozzles where the flow is radially pinched. This jet morphology resembles optical radio jets such as 3C273 and may apply in particular to compact symmetric sources (in light of the early time-evolution in the simulation).

The discovery of GRB supernovae with core-collapse of massive stars presents a novel setting for simulations of jets: those punching through a remnant stellar envelope[358]. They have in common with extragalactic jets the setting of light

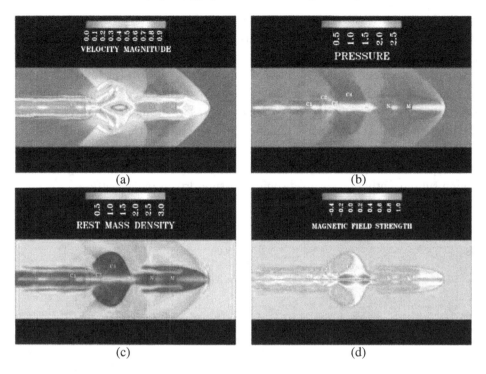

Figure 9.2 Shown is the morphological evolution of a light relativistic magne-tized jet propagating into an unmagnetized environment of high density. The results are obtained by parallel computation on the covariant hyperbolic equations of magnetohydrodynamics for a purely toroidal magnetic field. Shown are coordi-nate distributions of (a) V, (b) P, (c) r, and (d) B at time $t/\sigma_{jet} = 34.47$, where σ_{jet} is the radius of the jet. Colors vary linearly from blue to red. The jet aperture has boundary conditions $(\Gamma, M, H, P, r) = (2.46, 1.67, 0.46\hat{f}(\sigma), 0.10, 0.20)$, which are out of radial force-balance. Here, $f(\sigma) = \sigma \cos[(\sigma/\sigma_{jet})\pi/2]$, $\hat{f} = f/\|f\|$. The environment satisfies $(P, r) = (0.10, 1.00)$. The simulation shows an early stage of a jet with a notable confinement of enhanced pressure to the axis. The on-axis distributions of pressure and rest mass density of the jet vary by a factor of 190 and 45. The composite Mach disk C1-4 forms by pinch of the toroidal magnetic in combination with the formation of backflow. Similar backflow is at the head of the jet. A radial oscillation in the terminal Mach disk M produces a propagating $[v = 0.16 \pm 0.04c]$ supersonic nozzle N $(M = 1.28 \pm 0.03$ in the comoving frame of the nozzle) with pressure contrast 6.14 and rest mass density contrast 3.38 as a persistent feature – a bright knot – in between the stagnation point S and the Mach disk M. This defines a characteristic SNM morphology. A repeat of the radial oscillation in the terminal shock is observed at $t/\sigma_{jet} = 39.06$, which produces a second nozzle (not shown) ahead of N. These simulations were performed on the IBM SP2 Parallel Computer at the Cornell Theory Center. (Reprinted from [555])

magnetized relativistic outflows into regions with a negative density gradient, which at the center are higher in density than the jet. An interesting challenge is detailed modeling of the gamma-ray emissions process taking place through dissipation of kinetic energy in internal shocks[451, 452]. In frame of the center of mass, the collision of two relativistic outbursts produced by an intermittent source becomes the collision of two relativistic jets with modest relative Lorentz factor. Figure (9.3) shows a simulation of two such jets, each with Lorentz factor 1.5, producing a nearly steady-state central region of high-energy density with appreciable amplification of the transverse magnetic field. This high-density region subsequently creates a subsonic, pressure-driven radial outflow.

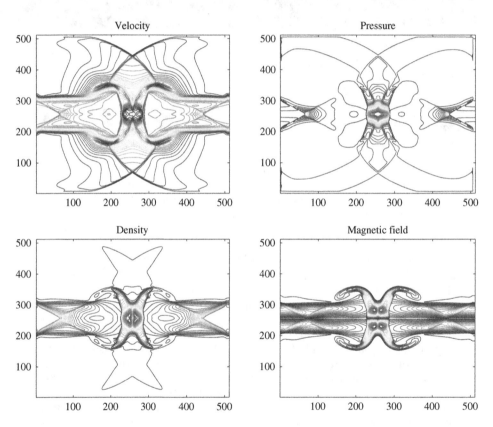

Figure 9.3 The morphological evolution of a the head-on collision of two heavy relativistic magnetized jets ("kissing jets") with rotational symmetry. The result represents the view in the frame of the center of mass of two ejecta from an intermittent source, wherein a fast-moving second overtakes a slow-moving first ejection. The collision (a) produces a slowly growing high-energy density region about a central stagnation point. Subsequently (b), it creates a subsonic pressure-driven radial outflow. These interactions provide sites for high-energy emissions of charged particles, such as in the internal shock model for GRBs.

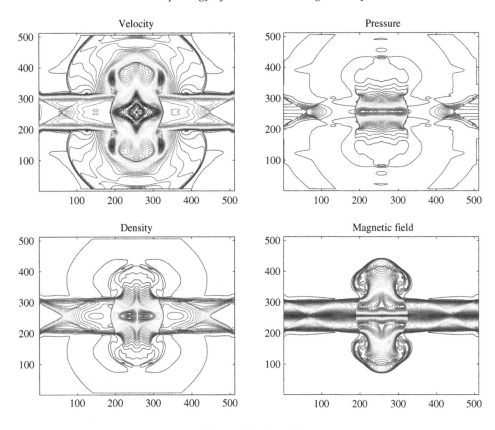

Figure 9.3 (cont.)

Exercises

1. Show that the rank-one update to the characteristic matrix of U(1) magnetohy-drodynamics generalizes to a rank-N update in SU(N) Yang–Mills magnetohy-drodynamics. Show further that the modified Friedrichs–Lax symmetrization procedure likewise carries over to SU(N).

2. For a jet with Lorentz factor Γ, calculate the *minimal* sound crossing time as a function of the radius of the jet. What happens in the limit as Γ approaches infinity?

3. Consider the condition of infinite conductivity in the laboratory frame $E + v \times B = 0$, where v denotes the three-velocity of the fluid. Derive an evolution equation for the magnetic field B by eliminating the electric field E using Maxwell's equations. What is the rank of this evolution equation? Devise a numerical scheme that maintains divergence-free magnetic fields for both smooth and shocked flows.

4. Consider the four-variant hyperbolic formulation of Faraday's equations in (9.8). Show that the shock capturing method of Section 9.7 carries over in the case of arbitrary-curved spacetime backgrounds (general relativistic MHD).

5. Show that the extrinsic curvature tensor of a smooth hypersurface is symmetric in its two indices.

6. The simulation shown in Figure 9.2 uses transverse magnetohydrodynamics. Show that the cylindrically symmetric flow in a nozzle is described by the Bernoulli equation $H = f^*\Gamma = \text{const.}$ and continuity $\Phi = ru^z A = \text{const.}$, where $A(r)$ denotes the local cross-section of the nozzle. Here $f^* = f + kr^2$ with specific enthalpy f as described for a polytropic fluid, where $k = h/r$ is a constant along streamlines set by the ratio of the transverse magnetic field-strength h relative to the rest mass density r.

7. Show that the bifurcation of the contact discontinuity into slow magnetosonic waves and alfven waves is an inertial effect. Show that in the force-free limit corresponding to vanishing inertia of the fluid, this bifurcation does not

occur, and that the slow magnetosonic wave and the Alfven wave coincide, and become luminal. Conclude that the fast magnetosonic wave also becomes luminal.

8. Consider a *sonic* nozzle, wherein the Mach number is reduced to $M = 1$ at the location of smallest cross-section A_s. In the non-relativistic regime, show that the Mach number and the cross-sectional area A are related by

$$\left(\frac{A}{A_s}\right)^{1/2} = \left(\frac{2}{\gamma+1}\right)^{\frac{\gamma+1}{4(\gamma-1)}} M^{-1/2} \left(1 + \frac{\gamma-1}{2} M^2\right)^{\frac{\gamma+1}{4(\gamma-1)}}. \tag{9.102}$$

What is the large M limit? Show that the same relation generalizes to relativistic fluids with sound speed $a_s = \tanh \lambda_s$ according to

$$\left(\frac{A}{A_s}\right)^{1/2} = \left(\frac{r}{r_s}\right)^{-1/2} \left(\frac{f_s^2}{f^2} \cosh^2 \lambda_s - 1\right)^{-1/4} \sinh^{1/2} \lambda_s. \tag{9.103}$$

In the asymptotic limit of large Mach number, $M = (r/r_s)^{-(\gamma+1)/2}(A_s/A)$ in the non-relativistic regime, and $a_s \to (\gamma-1)^{1/2}$ with $f = 1 + [f_s a_s^2/(\gamma - 1)](r/r_s)^{\gamma-1}$, $f_s = 1 - [a_s^2/(\gamma-1)]^{-1}$ in the relativistic regime. Derive the bounds

$$(\text{rel.}) \left(\sqrt{\gamma-1} \frac{A_s}{A}\right)^{1/(2-\gamma)} \leq \frac{r}{r_s} \leq \frac{A_s}{A}\sqrt{\frac{\gamma-1}{\gamma+1}} (\text{non-rel.}). \tag{9.104}$$

10

Nonaxisymmetric waves in a torus

"I cannot do't without counters."
William Shakespeare (1564–1616) *The Winter's Tale*, IV: iii.36.

Waves are common in astrophysical fluids. They define the morphology of outflows, which are related to accretion disks surrounding compact objects. Waves often appear spontaneously, in response to instabilities commonly associated with shear flows. The canonical example of a shear-driven instability is the Kelvin–Helmholtz instability. Even in the absence of shear, stratified flows with different densities can become unstable in the presence of acceleration and/or gravity – the Rayleigh–Taylor instability. Such instabilities do not fundamentally depend on compressibility, and hence they are appropriately discussed in the approximation of incompressible flows. In rotating fluids, instabilities represent a tendency to redistribute angular momentum leading towards a lower energy state. These, likewise, can be studied in the limit of incompressible flows.

A torus around a black hole is a fluid bound to a central potential well. The fluid in the torus is a rotational shear flow, which is generally more rapidly rotating on the inner face than on the outer one. In particular, when driven by a spin-connection to the black hole, the inner face develops a super-Keplerian state, while the outer one develops a sub-Keplerian state by angular momentum loss in winds. The induced effective gravity – centrifugal on the inner face and centripetal on the outer face – allows surface waves to appear very similar to water waves in channels of finite depth. As a pair of coupled surface waves, these interact by exchange of angular momentum. This leads to growth of retrograde waves on the inner face and growth or prograde waves on the outer one.

Papaloizou and Pringle[409] pointed out that nonaxisymmetric waves modes can thus arise on the inner and the outer face of a strongly differentially torus in the limit of infinite slenderness. This limit is not relevant in any astrophysical system. We here describe the formation of nonaxisymmetric instabilities in tori of finite slenderness.

10.1 The Kelvin–Helmholtz instability

The Kelvin–Helmholtz instability describes the instabilities arising in planar shear between two incompressible flows subject to gravity[597, 109].

A two-layer stratified fluid consists of a flow with density and horizontal velocity $(\rho_1, U_1 \neq 0)$ on top of a fluid flow at rest with $(\rho_2, U_2 = 0)$ as illustrated in Figure (10.1). Let both be subject to an external gravitational acceleration g.

A two-dimensional incompressible and irrotational flow can be described by a velocity potential ϕ: $(u, v) = (\phi_x, \phi_y)$. We denote the vertical perturbation of the interface between the two fluids by $\eta(x, t)$. We then have

$$\Delta\phi = 0 \text{ in } y > \eta, \quad \phi \to 0 \text{ as } y \to \infty, \tag{10.1}$$

$$\Delta\phi = 0 \text{ in } y < \eta, \quad \phi \to 0 \text{ as } y \to -\infty. \tag{10.2}$$

The interface between the two layers satisfies a kinematic boundary condition. Consider a particle at the interface, moving from $A_1 = (x_1, \eta_1)$ at $t = t_1$ to $A_2 = (x_2, \eta_2)$ a moment later at $t = t_2$. This particle has a vertical displacement

$$(t_2 - t_1)v \simeq \eta(x_2, t_2) - \eta(x_1, t_1) = \eta_t(t_2 - t_1) + \eta_x(x_2 - x_1). \tag{10.3}$$

In the limit as $t_2 - t_1$ becomes small, and noting that $(t_2 - t_1)u \simeq x_2 - x_1$, we are left with

$$\eta_t + u\eta_x = v. \tag{10.4}$$

Across the interface, therefore, we have the pair of kinematic surface conditions

$$(\eta_t + u\eta_x - v)^{\pm} = 0. \tag{10.5}$$

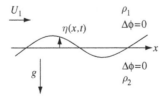

Figure 10.1 The Kelvin–Helmholtz instability describes the growth of a perturbation of an interface in a shear flow. Shown is a fluid with density and velocity (ρ_1, U_1) moving on top of another fluid at rest with density ρ_2. The instability is due to the Bernoulli effect and conservation of mass: a positive deflection $\eta(x, t)$ introduces a lower effective cross-section above and hence an enhanced velocity with reduced pressure. This stimulates growth of η, which may be stabilized by gravity.

Particles at the interface remain at the interface. The Bernoulli equation expresses a conserved energy in the Euler equations of motion for irrotational flow $(u_y - v_x = 0)$,

$$u_t + uu_x + vu_y = -P_x/\rho, \quad v_t + uv_x + vv_y = -P_y/\rho - g. \tag{10.6}$$

It is described by the integral

$$C(t) = \phi_t + \frac{1}{2}(u^2 + v^2) + \frac{P}{\rho} \tag{10.7}$$

which depends only on time. In the case at hand, the asymptotic boundary conditions at $\pm\infty$ impose $C(t) \equiv 0$. With pressure continuity across the interface, there obtains a single jump condition

$$\left[\rho\phi_t + \frac{1}{2}\rho(u^2 + v^2) \right] = 0. \tag{10.8}$$

Next, we linearize the boundary conditions (10.5) and (10.8), and use the harmonic ansatz $\phi = \delta\phi e^{i(kx - \omega t)}$ and $\eta = \eta e^{i(kx - \omega t)}$. This gives

$$\begin{pmatrix} \rho_1(-i\omega + ikU_1) & i\omega\rho_2 & (\rho_1 - \rho_2)g \\ k & 0 & -i\omega + ikU_1 \\ 0 & -k & -i\omega \end{pmatrix} \begin{pmatrix} \delta\phi_+ \\ \delta\phi_- \\ \eta \end{pmatrix} = \begin{pmatrix} 0 \\ 0 \\ 0 \end{pmatrix}. \tag{10.9}$$

Nontrivial solutions obtain when the matrix in (10.9) vanishes,

$$k\left[(\rho_1 + \rho_2)\omega^2 - 2k\rho_1 U_1 \omega + k^2 U_1^2 \rho_1 + gk(\rho_1 - \rho_2) \right] = 0. \tag{10.10}$$

For $k \neq 0$, (10.10) reduces to

$$(\rho_1 + \rho_2)(\omega/k)^2 - 2\rho_1 U_1(\omega/k) + U_1^2 \rho_1 + g(\rho_1 - \rho_2)/k = 0, \tag{10.11}$$

which defines the following dispersion relation $\omega = \omega(k)$:

$$\frac{\omega}{k} = U_1 \left(\frac{\rho_1}{\rho_1 + \rho_2} \pm \sqrt{-\frac{\rho_1\rho_2}{(\rho_1 + \rho_2)^2} + \frac{\rho_2 - \rho_1}{\rho_2 + \rho_2} \frac{c^2}{U_1^2}} \right) \tag{10.12}$$

where $c^2 = g/k$.

The Kelvin–Helmholtz instability describes the response to shear flow. In the presence of gravity, we have a critical wavenumber k_{KH} beyond which the interface becomes unstable,

$$k_{KH} = \frac{g(\rho_2^2 - \rho_1^2)}{\rho_1\rho_2 U_1^2}. \tag{10.13}$$

The Rayleigh–Taylor instability describes the response to a stratified fluid with different densities in the presence of gravity. Consider the case of $U_1 = 0$ and $\rho_1 \neq \rho_2$. Then for $\rho_1 > \rho_2$, (10.12) describes a heavy fluid on top of a light fluid, which is unstable,

$$\frac{\omega}{k} = \pm ic\sqrt{\frac{\rho_1 - \rho_2}{\rho_1 + \rho_2}}. \tag{10.14}$$

Alternatively, a light fluid on top of a heavy fluid is stable.

10.2 Multipole mass-moments in a torus

The effect of shear on the stability of a three-dimensional torus of incompressible fluid around a central potential well can be studied about an unperturbed, Newtonian angular velocity of the form

$$\Omega(r) = \Omega_a \left(\frac{a}{r}\right)^q \quad (3/2 < q < 2) \tag{10.15}$$

where the index of rotation q is bounded beteen the Keplerian value 3/2 and Rayleigh's stability criterion $q = 2$.

We consider irrotational perturbations to the underlying flow (vortical if $q \neq 2$) as initial conditions. In the inviscid limit, these perturbations remain irrotational by Kelvin's theorem. We shall expand the harmonic velocity potential of these perturbations in cylindrical coordinates (r, θ, z),

$$\phi = \Sigma_n a_n(r, \theta, t)z^n, \quad \Delta\phi = 0. \tag{10.16}$$

The equations of motion can conveniently be expressed in a local Cartesian frame (x, y, z) with the Newtonian angular velocity $\Omega_a = M^{1/2}a^{-3/2}$ of the torus of radius $r = a$ about a central mass M. These Cartesian and cylindrical coordinates are related by $x = r - a$, $\partial_x = \partial_r$ and $\partial_y = r^{-1}\partial_\theta$. We can readily switch between these two coordinate systems in coordinate invariant expressions. Infinitesimal harmonic perturbations of the form $e^{im\theta - i\omega' t}$ of frequency ω' as seen in the corotating frame at $r = a$ satisfy the linearized equations of momentum balance. For an azimuthal quantum number m and on the equatorial plane $(z = 0)$, these equations for the x and y velocity perturbations (u, v) are, in the notation of P. M. Goldreich, J. Goodman and R. Narayan[235]

$$-i\sigma u - 2\Omega v = -\partial_r(h + \psi),$$
$$-i\sigma v + 2Bu = -ik(h + \psi), \tag{10.17}$$

where h denotes a perturbation of the unperturbed enthalpy h^e, satisfying

$$\partial_r H^e = \Omega^2 r - Mr^{-2} \quad (z = 0), \tag{10.18}$$

Ψ denotes a perturbation to an external potential, $2B = (2-q)\Omega$, and $k = m/r$. The local frequency $\sigma(x) = \omega' - m\delta\Omega(x)$ $(\delta\Omega(x) = \Omega(x) - \Omega_a)$ is associated with the Lagrangian derivative $D_t(x) = -i\sigma(x) + u(x)\partial_x$. For a narrow torus, we note that (10.18) reduces to a quadratic $H^e(x) = (2q-3)\Omega_a^2(b^2 - x^2)/2(-b \le x \le b)$[235]. In what follows, we focus on tori of arbitrary width using the exact expression (10.18).

10.3 Rayleigh's stability criterion

Rayleigh's stability criterion refers to the observation that a revolving fluid is stable against azimuthally symmetric perturbations if and only if its specific angular momentum increases outwards. A Rayleigh stable state explicates the notion that it is "cheaper" to store angular momentum at larger radii than at smaller radii around a given potential well. The stability criterion therefore corresponds to a positive gradient in the specific angular momentum $j = \Omega r^2$, i.e.

$$(j^2)_r = r^3 \left(4\Omega^2 + r\frac{d\Omega^2}{dr}\right) = r^3 \kappa > 0, \qquad (10.19)$$

where $\kappa^2 = 4\Omega^2 + r d\Omega^2/dr$. See also C. Hunter[272].

10.4 Derivation of linearized equations

We derive (10.17) as follows. Consider the Euler equations of motion for a three-dimensional incompressible fluid with specific enthalpy H in the presence of an external potential Ψ,

$$\mathbf{u}_t + \mathbf{u} \cdot \nabla \mathbf{u} = -\nabla H - \nabla \Psi. \qquad (10.20)$$

In cylindrical coordinates (r, θ) with frame $(\mathbf{i}_r, \mathbf{i}_\theta)$,

$$\mathbf{u} = \mathbf{i}_r U + \mathbf{i}_\theta V, \quad \nabla = \mathbf{i}_r \partial_r + r^{-1}\mathbf{i}_\theta \partial_\theta \qquad (10.21)$$

(10.20) becomes

$$U_t + UU_r + r^{-1}VU_\theta - r^{-1}V^2 = -(H_r + \Psi_r),$$
$$V_t + Ur^{-1}(rV)_r + r^{-1}VV_\theta = -r^{-1}(H_\theta + \Psi_\theta). \qquad (10.22)$$

Consider a perturbation (u, v, h, ψ) about an equilibium state $(0, V, H, \Psi)^e$ of a uniformly revolving flow $(\partial_\theta = 0)$. To linear order, we have $\delta H = 2\Omega v$, and

$$u_t + r^{-1}V^e u_\theta - 2\Omega v = -h_r - \psi_r,$$
$$v_t + r^{-1}V^e v_\theta + ur^{-1}(rV^e)_r = -r^{-1}(h_\theta + \psi_\theta). \qquad (10.23)$$

With $V^e = \Omega r$ and $\kappa^2/2\Omega = 2\Omega + r\Omega_r$, the linearized perturbation equations may be written as

$$u_t + \Omega u_\theta - 2\Omega v = -h_r - \psi_r,$$
$$v_t + \Omega v_\theta + 2Bu = -r^{-1}(h_\theta + \phi_\theta), \tag{10.24}$$

where

$$2B = (2 - q)\Omega. \tag{10.25}$$

A frequency ω hereby corresponds to a frequency ω' expressed with respect to corotating coordinates, subject to $\omega = \omega' + \Omega_a$. For a harmonic perturbation $\propto e^{-i\omega t + m\theta}$, we have

$$u_t + \Omega u_\theta = (-i\omega + m\Omega)u = -i\sigma u, \tag{10.26}$$

where

$$\sigma = \omega' - m\delta\Omega, \quad \delta\Omega = \Omega - \Omega_a \tag{10.27}$$

in the notation of Goldreich, Goodman and Narayan[235]. In this notation, therefore, we arrive at (10.24), upon noting that $r^{-1}\partial_\theta = ik$ with $k = m/r$. In the ansatz (10.16), the equation of motion for the z-component w of the velocity of the fluid satisfies

$$-i\sigma w = -\partial_z h. \tag{10.28}$$

Reflection symmetry about the equatorial plane ensures that this third equation of motion decouples from (10.24).

In earlier linearized treatments[235], variations in $2B$ across the torus are neglected. This limits the application to narrow tori defined by $h^e(x_\pm) = 0$, which is of no immediate astrophysical relevance. For wide tori, we here include $(2B)_x = -(q/r)2B$. The equations of motion (10.24) hereby are

$$(\partial_r^2 + r^{-1}\partial_r - m^2 r^{-2})a_0 = qr^{-1}a_0' \tag{10.29}$$

for an azimuthal mode number m. Solutions which are symmetric about the equatorial plane hereby have the velocity potential

$$\phi = a_0 - \frac{qz^2}{2r}\partial_r a_0 + O(z^4), \quad a_0 = r^{p+} + \lambda r^{p-}, \tag{10.30}$$

where

$$p_\pm = q/2 \pm (q^2/4 + m^2)^{1/2}, \quad \lambda = \text{const.} \tag{10.31}$$

10.5 Free boundary conditions

The boundary of the torus is an interface with vanishing specific enthalpy

$$H = 0. \tag{10.32}$$

In the dynamical case, (10.32) defines a Lagrangian boundary condition in terms of a two-point boundary condition in the equatorial plane

$$0 = D_t H = -i\sigma h + u H_x^e \quad (x = x_\pm, z = 0), \tag{10.33}$$

where h denotes the perturbation of the enthalpy about its equilibrium H^e according to (10.18). The second equation in (10.24) gives $h = i\sigma\phi - \psi + ik^{-1}2B\phi_x$, whereby (10.33) gives[235]

$$k(\sigma^2\phi + i\sigma\psi) + (2B\sigma + kH_x^e)\phi_x = 0. \tag{10.34}$$

This holds at both zeros $x_\pm$ of $H^e(x_\pm) = 0$ in (10.18), which can be determined by numerical evaluation.

In the absence of a potential ψ, the stability of the torus is described in terms of a critical rotation index for each azimuthal quantum number m. The boundary conditions (10.34) become

$$\epsilon\sigma^2 + 2B\sigma + kH_x^e = 0 \quad (\epsilon = k\phi/\phi_x). \tag{10.35}$$

In the limit of small σ, (10.35) becomes linear in σ. This corresponds to the slender torus approximation $b \ll a$ of Papaloizou and Pringle and to the shallow water wave limit $kb \ll 1$. About $\omega' = 0$, this obtains the critical rotation index

$$q = \sqrt{3} \tag{10.36}$$

for all m[409]. This is easy to see. Dropping the quadratic term in (10.35), the narrow torus limit gives

$$2B\sigma + kH_x^e = 0, \quad H^e = (2q-3)\Omega_a^2(b^2 - x^2)/2. \tag{10.37}$$

About $\omega' = 0$, we have

$$\sigma = \omega' - m(\Omega - \Omega_a) = \frac{mq}{r}\Omega_a b \quad (x = b). \tag{10.38}$$

Together with $k = m/r$, (10.37) reduces to

$$(2-q)q - (2q-3) = 0 \tag{10.39}$$

independent of m, whereby (10.36) follows.

10.6 Stability diagram

Given the potential (10.30–10.31), the boundary conditions (10.35) define an eigenvalue problem in ω' for wave-modes of mode number m in tori of arbitrary width. Suppose we choose an outer torus boundary $x_+ = b/a$, leaving the associated inner boundary x_- determined by numerical root-finding of $H^e(x_-) = 0$ as a function of q. We can solve (10.35) simultaneously for (ω', λ) at $x = x_\pm$ for a given value of q. Eigenvalues ω' are distinct and real, or appear in pairs of complex conjugates. Hence, double zeros of ω' define a transition between stable and unstable wave-modes. This introduces critical values of $q = q_c(b/a; m)$ associated with double zeros of ω'. We can solve for these critical curves using numerical continuation methods of H. B. Keller[292].

Numerical continuation of curves of critical stability is most conveniently pursued on a single equation, following elimination of λ in (10.35). By (10.31) and the definition of $\epsilon = k\phi/\phi_x$ in (10.35), we have

$$\lambda = r^D \frac{1 - \epsilon n_+}{\epsilon n_- - 1}, \quad n_\pm = p_\pm/m. \tag{10.40}$$

where $D = \sqrt{q^2 + 4m^2}$. According to (10.35), we also have

$$\epsilon = \frac{N}{\sigma^2}, \quad N = -2B\sigma - kH^e_x. \tag{10.41}$$

Since λ is a constant, (10.40) holds at both boundary points $x_\pm$. Using (10.41), elimination of λ in (10.40) leaves a single fourth-order equation in ω',

$$G(\omega', q) := \alpha(\sigma^2_+ \sigma^2_- + N_- N_+ n_- n_+) + \beta_1 \sigma^2_+ N_- + \beta_2 \sigma^2_- N_- = 0, \tag{10.42}$$

where

$$\alpha = r^D_- - r^D_+, \quad \beta_1 = (r^D_+ n_- - r^D_- n_+), \quad \beta_2 = (r^D_+ n_+ - r^D_- n_-). \tag{10.43}$$

The stability curves are defined by the simultaneous solutions

$$G(\omega', q) = 0, \quad \partial_{\omega'} G(\omega', q) = 0. \tag{10.44}$$

We solve for the real roots (ω', q) (10.44) using the Newton–Raphson method. Doing so by continuation on the slenderness ratio b/a obtains the stability curves $q_c(b/a; m)$ for each azimuthal mode number m.

10.7 Numerical results

Figure (10.2) shows the numerical solution to (10.44) by continuation. Quadratic fits to the stability curves are

$$
q_c(\delta, m) = \begin{cases}
0.27(b/0.7506a)^2 + 1.73 \ (m = 1) \\
0.27(b/0.3260a)^2 + 1.73 \ (m = 2) \\
0.27(b/0.2037a)^2 + 1.73 \ (m = 3) \\
0.27(b/0.1473a)^2 + 1.73 \ (m = 4) \\
0.27(b/0.1152a)^2 + 1.73 \ (m = 5) \\
0.27(m\delta/0.56a)^2 + 1.73 \ (m > 5)
\end{cases}
\tag{10.45}
$$

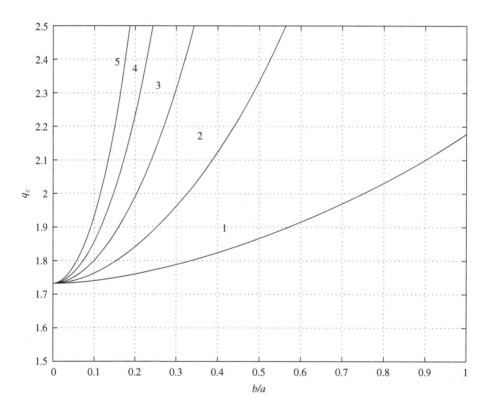

Figure 10.2 Diagram showing the neutral stability curves for the nonaxisymmetric buckling modes in a torus of incompressible fluid for finite slenderness ratios b/a, where b and a denote the minor and major radius of the torus, respectively. Curves of critical rotation index q_c are labeled with azimuthal quantum numbers $m = 1, 2, ..$, where instability sets in above and stability sets in below. (Reprinted from[561]. ©2002 The American Astronomical Society.)

Instability sets in above these curves, stability below. At the Rayleigh value $q_c = 2$ for critical stability of $m = 0$, we have

$$b/a = 0.7506, \ 0.3260, \ 0.2037, \ 0.1473, \ 0.1152, \dots, \ 0.56/m. \qquad (10.46)$$

These results show the creation of gravitational radiation in response to the spontaneous formation of multipole mass-moments in a torus which is strongly differentially rotating and sufficiently slender. The $m = 1$ mode produces a "*black hole-blob binary*" and the $m = 2$ mode produces a "*blob-blob binary*" system bound to the black hole. Both radiate at essentially twice the Keplerian velocity, as shown in Figure (10.3). Higher-order mass-moments define other lines of

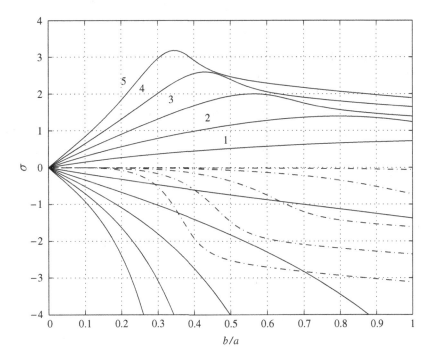

Figure 10.3 Frequency diagram of the pair of waves in a buckling mode on the neutral stability curves of a torus of incompressible fluid. The waves on the outer face are prograde (positive σ curves, labeled for each azimuthal quantum number $m = 1, 2, \cdots, 5$), whereas the waves on the inner face are retrograde (negative σ). The dot–dashed lines refer to the frequency ω' as seen in the corotating frame with the Newtonian angular velocity $\Omega_a = M^{1/2}/a^{3/2}$ of the torus at major radius $r = a$, where the highest (lowest) curve refers to $m = 1 (m = 5)$. Note that up to $b/a = 0.3$, ω' remains close to zero. Hence, the observed frequency of the gravitational radiation as seen at infinity is close to $m\Omega_a$ for low m. (Reprinted from[561].©2002 The American Astronomical Society.)

gravitational waves. In the presence of a spin-connection to the angular momentum of the central black hole, these emissions are long-lasting for the lifetime of rapid spin of the black hole. The torus hereby acts as a catalytic converter of black-hole spin energy into gravitational radiation.

10.8 Gravitational radiation-reaction force

A quadrupole buckling mode emits gravitational radiation at the angular frequency $\omega = 2(\Omega_a + \omega') \simeq 2\Omega_a$ (Figure 10.3). It describes an internal flow of energy and angular momentum from the inner to the outer face of the torus, in which total energy and angular momentum is conserved. The emitted gravitational radiation is therefore not extracted from the kinetic energy of this pair of waves. This contrasts with radiation from single surface waves of frequency $0 < \omega < m\Omega_T$ by the Chandrasekhar–Friedman–Schutz (CFS) instability. It may be noted that CFS instability is equivalent to a positive entropy condition $\delta S > 0$ in the first law of thermodynamics $-\delta E = \Omega_T(-\delta J) + T\delta S$ for a torus at temperature T, upon radiation of waves with specific angular momentum $\delta J/\delta E = m/\omega$ to infinity. See B. F. Schutz[484] on the entropy condition in the Sommerfeld radiation condition. The back-reaction of gravitational wave-emissions on the buckling mode can be assessed as follows.

The back-reaction of gravitational radiation consists of dynamical self-interactions and radiation-reaction forces, as described by K. S. Thorne[529, 530], S. Chandrasekhar and F. P. Esposito[111], and B. F. Schutz[484]. For slow-motion sources with weak internal gravity (e.g. a torus with low mass relative to the black hole) the latter can be modeled by the Burke–Thorne potential in the $2\frac{1}{2}$ post-Newtonian approximation

$$\Phi_{BT} = \frac{1}{5} x_j x_k \left(I^{jk} - \frac{1}{3} I\delta^{jk} \right), \tag{10.47}$$

where

$$I_{jk} = \int \Sigma x_j x_k \, dx dy \tag{10.48}$$

denotes the second-moment tensor of matter with surface density Σ. This intermediate order does not introduce a change in the continuity equation (as it does in the second-order post-Newtonian approximation[111, 484, 487]). In cylindrical coordinates $(r, x = r\cos\theta, y = r\sin\theta)$, and for harmonic perturbations $\eta = \eta e^{2i\theta - i\omega t}$ of the wave amplitude, we have, in the approximation of a constant surface density Σ,

$$I_{x_i x_j} = \Sigma \int_0^{2\pi} \int_{a+x_- + \eta_-}^{a+x_+ + \eta_+} x_i x_j \, dx dy. \tag{10.49}$$

Explicitly, $I_{xx} = (\pi\Sigma/2)[(a+x_+)^3\eta_+ - (a+x_-)^3\eta_-]$, which determines

$$\Phi_{BT} = \frac{1}{5}(x^2 I_{xx} + y^2 I_{yy} + 2xy I_{xy}) = \frac{1}{5}z^2 I_{xx}. \qquad (10.50)$$

Here, $z = x + iy$, $I_{xx} = I_{xx}e^{-i\omega t}$ and z^2 comprises $e^{2i\theta}$–combined, $\Phi = \Phi e^{2i\theta - i\omega t}$. The harmonic time dependence $e^{-i\omega t}$ derives from the integral boundaries in (10.50) and hence applies to all components of the moment-of-inertia tensor.

The linearized radiation-reaction force derives from the fifth time derivative, i.e. $\Phi = -i\omega^5\Phi_{BT}$ in the stability analysis of the previous section, supplemented with the kinematic surface conditions

$$-i\sigma\eta = \phi_x \qquad (10.51)$$

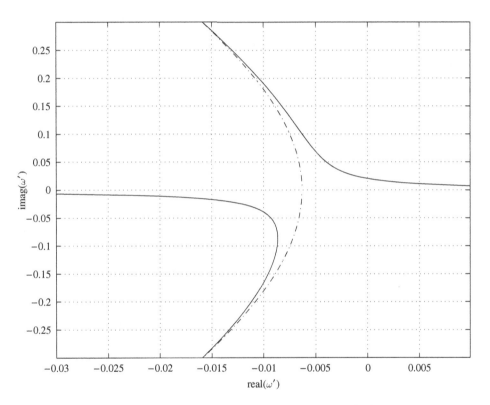

Figure 10.4 Complex frequency diagram of the frequency ω' of the quadrupole moment in the torus in response to the radiation-reaction force. The results are shown for a canonical value $b/a = 0.2$ and $\beta = 2 \times 10^{-4}$, corresponding to a torus mass of about 1% of the black hole. The dot–dashed curves are the asymptotes for $\beta = 0$. The results show that gravitational radiation-reaction forces contribute to instability of the quadrupole buckling mode. Similar results are found for modes $m \neq 2$, including $m = 1$. (Reprinted from[561]. ©2002 The American Astronomical Society.)

on the inner and outer boundaries $x = x_\pm$. Explicitly, we have

$$i\sigma_\pm = i\beta(1 + x_\pm)^2 K(x_-, x_+),$$ (10.52)

were

$$K(x_-, x_+) = (1 + x_+)^3 \phi_x(x_+) - (1 + x_-)^3 \phi_x(x_-),$$ (10.53)

and

$$\beta = \frac{1}{10} \pi a \Sigma (\omega a)^5.$$ (10.54)

A value $\beta = 10^{-4}$ is typical for a torus of mass $0.1 M_\odot$ and a radius $a = 3M$. Figure (10.4) shows the destabilizing effect of $\beta = 2 \times 10^{-4}$.

Exercises

1. By inspection of Figure (10.4), estimate the phase velocity of the $m = 1$ and $m = 2$ modes. Is gravitational radiation by the $m = 1, 2$ modes at exactly twice the angular velocity of the torus? Show that, however, the gravitational-wave luminosity of the torus due to its $m = 1$ multipole mass-moment is anomalously small.

2. Nonlinearities in wave-motion of finite amplitudes introduce coupling between the various wave-modes. What implications may this have for the gravitational-wave spectrum of the torus?

3. The presented perturbations are buckling modes, associated with the same sign of the radial velocity at the inner and the outer face. In contrast, two-dimensional incompressible vortical modes are defined by $\sigma \Delta \psi = k(2B)_x \psi$ in terms of the stream function ψ ($u = \psi_y$ and $v = \psi_x$). Derive this equation. These vortical modes are generally singular with divergent azimuthal velocities when $\psi \neq 0$ at the turning point $\sigma = 0$, although of finite net azimuthal momentum (ψ remains continuous). Elaborate a numerical approach to find these vortical eigenmodes.

11

Phenomenology of GRB supernovae

"Since you are now studying geometry and trigonometry, I will give
you a problem. A ship sails the ocean. It left Boston with a cargo of
wool. It grosses 200 tons. It is bound for Le Havre. The mainmast is
broken, the cabin boy is on deck, there are 12 passengers aboard, the
wind is blowing East-North-East, the clock points to a quarter past three
in the afternoon. It is the month of May. How old is the captain?"
Gustave Flaubert (1821–80), in a letter to his sister Cavoline, 1843.

Discovery of GRBs. Gamma-ray bursts were serendipitously discovered by the
nuclear test-ban monitoring satellites Vela (US), (Figure 11.1) and Konus (USSR).
Soon afterwards, it became clear that these events were not thermonuclear experiments of terrestrial origin, but rather a new astrophysical transient in the sky. These
data were first released in 1973 by R. Klebesadel, I. Strong and R. Olson[296]
and in 1974 by E. P. Mazets, S. V. Golenetskii and V. N. Ilinskii[368]. The first
detection of a gamma-ray burst in the Vela archives is GRB 670702 (Figure 11.2).
In the footsteps of Vela and Konus, a number of other gamma-ray burst detection
experiments and missions were conducted[12]: *Apollo* 16, *Helios* 2, *HEAO*-1,
International Sun Earth Explorer 3, *Orbiting Geophysical Observatory* 3 and 5,
Orbiting Solar Observatory 6–8, *Prognoz* 6–7, *Pioneer Venus Orbiter* (1978–92),
Konus and *SIGNE* on *Venera* 11–12 and *Wind, Transient Gamma-ray Spectrome-
ter* (TGRS) on *Wind, SIGNE* 3, *Solar Maximum Mission* (1980–89), *Solrad* 11AB,
MIR Space Station, *GINGA, WATCH* and *SIGMA* on *GRANAT* and *EURECA*,
and *Ulysses*.

The BATSE Catalog. Gamma-ray bursts come in two varieties – short and long –
whose durations are broadly distributed around 0.3 s and 30 s, respectively, in the
BATSE data of C. Kouveliotou (1999) *et al.*[305, 401] (Figure 11.3). The *Burst
and Transient Source Experiment* (BATSE[190, 401], launched in 1991, and shown
in Figure 11.3) confirmed the isotropic distribution in the sky[262]. Its unprece-
dented sensitivity unambiguously revealed a deficit in faint burst in a number versus
intensity distribution different from a $-3/2$ powerlaw. C. A. Meegan *et al.*[369]

Figure 11.1 The Vela satellite. (Courtesy of NASA Marshall Space Flight Center, Space Sciences Laboratory.)

hereby showed that they are cosmological in origin. The cosmological origin implies isotropic equivalent luminosities on the order of 10^{51} erg s^{-1}.

The cosmological origin of GRBs is further supported by a non-Euclidean distribution, given a $< V/V_{max} >$[482] of 0.334 ± 0.008[481], substantially less than the Euclidean value 1/2[415]. For short and long bursts, $< V/V_{max} >= 0.385 \pm 0.019$ and $< V/V_{max} >= 0.282 \pm 0.014$, respectively, both distinctly less than 1/2[290]. Short bursts might be disconnected from star-forming regions, and might be produced by black-hole–neutron-star coalescence[404], possibly associated with hyperaccretion on to slowly rotating black holes. Evidence to this scenario is not yet conclusive[242].

Long GRBs represent highly non-thermal gamma-ray emissions, ranging from a few keV up to tens of GeV. These emissions show spectral evolution from hard-to-soft[397, 193, 435]. The GRB-emissions over the BATSE energy range

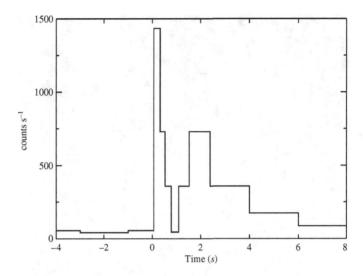

Figure 11.2 The light curve of GRB 670702, the first GRB detected by the Vela satellites (Klebasadel & Olson, Courtesy of NASA Marshall Space Flight Center, Space Sciences Laboratory.)

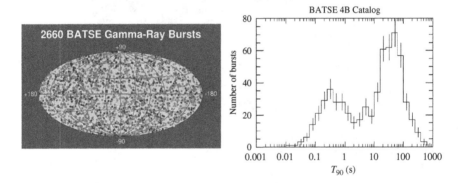

Figure 11.3 (*Left*) The isotropic angular distribution, shown in galactic coordinates, of GRBs in the BATSE 4B catalog indicates a cosmological origin of GRBs. (*Right*) The bimodal distribution of durations of short GRBs (T_{90} about 0.3 s) about long GRBs (T_{90} about 30 s) in the 4B Catalog, based on integrated lightcurves over all four channels ($E > 20$ keV). (Courtesy of NASA Marshall Space Flight Center, Space Sciences Laboratory.)

of from 30 keV to 2 MeV can be fitted by a Band spectrum[27] in terms of three parameters, consisting of low- and high-energy powerlaws connected by an exponential. The peak energies thus estimated show a broad distribution around 200 keV. Gamma-ray burst lightcurves often show rapid time variability, which reveals a compact source (with short timescale variability[429]).

The nonthermal gamma-ray emissions are well described by shock-induced dissipation of kinetic energy in ultrarelativistic plasmas by M. J. Rees and

P. Mészáros[451, 452]. A small baryon content suffices to convert the initially baryon-free radiation into kinetic energy with high Lorentz factor, described by A. Shemi and T. Piran[493]. These baryon-poor plasmas, in turn, can dissipate their energy in radiation by developing shocks internally[452] due to time-variability at the source, or in shocks upon interaction with the environment[451]. This *relativistic fireball shock model* grew out of an earlier *fireball* model [107, 402, 236, 493].

The modeling of GRBs by dissipation of kinetic energy in relativistic plasmas provides dramatic predictions for *lower* energy emissions, contemporaneous or subsequent to the GRB itself. This development serves to exemplify one of the few instances in which theory explaining contemporary observations defines important future observations. Theory further serves to point towards underlay correlations in the gamma-ray emissions which hitherto appeared as independent features, e.g.[167].

The observed high peak luminosities and time variability led B. Paczyński and J. E. Rhoads[406] to pose the existence of ultrarelativistic ejecta from a compact source. By appealing to an analogy to supernova remnants and radio galaxies, these authors predicted the existence of subsequent low-energy radio emissions as these ejecta deccelerate against the instellar medium. J. I. Katz[288, 289] independently predicted a broad spectrum of subsequent lower energy X-ray and radio synchrotron emissions from the debris of relativistic magnetized blast waves. M. J. Rees and P. Mészáros[449] derived predictions for contemporaneous lower-energy emissions in x-rays down to optical/UV in their model of relativistic plasmas decelerating against the intersteller medium. Late-time X-ray, optical and lower-energy emissions have been considered by P. Mészáros and M. J. Rees[354], and, in X-rays, by M. Vietri[575].

BeppoSax[436]: GRB afterglows and distances. The statistical view on the GRB landscape changed with the discovery by E. Costa *etal.*[135] of an X-ray afterglow (2–10 keV), (Figure 11.4) to GRB 970228 by the Italian-Dutch satellite BeppoSax, launched in 1996. This BeppoSax detection provided accurate localization, enabling J. van Paradijs[547] to point the Isaac Newton Telescope and the William Herschel Telescope in their detection of the first optical afterglow during X-ray observations of GRB 970228 Figure (11.4). The X-ray afterglows to GRB 970228 were also seen by A. Yoshida *etal.* [612] in observations by the Japanese satellite ASCA and by F. Frontera *etal.* [209] in observations by the German satellite ROSAT. This gamma-ray burst was also seen by K. Hurley *etal.*[274] in observations by Ulysses.

These lower-energy X-ray and optical afterglow emissions agree remarkably well with the previously mentioned predictions by the fireball model[598, 226, 455, 426, 427]. Even lower-energy, radio-afterglow emissions have been

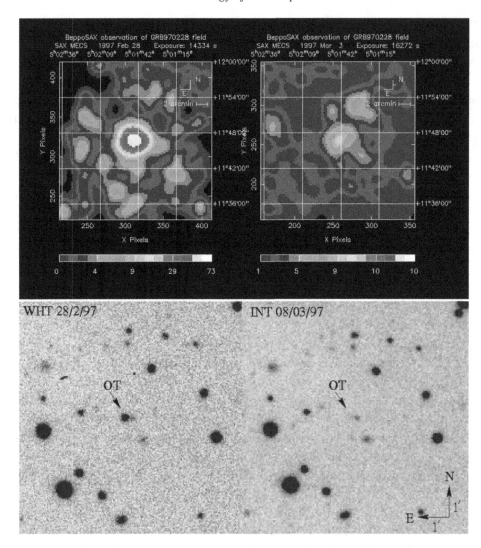

Figure 11.4 (*Top*) The X-ray source 1SAXJ0501.7 + 1146 in the error box of GRB 970228 detected by the BeppoSax Medium Energy Concentrator Spectrometer (2–10 keV). It represents an X-ray afterglow to GRB 970228, given a chance coincidence of 10^{-3}. Color refers to counts s^{-1} (white: $31\,s^{-1}$, green: $6\,s^{-1}$, grey: 0–$1\,s^{-1}$). The X-ray flux faded by a factor of 20 in 3 days. (Reprinted with permission from[135]. ©1997 Macmillan Publishers Ltd.) (*Bottom*) Follow-up identification of an optical transient by comparison of an early exposure by the William Herschel Telescope (WHT) and a late time 2.5 ks exposure by the Isaac Newton Telescope (INT). The optical decay is evident relative to the constant luminosity of a nearby faint M dwarf. (Reprinted with permission from[547]. ©1997 Macmillan Publishers Ltd.)

discovered by Frail, *et al.*[175] in GRB970228, as well as in a number of other cases[198, 310]. When present, these emissions can provide quantitative constraints on the fireball model[574, 195]. However, radio afterglows are not always observed (GRB970228[196]) while, if observed, their association to a fireball is not always unambiguous (GRB991216[194]). Ambiguities may arise as a result of combined radio afterglows from the deceleration of highly beamed ultrarelativistic baryon-poor outflows superimposed on subrelativistic unbeamed supernova ejecta. These processes have discrepant rates of late decline (the former being faster than the latter[342]). The reader is further referred to reviews by Piran[426, 427] and Mészáros[353].

No less significant than the afterglow phenomenon is the direct distance determination to the BeppoSax burst GRB 970508[433, 14]. Rapid follow-up by M. R. Metzger *et al.*[371] to the optical afterglow emission[73, 161, 501, 413, 106] provided an optical spectrum with absorption lines in FeII and MgII – redshifted at $z = 0.835$ in a star-forming dwarf galaxy[66]. A radio afterglow was discovered by D. A. Frail *et al.*[195]. In other cases, spatially coincident galaxies have been identified after the GRB-afterglow event. Notably, the *Hubble Space Telescope* revealed a galaxy[210] with redshift $z = 0.695$[67] in the error box of GRB 970228 (Figure 11.4). GRB 970228 appeared to be radio-quiet[496]. These redshift determinations formally provide a lower limit to the redshift of GRBs. The low probability of foreground galaxies, however, suggests that the redshift is that of a host galaxy. These redshifts are typically found to be of order 1, providing direct evidence of the cosmological origin of long GRBs such as those listed in Table 11.1. It confirms earlier suggestions on the cosmological origin by B. Paczyński[403]. Short GRBs, in contrast, do not appear to feature any afterglow emissions, which prohibits direct redshift identifications. Their cosmological origin remains based on an isotropic distribution and a $< V/V_{max} > < 1/2$.

Beyond fireballs: relativistic beamed ejecta. Recent indications of linear polarization in GRB 021206 suggests evidence of polarization in the gamma-ray emissions[126]. Various explanations have been proposed:

1. A certain amount of polarization can be attributed to synchrotron radiation[427, 434, 353, 428, 323]. Magnetic fields may represent an essential element in the creation of ejecta or outflows by long-lived inner engines. These outflows should then be beamed, or at least highly anisotropic. This becomes apparent in achromatic breaks in lightcurves (geometrical beaming). D. A. Frail *et al.*[196] infer a beaming factor of the observed population – clustered around a redshift of about 1 – around 500[196]. This defines a reduction of the isotropic equivalent energy in gamma-rays to a true GRB energy of about 3×10^{50} erg.

Table 11.1 *A redshift sample of thirty-three gamma-ray bursts.*

GRB	Redshift z	Photon flux (b)	Luminosity (c)	θ_j (d)	Instrument
970228	0·695	10	2.13×10^{58}		SAX/WFC
970508	0·835	0·97	3.24×10^{57}	0·293	SAX/WFC
970828	0·9578	1·5	7.04×10^{57}	0·072	RXTE/ASM
971214	3·42	1·96	2.08×10^{59}	> 0·056	SAX/WFC
980425	0·0085	0·96	1.54×10^{53}		SAX/WFC
980613	1·096	0·5	3.28×10^{57}	> 0·127	SAX/WFC
980703	0·966	2·40	1.15×10^{58}	0·135	RXTE/ASM
990123	1·6	16·41	2.74×10^{59}	0·050	SAX/WFC
990506	1·3	18·56	1.85×10^{59}		BAT/PCA
990510	1·619	8·16	1.40×10^{59}	0·053	SAX/WFC
990705	0·86			0·054	SAX/WFC
990712	0·434	11·64	7.97×10^{57}	> 0·411	SAX/WFC
991208	0·706	11·2*	2.48×10^{58}	< 0·079	Uly/KO/NE
991216	1·02	67·5	3.70×10^{59}	0·051	BATSE/PCA
000131	4·5	1·5*	3.05×10^{59}	< 0·047	Uly/KO/NE
000210	0·846	29·9	1.03×10^{59}		SAX/WFC
000301C	0·42	1·32*	8.37×10^{56}	0·105	ASM/Uly
000214	2·03				SAX/WFC
000418	1·118	3·3*	2.27×10^{58}	0·198	Uly/KO/NE
000911	1·058	2·86	1.72×10^{58}		Uly/KO/NE
000926	2·066	10*	3.13×10^{59}	0·051	Uly/KO/NE
010222	1·477				SAX/WFC
010921	0·45				HE/Uly/SAX
011121	0·36	15·04*	6.63×10^{57}		SAX/WFC
011211	2·14				SAX/WFC
020405	0·69	7·52*	1.58×10^{58}		Uly/MO/SAX
020813	1·25	9·02*	8.19×10^{58}		HETE
021004	2·3				HETE
021211	1·01				HETE
030226	1·98	0·48*	1.35×10^{58}		HETE
030323	3·37	0·0048*	4.91×10^{56}		HETE
030328	1·52	2·93*	4.31×10^{58}		HETE
030329	0·168	0·0009*	7.03×10^{52}		HETE

(a) Compiled from S. Barthelmy's IPN redshifts and fluxes (http://gcn.gsfc.nasa.gov/gcn/) and J. C. Greiner's catalog on GRBs localized with WFC (BeppoSax), BATSE/RXTE or ASM/RXTE, IPN, HETE-II[260] or INTEGRAL (http://www.mpe.mpg.de/jcg/grbgeb.html).

(b) in $cm^{-2}s^{-1}$.

(c) Photon luminosities in s^{-1} derived from the measured redshifts and observed gamma ray fluxes for the cosmological model of Porciani and Madau[439].

(d) Opening angles θ_j in the GRB emissions refer to the sample listed in Table I of Frail *et al.*[196]. (∗) Extrapolated to the BATSE energy range 50–300 keV using the formula given in Appendix B of Sethi and Bhargavi[488].

2. Polarization of gamma-rays can be attributed to inverse Compton scattering of low-energy circumburst radiation[492]. Upscattering is envisioned to take place by ultra-relativistic ejecta from a GRB inner-engine[148, 141, 142, 323, 326].
3. Polarization is a consequence of scattering of gamma-rays against a surrounding baryon-rich wind[168]. This model is particularly attractive, as it supports wide-angle low-luminosity emissions consistent with GRB 980425. It predicts that polarization is potentially strong over a wide range of viewing angles.

These three mechanisms are to some extent non-exclusive. Either one of them is effective in creating polarization, and is conceivably relevant in a particular burst given a particular viewing angle. Gamma-ray bursts are notoriously diverse in their durations and intermittent behavior, whereby at any one given epoch, one of these might dominate. Polarization measurements alone are probably not sufficient to uniquely identify any of these scenarios.

The supernova connection. Long GRBs are a now recognized as a subpopulation of Type Ib/c supernovae. The evidence includes GRB 980425/SN1998bw shown in Figure 11.5 [224, 514, 580], GRB 030329/SN2003dh[506, 265] shown in Figure 11.6, and an excess bump in the optical after about 1 week in the afterglow emissions[69, 310, 457]. There are now four GRB-supernova

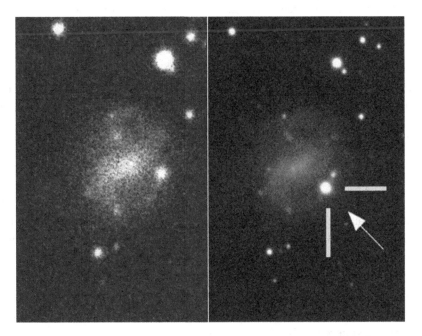

Figure 11.5 Shown is the optical identification of the supernova associated with GRB 980425. (*Left*) The Digital Sky Survey (DSS) image prior to GRB 980425. (*Right*) the R band image by the New Technology Telescope (NTT). (Reprinted with permission from[224]. ©1998 Macmillan Publishers Ltd.)

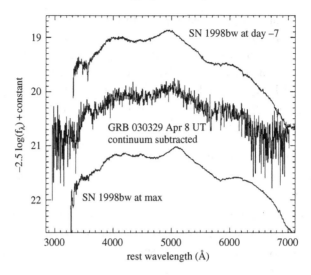

Figure 11.6　The optical spectrum of the Type Ic SN2003dh associated with GRB 030329 is very similar to that of the Type Ic SN1998bw of GRB980425 1 week before maximum; GRB 030329 displayed a gamma-ray luminosity of about 10^{-1} below typical at a distance of $z = 0.167$, $D = 800\,\mathrm{Mpc}$, whereas GRB 980425 was observed at an anomalously low gamma-ray luminosity (10^{-4} below typical) and small distance ($z = 0.008$, $D = 37\,\mathrm{Mpc}$). At the same time, their supernovae were very luminous with inferred ^{56}Ni ejecta of about $0.5 M_\odot$. (Reprinted with permission from[506]. ©2003 The American Astronomical Society.)

associations known, including GRB 021211/SN2002lt ($z = 1.0060$)[154, 155] and GRB 031203/SN2003lw ($z = 0.1055$)[512, 525, 361, 221]. Afterglow emissions to GRB 030329 include optical emissions[442] with intraday deviations from powerlaw behavior[541], possibly reflecting an inhomogeneous circumburst medium or latent activity of the inner engine[113, 442]. Retrospectively, an early indication of a supernova may be found in the late-time optical lightcurve of GRB 970228[456, 223, 457].

The supernova association is consistent with the identification of an underlying host galaxy, notably to GRB970228 by K. C. Sahu, M. Livio, L. Petro *et al.*[469], to GRB970508 by J. S. Bloom, S. G. Djorgovski, Kulkarni, S. R. *et al.*[66], to GRB980326 by P. J. Groot, T. J. Galama , P. M. Vreeswijk *et al.*[240, 69], and to GRB980703 by S. G. Djorkovski, S. R. Kulkarni, J. S. Bloom *et al.*[160]. More precisely, a number of GRBs are observed in association with star-forming regions[68]. When present, radio emissions may provide valuable information on an underlying supernova[70].

The association to supernovae indicates a correlation to the cosmic star formation rate. (And might be used conversely to infer the star formation rate at high redshift[478].) This, in turn, implies a true-to-observed event rate of about

450[570], consistent with the geometrical beaming factor of[196]. The true-but-unseen GRB event rate corresponds to a local event rate of about one per year within a distance of 100 Mpc. It defines a relatively small branching ratio of less than 1% of Type Ib/c supernovae into GRBs[439, 539, 471].

SN1998bw, associated with GBR980425, happened to be unusually close, allowing for detailed study of the supernova properties. SN1998bw is aspherical, representing a true kinetic energy of about 2×10^{51} erg as calculated by P. J. Höflich, Wheeler and Wang[268]. All core-collapse SNe are strongly nonspherical[267], as in the Type II SN1987A[266] and in the Type Ic SN1998bw[268], based, in part, on polarization measurements and direct observations. Observed is a rotational symmetry with axis ratios of 2 to 3 in velocity anisotropy. This generally reflects the presence of rotation in the progenitor star and/or in the agent driving the explosion.

Type Ib/c SNe tend to be radio-loud[539], as in SN1990B[546, 112, 586, 47]. This includes GRB 980425/SN1998bw as observed by S. Kulkarni *et al.*[313] and K. Iwamoto[279] as the brightest Type Ib/c radio SN at a very early stage[585]. No such supernova radio-signature appears to be present in GRB 030329 (but see Willingale *et al.*[600]). Radio emissions in these SNe are well described by optically thick (at early times) and optically thin (at late times) synchrotron radiation of shells expanding into a circumburst medium of stellar winds from the progenitor star[341].

Furthermore, some of these GRB supernovae might feature bright X-ray emission lines. Tentative evidence includes GRB 970508 by L. Piro[432], GRB 970828 by A. Yoshida[613], GRB 991216 by L. Piro *et al.*[434], GRB 000214 by A. Antonelli *et al.*[17] and GRB 011211 by J. N. Reeves *et al.*[454]. The aforementioned X-ray line-emissions in GRB 011211 might be excited by high-energy continuum emissions of much larger energies[229] in various scenarios [322, 325]. For the Type Ib/c supernova association with GRBs, this led S. E. Woosley, Eastiman and Schmidt, to suggest the presence of a new explosion mechanism[611] in various Scenarios[322, 325]. At present, the observational evidence for X-ray lines is not universally accepted. The upcoming *Swift* mission is expected to put this issue "under the microscope."

The *astronomical* mystery of long GRBs is solved through their association to supernovae, providing a link to stellar evolution[608]. They probably represent the explosive endpoint of binary evolution of massive stars[404]. It confirms the earlier suggested association to supernovae by Stirling Colgate, except that the observed gamma-rays are produced not by shocks in the expanding remnant stellar envelope but by dissipation of kinetic energy in an ultrarelativistic jet in internal or external shocks[451, 452]. G. E. Brown *et al.* propose that their remnants may be found in some of the current soft X-ray binaries in our galaxy[87].

The mystery of the *physical mechanism* producing long GRBs – ultrarelativisitic baryon-poor jets *in* aspherical supernovae of massive stars – poses a challenge which could guide us to new and "unseen" phenomena, perhaps also new or untested physics.

11.1 True GRB energies

In a number of cases, GRBs display achromatic breaks in their lightcurves such as GRB 990510[253] shown in Figure (11.7) and GRB991216[244]. This confirmed earlier indications of nonspherical jets in GRB 990123 observed by S. R. Kulkarni *et al.*[313, 311, 312] and Fruchter *et al.*[211]. Gamma-ray burst emissions are either limited to two cones or are highly anisotropic (in two directions). The latter either takes the form of outflows with anisotropic emissions inside a cone ("structured jets"[169, 616, 465, 411]), or a superposition of conical emissions and low-luminosity emissions over arbitrary angles[570] (Figure 11.8).

Achromatic breaks in the lightcurves indicate a transition between an ultrarelativistic phase and a relativistic or non-relativistic phase of a radiative front[459,

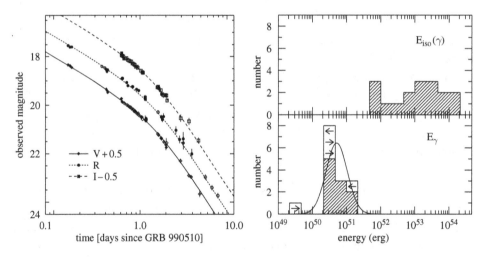

Figure 11.7 (*Left*) Optical lightcurves at V, I, R-bands observed in GRB 990510. The achromatic break in these lightcurves takes place around 1 day after the GRB, indicative of a geometric transition to a non-relativistic radiative front whose luminosity is opening-angle limited. The estimated powerlaw indices are -0.82 ± 0.02 before and -2.18 ± 0.05 after the break. (Reprinted with permission from[253].) ©The Astrophysical Journal. (*Right*). The distribution of apparent isotropic gamma-ray emissions in a sample of GRBs with individually measured redshifts and opening angles (*top*) and the true GRB-energies following a correction by the inferred spherical opening angle (*bottom*). Arrows refer to upper and lower limits. (Reprinted with permission from[196]. ©2001 The American Astronomical Society.)

505, 460, 473, 472]. In an ultrarelativistic phase, the observed luminosity is limited by a finite surface patch on the radiative front, whose angular size is equal to the reciprocal of its Lorentz factor. As the front propagates into the environment and gradually slows down, the observed patch grows in size until it reaches the physical angular size of the front. This transition introduces a break in the light curve, irrespective of color. As a function of the expected host environment, the time of transition defines the opening angle θ_j of the front, as reviewed in[428].

The true energy in gamma-rays from GRBs is given by the observed isotropic equivalent emission reduced the average beaming factor $1/f_b = 500$[196], where $f_b = \theta_j^2/2$. The true GRB-energies thus emitted in bipolar jets is on average 3×10^{50} (Figure 11.7, right window). The distribution of true-GRB energies is hereby also much narrower than the distribution of isotropic equivalent energies. This has been interpreted to reflect a standard energy reservoir[196].

An anticorrelation between the observed opening angle and redshift shown in Figure (11.9) points towards a deviation from conical outflows (alternative (a) in Figure (11.8)). It favors structured jets or strongly anisotropic outflows, i.e. alternatives (b) and (c) in Figure (11.8). The latter includes wide-angle GRB emissions which are extremely weak, as in GRB 980425. Given that the event rate of GRB 980425 at $D = 34$ Mpc is roughly consistent with one per year within $D = 100$ Mpc, these wide-angle emissions may also be standard. With alternative (c) in Figure (11.8), GRB980425 ($E_{\gamma,iso} \simeq 10^{48}$erg, $z = 0.0085$) is *not* necessarily anomalous unless calorimetry shows otherwise. GRB 030329 ($E_\gamma \simeq 3 \times 10^{49}$erg,

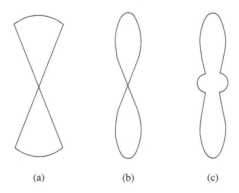

      (a)   (b)   (c)

Figure 11.8 Possible radiation patterns (not to scale) of beamed gamma-ray emissions: conical (a), structured (b) and strongly anisotropic accompanied by weak emissions over arbitrary angles (c). Both (b) and (c) give rise to an anticorrelation of observed opening angle with redshift. (c) allows all nearby events to be detected, irrespective of orientation.

$z = 0.167)[442]$ may be considered to be seen slightly off-axis in either alternative (b) or (c) in Figure (11.8).

11.2 A redshift sample of 33 GRBs

There is a rapidly growing list of GRBs with individually determined redshifts, based on a localizations by a number of different satellites. Table 11.1 lists thirty-three GRBs with redshift and the instrument in which the event was detected.

The sample of Table 11.1 is biased strongly towards low redshifts. Conical emissions introduce an orientation cut-off in any sample, regardless of the sensitivity of the instrument, whereas highly anisotropic emissions introduce an orientation cut-off which decreases with instrumental sensitivity ((b) or (c) in Figure 11.8). In the ideal limit of infinite sensitivity, all highly anistropic events are observed, and the sample becomes unbiased. In a flux-limited sample, we detect mostly events which are pointed towards us or those that are extremely close. The latter are thus apparent even at low intrinsic luminosities such as GRB980425. In alternative (c),

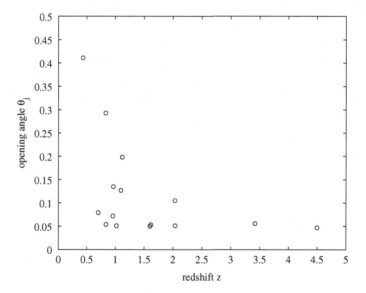

Figure 11.9 Shown is a plot of the opening angle θ_j of GRB emissions versus redshift z in the sample of Frail *et al.*[196], as derived from achromatic breaks in the GRB lightcurves. The results indicate an anticorrelation between θ_j and z. For standard GRB energies, this introduces a peak luminosity function of GRBs which is correlated with the beaming factor. This allows the beaming factor to be determined also in terms of the unseen-but-true GRB event rate to the observed GRB event rate, using the sample of 33 GRBs with individually measured redshifts shown in Table 11.1. (Reprinted with permission from[570]. ©2003 The American Astronomical Society.)

but not (a) or (c) in Figure 11.8, the true redshift distribution would be observed in the ideal case of a zero-flux limit.

11.3 True GRB supernova event rate

The observed redshift distribution of Table 11.1 can be contrasted with the underlying redshift distribution of the cosmological star-formation rate. The latter provides the redshift distribution of the true GRB event rate, up to an overall scaling factor. This comparison can be used to infer the orientation averaged GRB-luminosity function. To leading order, the intrinsic GRB-luminosity function can be assumed to be redshift-independent, neglecting any intrinsic cosmological evolution of GRB-supernova progenitors.

Assuming that the GRB luminosity function is redshift-independent, i.e. without cosmological evolution of the nature of its progenitors, consider a lognormal probability density for the luminosity shape function, with mean μ and width σ given by

$$p(L) = \frac{1}{(2\pi)^{1/2}\sigma L} \exp\left(\frac{-(\log L - \mu)^2}{2\sigma^2}\right),$$ (11.1)

where log refers to the natural logarithm and L is normalized with respect to $1\,cm^{-2}/s$. Optimal parameters of this model, assuming a flat Λ-dominated cold dark matter cosmology with closure energy densities $\Omega_\Lambda = 0.70$ and $\Omega_m = 0.30$, are (van Putten & Regimbau[570])

$$(\mu, \sigma) = (124, 3) \pm (2, -0.4).$$ (11.2)

This notation means that the estimated parameters can be either (122,3.4), (123,3.2), (125,2.8), or (126,2.6), but not (122,2.6), for instance. These results compare favorably with the expectations of Sethi and Bhargavi[488], who derive a lognormal luminosity function with $\mu = 129$ and $\sigma = 2$ from a different flux limited sample.

The observed redshift distribution and the redshift distribution predicted by the star formation rate are shown in Figure (11.10) in case of optimal parameters (11.2). The fraction of *detectable* GRBs as a function of redshift,

$$F(z) = \frac{dR_{detect}}{dR_{GRB}(z)} = \int_{L_{lim}(z)} p(L)dl,$$ (11.3)

shows a steep decrease in $F(z)$ as the luminosity threshold increases, making high-redshift GRBs less likely to be detected. The fluxes derived from our luminosity function in 50–300 keV have been extrapolated to the IPN range of 25–100 keV, assuming an E^{-2} energy spectrum and using the formula given in Appendix B of Sethi and Bhargavi[488]. The conversion factor from erg cm^{-2}/s

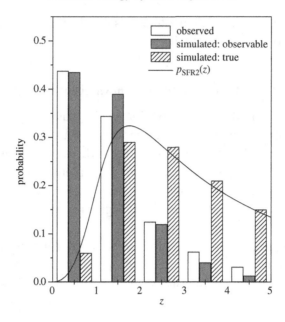

Figure 11.10 Three redshift distributions: the observed sample derived from Table 11.1 (white), the true sample assuming the GRB event rate is locked to the star formation rate (hachured), and the sample of detectable GRBs predicted by a lognormal peak luminosity distribution function (grey). The continuous line represents the cosmic star formation rate according to a Λ-dominated cold dark matter universe. (Reprinted from[570]. ©2003 University of Chicago Press.)

to photon cm^{-2}/s has been taken to be 0.87×10^{-7}, and the sensitivity threshold equal to 5 photon cm^{-2}/s[273].

For the optimal parameters (11.2), we find a true-to-observed GRB event rate

$$1/f_r = 450. \tag{11.4}$$

The factor $1/f_r$ is between 200–1200 in the error box of (11.2). This true-to-observed GRB event (11.4) is independent of the mechanism providing a broad distribution in GRB luminosities. Without further input, our results may reflect isotropic sources with greatly varying energy output, or beamed sources with standard energy output and varying opening angles.

The fraction $1/f_r$ is strikingly similar to the GRB-beaming factor $1/f_b$ of about 500 derived by Frail *et al.*[196]. We conclude that the GRB peak luminosities and beaming are strongly correlated. A strong correlation between peak luminosities and beaming is naturally expected in conical outflows with varying opening angles with otherwise standard energy output, as well as alternative (c) with standard geometry in Figure 11.8. We favor the latter in view of GRB980425/SN1998bw. This correlation implies an anticorrelation between *observed* beaming and distance

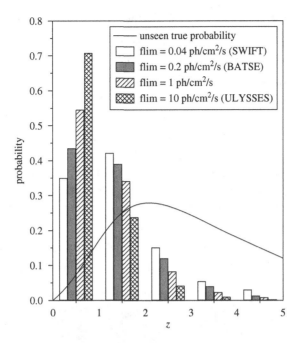

Figure 11.11 Simulation of the observed GRB redshift distribution as a function of flux limit, set by various instruments including the upcoming *Swift* mission. Here, the GRB event rate is locked to the SFR, using the best-fit lognormal peak luminosity distribution function. *HETE* − 2 thresholds are 0.21 (soft X-ray camera), 0.07 (wide-field X-ray monitor), and 0.3 (French Gamma Telescope) in units of cm^{-2}/s. (Reprinted from[570]. ©2003 The American Astronomical Society.)

such that leading-order $\theta_j z \sim$ const. Figure 11.9 shows that this anticorrelation holds approximately in the sample of Frail *et al.*[196].

The phenomenology of GRB supernovae can be summarized as follows.

1. They are cosmological in origin, and last tens of seconds.
2. GRBs represent shocked emissions of ultrarelativistic kinetic energy in magnetized, beamed baryon-poor outflows, with lower-energy after glow emissions in X-rays, optical and radio.
3. True GRB energies cluster around 3×10^{50} erg.
4. The true-to-observed GRB event rate is 450–500.
5. They are produced by Type Ib/c SNe with branching ratio $(2-4) \times 10^{-3}$ in association with star-forming regions.
6. Type Ib/c are aspherical and are typically radio-loud.
7. GRB-SNe show bright X-ray line-emissions in a number of cases.
8. GRB-SNe probably take place in compact binaries.
9. GRB-SNe remnants are probably black holes with a stellar companion.
10. GRB-SNe late-time remnants are probably soft X-ray transients.

11.4 Supernovae: the endpoint of massive stars

Stars have a finite lifetime, set by their ability to support thermal pressure by nuclear burning in their core. Hydrogen burns by fusion into He. The lifetime of a star on the main sequence during this initial stage of H-burning is hereby a function of the mass of the star,

$$T_{MS} \simeq 13 \left(\frac{M}{M_\odot} \right)^{-5/2} \text{Gyr.} \tag{11.5}$$

Note the steep decline in lifetime with the mass of the star. Following H-burning, He and its products are converted into heavier elements. Ultimately, the core of a star is depleted of fuel, leaving iron at its center. After cooling, the core collapses until it is supported by electron-degenerate pressure. Degenerate pressure will suffice as support against gravitational self-interaction, provided the mass of the star is less than about $\simeq 4M_\odot$. In this event, the remnant is a white dwarf.

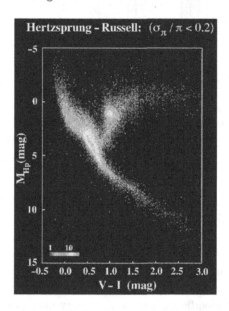

Figure 11.12 Hertzsprung–Russell diagram from the Hipparcos catalog, showing the zero-age main sequence stars (ZAMS, in their H-burning phase) on a diagonal in a color-magnitude diagram. This represents about 90% of the stars. The colors indicated correspond to a surface temperate range of about $1/T = 1/30000\text{K} - 1/3000\text{K}$. The increasing luminosity on the main sequence corresponds to an increasing stellar mass. The high-luminosity low-temperature branch on the upper right represents giants and supergiants, such as Betelgeuse (α-Ori), which are in their short-lived (tens of M yr) He-burning phase. A few objects in the lower left corner represent white dwarfs, notably Sirius B, as low-luminosity high-temperature compact objects. They reside on a narrow strip (only a few are sampled in the Hipparcos catalog), consistent with cooling by black body radiation[511].

Higher mass stars evolve differently, and are believed to produce supernovae following the collapse of the core. The details of the explosion mechanism are still not well understood, although a shock rebounce on the core is probably part of the process. The remnant in this case is a neutron star or black hole, which is produced promptly during collapse or as a result of the shock rebounce – possibly aided by additional accretion on to the remnant in the core. An in-depth review of supernova physics is given by[606].

The density of these compact objects in our galaxy can be determined from the Salpeter birthrate function[490]

$$\psi_s d(M/M_\odot) = 2 \times 10^{-12} (M/M_\odot)^{-2.35} d(M/M_\odot) \ \mathrm{pc}^{-3}/\mathrm{yr} \qquad (11.6)$$

applied to the volume $V_d = 1.3 \times 10^{11} \ \mathrm{pc}^3$. Stars of mass $1 M_\odot$ or larger have ages shorter than the age of the galaxy – about 12 Gyr. The population of stars in this mass range has reached equilibrium in birth and death rates. White dwarfs form from low-mass progenitors (about $1 - 4 M_\odot$), neutron stars from intermediate-mass progenitors (about $4 - 10 M_\odot$), and black holes from high-mass progenitors ($M > 10 M_\odot$). Integration gives densities $0.015 \ \mathrm{pc}^{-3}$ of white dwarfs, $0.002 \ \mathrm{pc}^{-3}$ of neutron stars and $0.0008 \ \mathrm{pc}^{-3}$ of black holes.

Ultimately, all stars die in collapse, as the nucleus runs out of nuclear fuel to provide thermal pressure support to the star. This may be rather uneventful, leaving a compact remnant in the form of a white dwarf, or explosive in the form of a supernova, leaving a remnant in the form of a neutron star or black hole. If the star is a member of a binary, the white dwarf may be accreting, in which case it could be induced to a final explosive burning phase leaving no remnant at all.

11.4.1 Classification of supernovae

Currently, the supernovae classification is entirely observational in the electromagnetic spectrum[594, 186, 187, 245]. It is preferred to do so at an early stage within about 1 month[222]. Broadly, supernovae fall into two groups: H-deficient supernovae (Type I) and H-rich supernovae (Type II). The supernovae classification is done according to spectra. Current reviews (e.g. Filippenko[186], Turatto[539, 540], Cappellaro[99], present the following picture of the various supernovae types as exemplified in Figures (11.13) and (11.14) by M. Turatto[539].

Spectral features reveal the presence of primary chemical elements H, He and Si. Additional important chemical elements are Ca, Ca, S and Mg, as well as $^{56}\mathrm{Ni}$, $^{56}\mathrm{Co}$ and $^{56}\mathrm{Fe}$. Heavier elements generally derive from the inner regions of the star. The latter three are closely related: $^{56}\mathrm{Ni}$ may decay into $^{56}\mathrm{Co}(\tau_{1/2} = 6.1\mathrm{d})$, and $^{56}\mathrm{Co}$ may decay into $^{56}\mathrm{Fe}(\tau_{1/2} = 77.1\mathrm{d})$, both transitions by electron capture

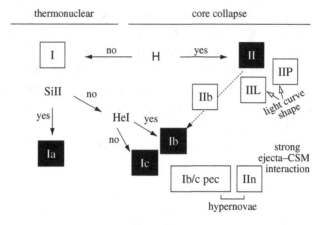

Figure 11.13 Classification scheme according to lines of various chemical elements of Type I-Ia, Ib-c (H-deficient), and different Type II (H-rich) supernovae. Type Ia are thermonuclear explosions of ^{12}C in white dwarfs, the other types are core-collapse events with distinct explosion mechanism(s). Type Ib/Ic and Type II are aspherical, wherein the Type Ib/c further may show anomalously high ejection velocites (about $0.1c$ in GRB011211). Type Ib/Ic hereby appear with isotropic equivalent kinetic energies in excess of 10^{52} erg[391] ("hypernovae"). Their true kinetic energy, corrected for asphericities, assume standard values of a few times 10^{51} erg (SN1998bw[268]). Type Ib/c are associated with "naked" stellar cores of initally massive stars as described by J. C. Wheeler[595], stripped of their H-envelope. Type II are associated with H-envelope retaining massive stars. The H-envelope in Type Ib/c is believed to be removed by winds (isolated type-WC Wolf-Rayet stars[608, 510], or through interaction with a companion star in a compact binary by mass-transfer[393, 609, 404]). Such interaction might also remove the He-envelope and/or might start late after core He-burning (Type Ic) (see[186, 87]). (Reprinted with permission from[539]. 2003 ©Springer-Verlag Berlin and Heidelberg).

or positron emission as discussed by H. A. Bethe, G. E. Brown and C. H Lee[53]. Consequently, detected Fe lines need not represent ejection of Fe itself, but rather decaying Ni[53]. In contrast, the lightest elements H and He represent ejecta from the outer layers of the progenitor star.

Supernova spectra consist characteristically of a thermal continuum and P-Cygni profiles – the sum produced by a spherically symmetric star and its stellar winds, the latter producing blue-shifted absorption in the direction of the observer.

11.4.2 Type Ia

Type Ia supernovae such as SN1987L and SN1987N are thermonuclear explosions of C-O white dwarfs with luminosities of about 10^{43} erg s^{-1} and total kinetic energies of about 10^{51} erg[392, 606]. These events are probably triggered by

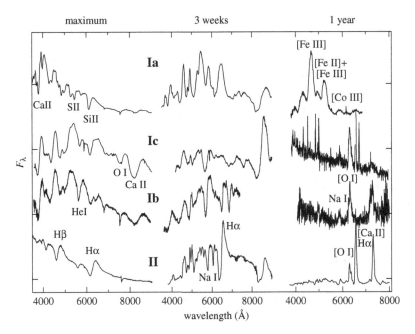

maximum 3 weeks 1 year

wavelength (Å)

Figure 11.14 Examples of spectral evolution of H-deficient Type Ia (SN1996X[471]), Type Ib (SN1999dn: left, center; SN1990I: right) and Type Ic (SN1994I[125]: left, center; SN1997B: right), and H-rich Type II (SN1987A[422]). Type Ia show deep absorption at SII (6150A), which is absent in Type Ib/c. At late times, note the excess Fe-features in Type Ia, which are attributed to downscattered ^{56}Ni. Type Ib shows a prominent HeI-absorption feature, otherwise absent in Ic. Notice further the OI emission feature, strong in Ic and weaker in Ib at late times. If Type II also take place binaries[494] (e.g. SN 1993J[9, 367], rather than isolated stars), then it may form a continuous class with Type Ib/Ic. This is suggested by the temporal evolution of SN1987K and SN1993J (see[186]), which displayed a gradual disappearance of Hα absorption and a gradual appearance of OI. Perhaps Type II and Ib/Ic are determined by binary separation. (Reprinted with permission from[539]. 2003 ©Springer-Verlag Berlin and Heidelberg.)

accretion from a binary companion. Type Ia SNe show a characteristic absence of H and presence of a deep SiII absorption line near 6150Å (blue-shifted from 6347Å and 6371Å), along with late-time lines of Ca and Fe[186]. Their spectra and lightcurves are remarkably consistent, showing a rather tight *Phillips relation*[421, 246, 247, 461, 462] between the width of the lightcurves and brightness. The Phillips relation may be used to normalize their lightcurves, thus making Type Ia of great interest as calibrated distance markers to cosmology (following corrections for extinction[462] within $z < 1$). They are radio-quiet. Type Ia may be found in elliptical and spiral galaxies alike[100].

SNIa are triggered when their mass reaches the limiting Chandrasekhar mass $1.38 M_\odot$. The explosion must start before reaching this threshold, for otherwise collapse to a neutron star would occur. No such remnants are identified in case of SNIa. This onset is attributed to C-burning, sufficiently rapid to counter neutrino cooling. At nuclear efficiencies of the order of 1%, a binding energy of about 10^{50} erg, or about $10^{-4} M_\odot$, can be readily overcome by burning a few percent of a solar mass $M_\odot$ in ^{12}C.

The SNIa peak luminosity is linked directly to the amount of nuclear ashes produced. Most of this consists of radioactive ^{56}Ni. It decays into a large amount of Fe, about $(0.4 - 1.2) \times M_\odot$ in SN1991T[504]. Different amounts of ^{56}Ni presumably introduce variations in the peak luminosity according to Arnett's rule. Smaller, though spectrally important, amounts are in Si, S, Ar, and Ca, or about $0.2 M_\odot$. The amount of stable elements ^{54}Fe and ^{58}Ni is $0.1 M_\odot$ or less. These observational constraints have led to the conjecture that the star must first pre-expand to avoid electron capture, before expanding rapidly by fusion at densities of about $10^7 - 10^8$ g cm^{-3}. This has led to extensive explorations of various two-step burning mechanisms (e.g. by including hydrodynamical aspects (instabilities, turbulence, pulsations) and aspherical burning).

The decay

$$^{56}Ni + e^- \rightarrow {}^{56}Co + \nu + 3.0 \times 10^{16} \text{ erg/g} \tag{11.7}$$

$$^{56}Co + e^- \rightarrow {}^{56}Fe + \nu + 6.4 \times 10^{16} \text{ erg/g} \tag{11.8}$$

provides additional late-time energy of about 1.1×10^{50} erg $0.6 M_\odot^{-1}$. This energy output matches well with the observed optical light curve for up to a few months[130, 131, 80, 366, 46, 535, 101, 375, 593]. Their progenitors are low-mass stars, broadly of less than $10 M_\odot$ and, hence, relatively old stars. Detailed modeling of Type Ia is pursued by various groups, e.g.[252].

11.4.3 Type Ib/c

Like SNIa, supernovae of Type Ib/c lack H lines. They are associated with core collapse of H-envelope stripped stars of initially large mass[595].

SNIb/c lack in SII absorption lines. The observational difference between SIb/c is in He abundance: SNIb show strong HeI absorption lines around 5876Å, which are otherwise weak in SNIc. SNIc have been found with HeI around 10830Å in SN1994I. The HeI lines could be associated with gamma ray emission from the decay of ^{56}Ni and ^{56}Co.

Type Ib/c SNe appear to occur only in spiral galaxies[102]. They may be radio-loud, such as the event SN1998bw. As an absorption feature, Hα has been observed in the SNIb 1983N[125] and SN1984L (in Filippenko[186]), which may be attributed to the host environment.

The envelope stripping prior to the supernova event is believed to be due to a common envelope phase with a binary companion[390, 539]. They therefore could be rapidly rotating, due to transfer of orbital angular momentum during the common envelope phase. The supernova mechanism could hereby be rotationally powered by a compact and rapidly rotating core, possibly in the form of a black hole formed in the process of core collapse.

11.4.4 Type II and IIb

Type II and IIb supernovae are produced by stars which have retained their H-envelope, believed to be in a $10\text{--}15M_\odot$ mass range. The most exciting and revealing event is SN1987A in the LMC. Figure (11.15) shows the light curve of the neutrino emissions from this event.

The Type II have been subdivided according to shape of their optical lightcurves[186], e.g. those featuring a plateau (IIP) or a linear light curve (IIL) followed by an exponential decline attributed to the decay ^{56}Co into ^{56}Fe. Both are believed to have progenitors with an H envelope of more than one solar mass. They may be radio-loud, and an observed UV excess is attributed to Compton scattering of photospheric radiation by high-speed electrons in the shock-heated circumstellar medium[186, 539]. Type IIb are similar to Ib/c at late times, notably SN1993J[186, 367] in M81 is Type IIb, showing an early blue continuum, broad H and HI at 5876Å. At later times, it showed stronger He I $\lambda\lambda 5876, 6678, 7065$ similar to Ib. Others show particularly narrow He I emission lines and Na I absorption lines (Type IIn), which correspond to low-expansion velocities of about $1000\,\text{km s}^{-1}$. Type II may further show unresolved forbidden lines in O and Fe.

Type II SNe are *envelope-retaining*, rather than envelope-stripped SNe (Ib/c). This suggests the following Nomoto–Iwamoto–Suzuki sequence[390, 539]

$$\text{IIP} \rightarrow \text{IIL} \rightarrow \text{IIb} \rightarrow \text{Ib} \rightarrow \text{Ic} \tag{11.9}$$

in the order of decreasing H-mass envelope.

In principle, the physics of core collapse can be probed using neutrinos, as in 1987A, and gravitational radiation. Detection of a neutrino burst from 1987A dramatically confirmed the theory of core collapse in Type II SNe as that associated with the formation of matter at nuclear densities, which may have been a neutron or nucleon star[55, 263]. The latter was probably an object in transition to a black hole, since no remnant appears observable at present.

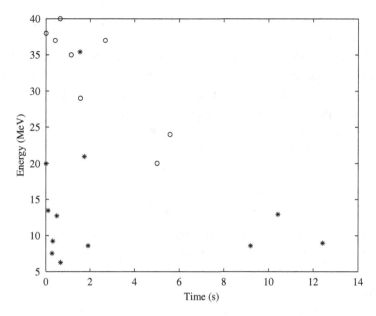

Figure 11.15 Light curve of MeV neutrino emission in the Type II event SN1987A in the LMC, compiled from Kamioka (stars) and IMB (circles) as listed in[95]. These emissions provide direct evidence of core collapse to supranuclear densities. Note the 10 s timescale of the neutrino burst, and the decay in energy by a factor of about 4. Because there appears to be no neutron star remnant, the neutrino emission is related to matter at nuclear densities in transition to a black hole. The duration of the burst is consistent with the diffusion timescale of neutrinos from a nucleon star[95, 86], as well as the free-fall timescale matter in core collapse. If the latter were rotating, the collapsing matter would briefly form a torus at nuclear densities.

Because of its proximity, the progenitor star of 1987A was identified, i.e. a blue giant B3 I (Sk-69 202). Its explosion energy was $E_k = 1 \times 10^{51}$ erg with an ejection of $0.07M_\odot$ in ^{56}Ni[186, 539]. Progenitor masses of other nearby events are known in case of SN1993J (13–$20M_\odot$[367]) SN1999gi ($< 9^{+3}_{-2}M_\odot$[539]) and SN1999em ($< 12\pm1M_\odot$[539]). This supports the notion that Type II progenitors are probably less massive than Type Ib/c progenitors.

11.5 Supernova event rates

Current observations of supernovae in ellipticals (Type Ia) and spirals (Type Ia, Type Ib/c and Type II) show the event rates[540]

$$\dot{N}(\text{Type Ia}) = 0.27, \quad \dot{N}(\text{Type Ib/c}) = 0.11, \quad \dot{N}(\text{Type II}) = 0.53 \qquad (11.10)$$

in units of $10^{-11}M_\odot$ 100 yr$^{-1}(H/75)^2$. In particular, the event rate of Type Ib/c is approximately 20% of Type II.

11.6 Remnants of GRB supernovae

Type II and Type Ib/c supernovae (all or most) probably take place in binaries. These core-collapse events are believed to produce neutron stars and black holes as their remnants. If the binary remains intact, the remnant will be a binary surrounded by a supernova remnant. A notable Type II-Ib supernova with binary companion is SN1993J[9, 367]. The binary association is further supported by the strong asphericity in the explosion mechanism.

These indications suggest that GRB supernovae are likewise taking place in binaries. Indeed, a rotationally driven explosion mechanism could naturally derive

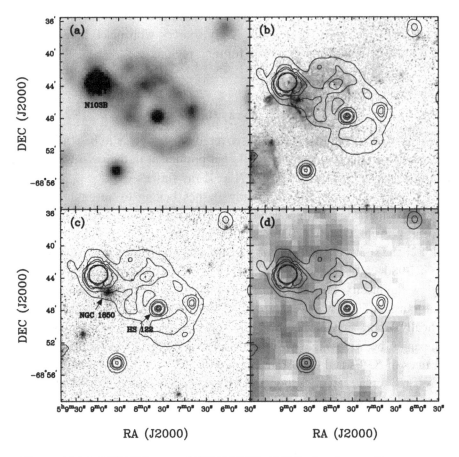

Figure 11.16 ROSAT image of RX J050736-6847.8, showing an X-ray supernova remnant around a point source (a). The remnant is also shown in a Curtis Schmidt $H\alpha$ image (b), a Digital Sky Survey Image (c), and an HI image (d), each overlaid by X-ray contours. (Reprinted with permission from[121]. ©2000 The University of Chicago Press.)

its angular momentum from the companion star, assuming spin-up of the progenitor star prior to core collapse by tidal coupling to orbital motion.

The end result of a GRB supernova is hereby a black hole with a stellar companion surrounded by a supernova remnant. These remnants may appear as black hole binaries in supernova remnants. Gamma ray burst supernovae are believed to produce a soft X-ray transient[87]. A particularly striking example of an X-ray binary surrounded by a supernova remnant is RX J050736-6847.8[121] which may be harboring a black hole (Figure 11.16). It remains an open observational question to ascertain if such is a remnant of a GRB supernova.

Searches for GRB remnants may therefore focus on aspherical remnants of beamed outflows[239, 25], late-time spectral peculiarities produced by low-luminosity activity of a remnant inner engine[448], chemical abundances in SNRs simular to the α-nuclei found in the companion star of the soft X-ray transient GRO J1655-40[277], binary X-ray sources with black hole candidates in SNRs, and an association with star-forming regions.

11.7 X-ray flashes

In a recent development, BeppoSax discovered what appears to be a new class of bursts, similar in duration to long GRBs but prominent in their X-ray energetics. These were introduced by J. Heise[257] as X-ray flashes.

It is presently unclear whether X-ray flashes belong to an entirely separate class, or whether they form a continuous extension of the GRB phenomenon. The nearby event GRB021203 is sufficiently soft to be considered an X-ray flash[582] and showed tentative evidence for an association with SN2003lw[361]. Like GRB980425, it is a nearby event ($D = 453$) with very low burst energy and afterglow luminosity[499], also in the radio[497]. Only XRF 020903, also with optical transient[500], was weaker[470]. This "weak-nearby" relation is expected statistically, upon viewing nearby events off-axis, provided that XRF/GRBs are accompanied by wide-angle weak emisisons, in view of the fact that observed event rates of XRFs and GRBs are comparible.

A second, tentative connection to GRBs is based on the *Amati relation*[13], describing a positive correlation

$$\frac{E_{peak}}{100 \text{ keV}} \simeq \left(\frac{E_{iso}}{10^{52} \text{ erg}}\right)^{1/2} \tag{11.11}$$

in the prompt emission. While GRB 021203 and GRB 980425 appear to be exceptions to this relationship, most of the X-ray flashes and GRBs appear to satisfy this relation over a remarkably wide range of energies[315, 470].

11.8 Candidate inner engines to GRB/XRF supernovae

The various similarities and differences between GRBs, XRFs, and weak GRBs (GRB980425 and GRB031205[498, 477]) pose the challenge of finding a unified model, representing a common origin in the end point of massive stars. Such a theory should explain their various event rates, total energy output and spectra, their durations and pose observational tests. Let us look at two alternatives.

These three populations have the same inner or different inner engines, yet all form in core collapse of massive stars. Their distinct phenomenology is due to distinct different viewing angles, or to different driving mechanisms from their emissions. These alternatives can be tested through their unseen emissions in neutrinos and gravitational radiation, since these are largely unbeamed. Unification by viewing angle predicts that all three produce largely similar emissions in these as-yet unseen channels. Unification by branching of core collapse into different inner engines predicts possibly distinct emissions in these unseen channels.

At present, it appears that GRB980425 and GRB031205 are genuinely weak in their total energy output[498] If true, this challenges unification by viewing angle. Nor do they or the XRF/GRB021203 satisfy the Amaldi relation (11.11).

Yet, core collapse of massive stars is unlikely to produce the same inner engine in all cases. The rotational state of the inner engine is expected to depend on whether the progenitor is single, or lives in a binary. A compact binary tends to spin up the progenitor by tidal interaction, which contributes to the angular momentum in the newly formed compact object, a neutron star or black hole. Furthermore, the compact object generally receives a kick, as neutron stars do in Type II supernovae and as black holes should receive by the Bekenstein gravitational-radiation recoil mechanism in aspherical collapse[42]. Rotation and kick undoubtedly produce a continuum of inner engines (parametrized by mass M, angular momentum J and kick velocity K), whereby no two are the same. These kick velocities can reach large values, about $100\,\mathrm{km\ s^{-1}}$ for black holes or more for neutron stars, whereby the newly formed compact leaves the core prematurely before core collapse is completed. Kick velocities K hereby introduce a distribution of inner engines, from low-mass and rapidly moving to true gems: high-mass inner engines are at the center of the remnant envelope of the progenitor star.

The branching ratios of core collapse into these various inner engines define the relative, true event rates between the various observational outcomes, powered by their varying emissions and interactions with the remnant envelope. Evidently, kick velocities produced randomly predict true gems (small K) to be the most rare, and the most powerful if rapidly rotating.

Exercises

1. Show that $< V/V_{max} >= 1/2$ for any flux-limited sample of sources distributed uniformly in Euclidean space. Here, $V(r)$ denotes the volume of the sphere, whose radius r is given by the distance to the source. Infer that in an expanding universe, a cosmological distribution gives rise to $< V/V_{max} > < 1/2$.

2. Show that the GRB peak luminosities and beaming must be correlated, based on the true-to-observed rate (11.4) and the Type Ib/c supernova rate.

3. Calculate the free-fall timescale in a core-collapse event.

4. Determine the condition of *Roche lobe overflow* in binaries of stars, sufficiently compact to suppress the Newtonian gravitational barrier against mass transfer.

5. Determine the evolution of binary separation as a result of conservative mass transfer.

6. Determine the critical mass loss of a member of a binary, as in a supernova explosion, for the binary to unbind.

7. Calculate the local density of GRB remnants with observable supernova remnant. Assume that a supernova remnant remains coherent for about 5000 yrs. What is statistically the expected distance of the nearest GRB plus supernovae remnant?

12

Kerr black holes

"The black holes of nature are the most perfect macroscopic objects
there are in the universe: the only elements in their construction are our
concepts of space and time."[110][1]

Kerr derived the exact solution of rotating black holes as fundamental objects in general relativity[293]. These solutions show a rotating null surface surrounded by a differentially rotating spacetime, characterized by frame-dragging: surrounding particles with zero angular momentum are engaged with nonzero angular velocities. The Kerr solution shows the potential for storing a large fraction of its mass energy in angular momentum, about an order of magnitude larger than that in rapidly rotating neutron stars. These solutions are parametrized by mass M and angular momentum J_H (later generalized to include electric charge), and they satisfy the bound

$$J_H \leq GM^2/c. \tag{12.1}$$

Angular momentum of a spinning black hole couples with curvature in its surroundings. Through curvature-spin or curvature-angular momentum coupling, this points towards energetic interactions between the black hole and its surrounding particles. In this chapter we shall discuss these interactions from a first-principle point of view.

In isolation, rotating stellar black holes are stable and nonradiating. Interactions of the black hole with its environment are subject to the first law of thermodynamics[32, 254, 255]

$$\delta M = \Omega_H \delta J_H + T_H dS_H \tag{12.2}$$

for a black hole with angular velocity Ω_H. Here, and in this chapter we use geometrical units $(G = c = 1)$. The first term on the right-hand side of (12.2)

---

[1] Where we include conservation laws.

179

represents useful work performed by black-hole spin energy on environment. The entropy-creating term $T_H dS$ is *not* pro forma. Bekenstein[42, 43] proposed that this entropy is genuine, to be associated with the irreducible mass of the black hole, $S \propto A_H$, which led Hawking to propose that black holes radiate at a finite temperature T_H[254].

We discuss two interactions: gravitational spin–orbit interactions on particles with spin along the axis of rotation, and interactions with particles in the equatorial plane.

12.1 Kerr metric

Rotating black holes can be parametrized in terms of the dimensionless specific angular momentum a/M, where $a = J_H/M$, as shown in Table (12.1). The Kerr metric possesses a timelike and azimuthal Killing vector $(\partial_t)^b$ and $(\partial_\phi)^b$. Its line element in Boyer–Lindquist coordinates[76, 110, 534] is

$$ds^2 = -dt^2 + \frac{\rho^2}{\Delta} dr^2 + \rho^2 d\theta^2 + (r^2 + a^2) \sin^2 \theta d\phi^2$$

$$+ \frac{2Mr}{\rho^2} \left(dt - a \sin^2 \theta d\phi \right)^2, \tag{12.3}$$

where $\rho^2 = r^2 + a^2 \cos^2 \theta$ and $\Delta = r^2 - 2Mr + a^2$. The event horizon of the black hole is given by the outermost null surface, the largest root of $\Delta = 0$,

$$r_H = M + \sqrt{M^2 - a^2} = 2M \cos^2(\lambda/2). \tag{12.4}$$

In the same coordinates, (12.3) is often also expressed as[110]

$$ds^2 = -\frac{\rho^2 \Delta}{\Sigma^2} dt^2 + \frac{\Sigma^2}{\rho^2} (d\phi - \omega dt)^2 \sin^2 \theta + \frac{\rho^2}{\Delta} dr^2 + \rho^2 d\theta^2, \tag{12.5}$$

where

$$\omega = \frac{2aMr}{\Sigma^2} \tag{12.6}$$

denotes the frame-dragging angular velocity, where $\Sigma^2 = (r^2 + a^2) - a^2 \Delta \sin \theta$.

The specific angular momentum of a particle with velocity four-vector u^b and angular velocity $\Omega = u^\phi/u^t$ is given by

$$L = \chi^a g_{ab} u^b = g_{\phi t} u^t + g_{\phi\phi} u^\phi = g_{\phi\phi} u^t (\Omega - \omega), \tag{12.7}$$

where

$$\omega = -\frac{g_{\phi t}}{g_{\phi\phi}}, \quad g_{\phi t} = -\frac{2aMr}{\rho^2}, \tag{12.8}$$

Table 12.1 *Trigonometric parametrization of a Kerr black hole in geometrical units with $\hbar = 1$. Here, M denotes the mass energy at infinity, $a = J_H/M$ denotes the specific angular momentum, and M_{irr} denotes the irreducible mass.*

SYMBOL	EXPRESSION	COMMENT
λ	$\sin \lambda = a/M$	
r_H	$2M \cos^2(\lambda/2)$	
Ω_H	$\tan(\lambda/2)/2M$	
J_H	$M^2 \sin \lambda$	
E_{rot}	$2M \sin^2(\lambda/4)$	$\leq 0.29M$
M_{irr}	$M \cos(\lambda/2)$	$\geq 0.71M$
A_H	$16\pi M_{irr}^2$	
S_H	$A_H/4$	
T_H	$\cos \lambda / 8M \cos^2(\lambda/2)$	

and

$$g_{\phi\phi} = \sin^2 \theta \left[(r^2 + a^2) + \frac{2Mr}{\rho^2} a^2 \sin^2 \theta \right]$$

$$= \rho^{-2} \sin^2 \theta \left[(r^2 + a^2)^2 - \Delta a^2 \sin^2 \theta \right]. \qquad (12.9)$$

Zero angular momentum observers (ZAMOs) hereby rotate with the angular velocity ω. On the event horizon, ZAMOs corotate with the black hole,

$$\omega = \Omega_H. \qquad (12.10)$$

At large distances, $\omega \propto 1/r^{-3}$ as r approaches infinity. This shows that frame-dragging is differential in nature. It cannot be transformed away by a global change of angular velocity of the coordinate system, and hence it is a real physical effect in accord with Mach's principle in the neighborhood of the black hole illustrated in Figure (12.1).

The specific energy of a zero-angular velocity particle is

$$E = -\xi^a g_{ab} u^b = -g_{tt} u^t, \quad g_{tt}(u^t)^2 = -1, \qquad (12.11)$$

where

$$g_{tt} = \rho^{-2} \left(a^2 \sin^2 \theta - \Delta \right). \qquad (12.12)$$

It follows that zero-angular velocity particles exist only outside the *ergosphere*, i.e. outside the region $g_{tt} > 0$:

$$r_H < r < M + \sqrt{M^2 - a^2 \cos^2 \theta}. \qquad (12.13)$$

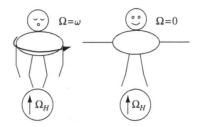

Figure 12.1 Frame-dragging around a rotating black hole with angular velocity Ω_H breaks the correspondence of zero angular momentum and zero angular velocity. *Left*: An observer corotating with the frame-dragging angular velocity ω is in a state of zero angular momentum, and experiences no centrifugal forces. *Right*: An observer fixed relative to the distant stars assumes a state of negative angular momentum, and experiences centrifugal forces. Frame-dragging is differential, stronger near the black hole and weaker at larger distances and, hence, not a choice of gauge.

Inside, they must be rotating at some finite fraction of the angular velocity of the black hole.

The effect of frame-dragging becomes explicit in a $3+1$ decomposition of the line-element[534]

$$ds^2 = -\alpha^2 dt^2 + h_{ij}(dx^i - \omega^i)(dx^j - \omega^j), \tag{12.14}$$

where (α, h_{ij}) is diagonal:

$$\alpha = \frac{\rho}{\Sigma}\sqrt{\Delta}, \quad h_{ij} = \begin{pmatrix} \frac{\rho^2}{\Delta} & 0 & 0 \\ 0 & \rho^2 & 0 \\ 0 & 0 & \tilde\omega^2 \end{pmatrix}, \tag{12.15}$$

$$\omega^\phi = \frac{2aMr}{\Sigma^2}, \quad \varpi = (\Sigma/\rho)\sin\theta. \tag{12.16}$$

The condition $L = 0$ expresses the geometrical property, that the velocity four-vector of ZAMOs is orthogonal to the azimuthal Killing vector $(\partial_\phi)^b$, i.e. orthogonal to the coordinate surface of constant (r, ϕ). The eigentime of these zero angular momentum particles evolves according to $ds/dt = \alpha$, where α is known as the redshift factor or lapse function. The four-dimensional volume element $\sqrt{-g} = \alpha\sqrt{h} = \rho^2 \sin\theta$. At large distances, (12.14) satisfies

$$ds^2 \simeq -\left(1 - \frac{2M}{r}\right)dt^2 + \left(1 + \frac{2M}{r}\right)dr^2 + r^2 d\theta^2 + r^2 \sin^2\theta d\phi^2$$

$$+ \frac{4Ma}{r}\sin^2\theta d\phi dt. \tag{12.17}$$

12.2 Mach's principle

Mach recognized that a zero-angular momentum state is defined by zero angular velocity relative to a surrounding dominant distribution of matter ("the distant stars"). A state of zero angular velocity relative to infinity defines a state of zero angular momentum to within our current experimental uncertainties. (This might change with the upcoming Gravity Probe B experiment, and might have been measured by nodal precession in the orbits of the LAGEOS and LAGEOS II Satellites[122, 123].)

Mach's principle can be extended by taking into account nearby compact objects such as black holes. The Kerr solution shows that zero angular momentum trajectories assume a state of corotation in the proximity of the horizon. Particles in the neighborhood are effectively "sandwiched" between the black hole horizon and infinity. The angular velocity of zero angular momentum particles will be between zero and that of the black hole, showing rotational shear in spacetime due to frame-dragging. In order to illustrate this departure from the familiar correspondence of zero angular momentum and zero angular velocity in flat spacetime, consider lowering an observer to the north pole along the axis of rotation of a Kerr black hole. In this process, frame-dragging acts on his/her arms and legs. If maintaining a state of zero angular momentum, the legs twist spontaneously while arms remain straight down. Posture is that of the Etruscan sculpture *Lady with the Mirror* shown in Figure (12.2). Resisting this by keeping legs straight, the subject becomes more attracted to the black hole by gravitational spin–orbit coupling, between positive angular momentum of the black hole and negative angular momentum of the legs. If tall (short), he/she will see the sky in slow (rapid) rotation given by minus the local frame-dragging angular velocity ω. If, on the other hand, eyes are fixed on to the distant stars, arms will lift spontaneously due to their nonzero angular momentum.

This illustrates curvature induced by black hole spin energy, as quantified by the Kerr metric.

12.3 Rotational energy

Black holes become luminous by suppressing or circumventing the canonical angular momentum barriers of radiation fields. There are several avenues to make this happen, "by hand" or otherwise. In response, the black hole evolves according to conservation of mass, angular momentum and electric charge. The most efficient process is adiabatic, described by $T_H dS_H = 0$, in which case

$$\dot{M} = \Omega_H \dot{J}_H, \tag{12.18}$$

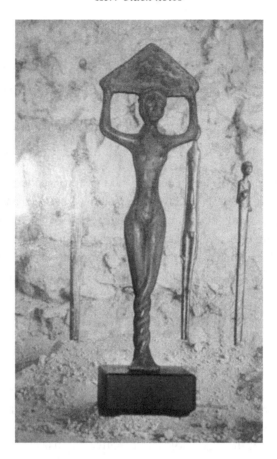

Figure 12.2 Etruscan sculpture *Lady with the Mirror*, an ancient symbol of fertility. The twisted legs represent the action of frame-dragging, upon suspension above the north pole of a rotating black hole, described by the Kerr metric. This ejects matter with high specific angular momentum by spin–orbit coupling along the axis of rotation.

where the dot refers to differentiation with respect to λ. Hence,

$$\dot{M} = \frac{1}{2M} \tan(\lambda/2) \left[2M\dot{M}\sin\lambda + M^2\cos\lambda\right], \qquad (12.19)$$

or

$$\left(1 - 2\sin^2(\lambda/2)\right)\frac{\dot{M}}{M} = \frac{1}{2}\cos\lambda\tan(\lambda/2). \qquad (12.20)$$

Integration gives

$$\frac{M_2}{M_1} = \frac{\cos(\lambda_1/2)}{\cos(\lambda_2/2)}. \qquad (12.21)$$

The relationship between the mass M of the spinning black hole and the so-called *irreducible* mass M_{irr} of the adiabatically related nonrotating black hole, therefore, is expressed by

$$M_{irr} = M \cos(\lambda/2). \tag{12.22}$$

The difference $E_{rot} = M - M_{irr}$ is the rotational energy – the maximal possible energy liberated from the black hole – given by

$$E_{rot} = 2M \sin^2(\lambda/4). \tag{12.23}$$

At $\lambda = \pi/2$, the rotational energy is about 29% of the mass of the black hole. This specific energy in rotation is far in excess of that in a rapidly spinning neutron star, which is limited to few percent of its mass-energy at best.

12.4 Gravitational spin–orbit energy $E = \omega J$

The Kerr metric shows that spin induces curvature. This is the converse of curvature–spin considered in Chapter 4. Consequently, spinning bodies couple to spinning bodies. Such interactions are commonly referred to as gravitomagnetic effects[534], by analogy to magnetic moment–magnetic moment interactions. To study this spin–orbit coupling in the Kerr metric, we focus on interactions along the axis of rotation.

Gravitational spin–spin interactions are such that antiparallel spin–spin orientations repel, while parallel spin–spin orientations attract. This can be illustrated by considering a balance, located on the north pole of a massive object M with angular velocity Ω_M (12.3). Equal weights will be measured of objects of the same mass and zero spin. Distinct weights will be measured in case of objects of the same mass and opposite spin: the object whose spin is parallel to that of the planet weighs less than the object whose spin is antiparallel to that of the massive object. Based on dimensional analysis, we expect a gravitational potential for spin aligned interactions given by

$$E = \omega J, \tag{12.24}$$

where ω refers to the frame-dragging angular velocity produced by the massive body and $\mathbf{J} = J\mathbf{e}_2$ is the angular momentum of the spinning object.

The nonzero components of the Riemann tensor of the Kerr metric can be expressed in tetrad 1-forms

$$\mathbf{e}_0 = \alpha dt, \quad \mathbf{e}_1 = \frac{\Sigma}{\rho}(d\phi - \omega dt)\sin\theta, \quad \mathbf{e}_2 = \frac{\rho}{\sqrt{\Delta}}dr, \quad \mathbf{e}_3 = \rho d\theta. \tag{12.25}$$

as[110]

$$R_{0123} = A$$

$$R_{1230} = AC$$

$$R_{1302} = AD$$

$$-R_{3002} = R_{1213} = -A3a\sqrt{\Delta}\Sigma^{-2}(r^2 + a^2)\sin\theta$$

$$-R_{1220} = R_{1330} = -B3a\sqrt{\Delta}\Sigma^{-2}(r^2 + a^2)\sin\theta$$ (12.26)

$$-R_{1010} = R_{2323} = B = R_{0202} + R_{0303}$$

$$-R_{1313} = R_{0202} = BD$$

$$-R_{1212} = R_{0303} = -BC,$$

where

$$A = aM\rho^{-6}(3r^2 - a^2\cos^2\theta),$$

$$B = Mr\rho^{-6}(r^2 - 3a^2\cos^2\theta),$$

$$C = \Sigma^{-2}[(r^2 + a^2)^2 + 2a^2\Delta\sin^2\theta],$$ (12.27)

$$D = \Sigma^{-2}[2(r^2 + a^2)^2 + a^2\Delta\sin^2\theta].$$

Notice that on-axis, where $\theta = 0$,

$$2A = -\partial_r\omega = \frac{2aM}{\rho^6}(3r^2 - a^2), \quad C = 1, \quad D = 2.$$ (12.28)

This brings about explicitly black hole spin-induced curvature components in the first three terms of (12.26) on the axis of rotation and, hence, an implied curvature–spin-connection.

This interaction bears out by inspection of (5.37), by considering orbital motion around the spin axis. Evaluated in an orthonormal tetrad, we have according to (12.28) a radial force

$$F_2 = JR_{3120} = JAD = -\partial_2\omega J.$$ (12.29)

The assertion of (12.24) follows from

$$E = \int_r^\infty F_2 ds = \omega J,$$ (12.30)

The result (12.30) may also be recognized, by considering the difference in total energy between particles that orbit the axis of rotation of the black hole with opposite spin. Let u^b denote the velocity 4-vector and $u^\phi/u^t = \Omega$ the angular velocities of either one of these,

$$-1 = u^c u_c = [g_{tt} + g_{\phi\phi}\Omega(\Omega - 2\omega)](u^t)^2.$$ (12.31)

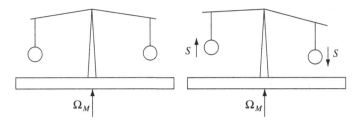

Figure 12.3 Gravitational spin–spin interactions are similar, except for sign, to magnetic moment–magnetic moment interactions: spinning bodies suspended are repelled when their spin is parallel and are attracted when it is antiparallel to the angular velocity of the underlying mass M. For massive bodies, this implies unequal weights as measured by a balance. Spinning particles with large specific angular momentum will be ejected to infinity in accord with the Rayleigh criterion, as spin–spin coupling overcomes the gravitational Coulomb attraction.

This normalization condition has the two roots

$$\Omega_\pm = \omega \pm \sqrt{\omega^2 - (g_{tt} + (u^t)^{-2})/g_{\phi\phi}}. \tag{12.32}$$

We insist that these two particles have angular momenta of opposite sign and equal magnitude

$$J_\pm = g_{\phi\phi} u^t (\Omega_\pm + \omega) = g_{\phi\phi} u^t \sqrt{\omega^2 - (g_{tt} + (u^t)^{-2})/g_{\phi\phi}} = \pm J. \tag{12.33}$$

This shows that u^t is the same for each particle. The total energy of the particles is given by

$$E_\pm = (u^t)^{-1} + \Omega_\pm J_\pm, \tag{12.34}$$

and hence one-half their difference

$$E = \frac{1}{2}(E_+ - E_-) = \omega J. \tag{12.35}$$

The curvature–spin coupling (12.30) is universal, and applies whether the angular momentum is mechanical or electromagnetic in origin[558].

12.5 Orbits around Kerr black holes

The motion of test particles is described by a Lagrangian $L = p^2/2m$ (Chapter 2). In the equatorial plane, this is described by three conserved quantities[30, 490]: a rest mass m, energy E and angular momentum L. Using the Killing vectors ∂_t and ∂_ϕ of the Kerr metric in the equatorial plane, the first gives

$$2L/m = -\left(1 - \frac{2M}{r}\right)\dot{t}^2 - \frac{4aM}{r}\dot{t}\dot{\phi} + \frac{r^2}{\Delta}\dot{r}^2 + \left(r^2 + a^2 + \frac{2Ma^2}{r}\right)\dot{\phi}^2, \tag{12.36}$$

and the constants of motion

$$mE = -p_t = \frac{\partial L}{\partial \dot{t}}, \quad mL = p_\phi = \frac{\partial L}{\partial \dot{\phi}}. \tag{12.37}$$

Combined, (12.36) and (12.37) give[490]

$$\frac{\dot{t}}{m} = \frac{(r^3 + a^2 r + 2Ma^2)E - 2aML}{r\Delta}, \quad \frac{\dot{\phi}}{m} = \frac{(r - 2M)L + 2aME}{r\Delta} \tag{12.38}$$

and

$$\left(\frac{dr}{d\lambda}\right)^2 + V(r, E, L) = 0, \tag{12.39}$$

where

$$r^3 V = -m^2 \left(E^2(r^3 + a^2 r + 2Ma^2) - 4aMEL - (r - 2M)L^2 - r\Delta \right). \tag{12.40}$$

or more explicitly[577]

$$m^{-2}V = -\frac{M}{r} + \frac{L^2}{2r^2} + \frac{1}{2}\left(1 - E^2\right)\left(1 + \frac{a^2}{r^2}\right) - \frac{M}{r^3}(L - aE)^2. \tag{12.41}$$

This is illustrated in Figure (12.4). Turning points correspond to $V = 0$. Using, for example, MAPLE, solutions of circular orbits can be found corresponding to $V = V'$ or, equivalently,

$$R = R'. \tag{12.42}$$

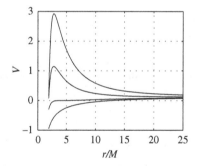 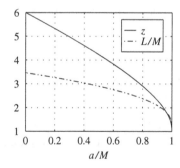

Figure 12.4 *Left*: The potential V as a function of r/M around a Kerr black hole of mass M and $a/M = 0.5$ for various values $L = iL_{ISCO}(i = 0, 1, 2, 3)$ of the specific orbital angular momentum of particles of energy $E = E_{ISCO}$. The ISCO values represent the constants of motion in the innermost stable circular orbit, where $V = V' = V'' = 0$. *Right*: The location $z = r_{ISCO}/M$ of the innermost stable circular orbit (continuous line) and the dimensionless specific angular momentum L/M (dot-dashed line) as a function of a/M.

These are[490]

$$E = \frac{r^2 - 2Mr \pm a\sqrt{Mr}}{r(r^2 - 3Mr \pm 2a\sqrt{Mr})^{1/2}}, \quad L = \pm \frac{\sqrt{Mr}(r^2 \mp 2a\sqrt{Mr} + a^2)}{r(r^2 - 3Mr \pm 2a\sqrt{Mr})^{1/2}}. \quad (12.43)$$

The plus and minus signs correspond, respectively, to corotating and counterrotating orbits, relative to the spin of the black hole. Of particular interest is the angular velocity

$$\Omega = \pm \frac{M^{1/2}}{r^{3/2} \pm aM^{1/2}} \quad (12.44)$$

of circular orbits around Kerr black holes.

The *innermost stable circular orbit* (ISCO) is the circular orbit for which $V'' = 0$, or

$$R'' = 0. \quad (12.45)$$

This defines the transition between stable ($R'' \le 0$) and unstable ($R'' > 0$) orbits. The solutions are due to J. M. Bardeen[30] (see further[490])

$$E = \sqrt{1 - \frac{2}{3z}}, \quad L = \frac{2M}{3\sqrt{3}} \left(1 + 2\sqrt{3z - 2}\right) \quad (12.46)$$

at a radius $z = r_{ISCO}/M$ in terms of $\hat{a} = a/M$:

$$z = 3 + Z_2 \mp [(3 - Z_1)(3 + Z_1 + 2Z_2)]^{1/2}, \quad (12.47)$$

where $Z_1 = 1 + (1 - \hat{a}^2)^{1/3} [(1 + \hat{a})^{1/3} + (1 - \hat{a})^{1/3}]$, and $Z_2 = (3\hat{a}^2 + Z_1^2)^{1/2}$. Notice that L/E decreases from $3\sqrt{3}/2M$ for $a = 0(z = 6)$ down to $L/E = 2M$ for $a = 0(z = 1)$. The specific angular momentum j of particles in stable circular orbits around black holes satisfy

$$j \ge GlM/c, \quad (12.48)$$

where $2/\sqrt{3} < l < 2\sqrt{3}$ is the specific orbital angular momentum $l = l(a/M)$ given by L/M in (12.46), corresponding to an extremal black hole ($a = M_H$, $z = 1$) and a nonrotating black hole ($a = 0$, $z = 6$).

12.6 Event horizons have no hair

The horizon of a black hole is a surface with no hair. This commonly refers to the notion that a black hole is uniquely described by its three parameters: mass, angular momentum and electric charge. These three quantities refer to conserved quantities with associated long-range interactions.

The event horizon has the unique property of *topological equivalence*, in that all its points are mutually identified. If we drop particles carrying ($\delta M, \delta J, \delta q$)

on to the black hole, the black hole evolves as described by the three parameters (M, J_H, q) with no memory of the point of intersection of the particle trajectory with the black hole event horizon. The horizon surface is topologically "no-hair." This destruction of information implies equivalence between horizon surfaces regardless of their past history. This uncertainty represents an entropy associated with the event horizon. This topological equivalence is beautifully illustrated by considering the evolution of the electric field as charged particles are dropped onto a Schwarzschild black hole[344, 134, 128, 250, 534]. Regardless of the initial condition and the trajectory of the particle, the final state of the electric field is that of electric charge delocalized uniformly over the horizon or, equivalently, a point charge at the center of the black hole. In what follows, we consider nonrotating black holes.

We are at liberty to envision the horizon surface partitioned into small black holes of radius l_p and mass M_p at the Planck scale, since any detailed structure of the horizon is hidden behind its infinite redshift. According to the Schwarzschild solution and the Heisenberg uncertainty relation $\Delta p \Delta x \simeq \hbar/2$, we have

$$\frac{2GM_p}{l_p} = c^2, \quad (M_p c) l_p = \hbar/2. \tag{12.49}$$

This defines the Planck length

$$l_p = \frac{G\hbar}{c^3}. \tag{12.50}$$

The number of ways the mass M of the black hole can be partitioned in Planck masses over the surface area A of the event horizon is about $N_p!$, where

$$N_p = \frac{A}{(2l_p)^2} \tag{12.51}$$

which gives the *Bekenstein–Hawking entropy*

$$S_H = \frac{kc^3}{4G\hbar} A, \tag{12.52}$$

where $k = 1.38 \times 10^{-16}$ erg K^{-1}(8.62×10^{-5} eV K^{-1}) denotes the Boltzmann constant. Here, we ignore logarithmic corrections in the definition of entropy from the number of permutations (essentially $N_p!$).

The Planck-sized black holes are restricted to "surfing" on the two-dimensional event horizon, i.e. a box of linear size set by its circumference $2\pi R_s$, where $R_s = 2M$ in case of a non-rotating black hole. The circumference gives the size of great circles and, hence, the lowest momenta of the surfing black holes. Their kinetic energy ΔE, therefore, satisfies $\Delta E(4\pi GM/c^2) = \hbar c/2$. The corresponding *Hawking temperature* is

$$kT = \frac{\hbar c^3}{8\pi GM},$$ (12.53)

where we take into account a two-dimensional thermal distribution with an energy $(1/2)kT$ in each direction. The reader will recognize that $TdS = dMc^2$.

The result (12.53) can be recognized to correspond to surface gravity, as seen by a distant observer. This refers to gently lowering a test particle attached to a long rope of constant length: the pull at infinity is defined to be the surface gravity in the limit as the test particle approaches the horizon. The energy-at-infinity of the test particle (per unit mass) is given by the redshift factor α. It therefore follows that

$$g_H = \lim_{r \to r_H} \frac{d\alpha}{ds} = \lim_{r \to r_H} \frac{\partial_r \alpha}{\partial_r s} = \lim_{r \to r_H} \frac{\sqrt{\Delta}}{\rho} \partial_r \left(\frac{\rho}{\Sigma} \sqrt{\Delta} \right) = \frac{r_H - M}{2Mr_H},$$ (12.54)

or

$$g_H = \frac{\cos \lambda}{8M \cos^2(\lambda/2)}.$$ (12.55)

With this identification, consistent with[542], the temperature (12.53) generalizes to[534]

$$kT = \frac{\hbar}{2\pi} g_H.$$ (12.56)

The above topological no-hair argument heuristically suggests that black holes are radiating objects: Schwarzschild black holes radiate a thermal spectrum of temperature

$$T = 6 \times 10^{-8} \left(\frac{M}{M_\odot} \right)^{-1} \text{K}.$$ (12.57)

In the above, we used Planckian black holes to discretize the area of the horizon. The notion of area discretization is central to Bekenstein's discretization of black holes and loop quantum gravity, which seeks to develop a theory of nonperturbative quantum gravity.

The no-hair theorem shows the evolution of black holes is completely described by the evolution of its mass, angular momentum and electric charge. Each of these are subject to the associated conservation laws.

In astrophysical situations, the no-hair property of black holes is augmented by the kick velocity K in the interaction with its environment.

12.7 Penrose process in the ergosphere

An instructive method for liberating rotational energy from a Kerr black hole has been invented by Roger Penrose[416]. Consider dropping a particle to close proximity of a rotating black hole. Allow the particle to break apart into two pieces, one half to be sent into the black hole and one half on an escape trajectory to infinity. The particle falling into the black hole spins down, so that escaping to infinity is the recipient of additional angular momentum by conservation of angular momentum. The particle falling into the black hole can further be put on a negative energy trajectory. The half escaping to infinity is the recipient of additional energy by conservation of energy. Under appropriate conditions, the latter thereby delivers more energy to infinity than provided to the original, single, piece from the start. The results are easily described in terms of conserved quantities on geodesics.

A key feature in the Penrose process is the existence of negative energy trajectories. These trajectories are limited to a finite region outside the black hole: the ergosphere. In this region, frame-dragging forces particles to rotate in the direction of the angular velocity of the black hole.

"Penrose's hand" may split the particle in the vicinity of the black hole, leaving two particles on a counterrotating trajectory with energy $E' < 0$ – falling into the black hole – and a corotating trajectory E'' – out to infinity, subject to

$$E = E' + E''. \tag{12.58}$$

At a turning point, where $\dot{r} = 0$, the energy mE of a particle satisfies[490]

$$E = \frac{2aML + \sqrt{L^2 r^2 \Delta + r\Delta}}{r^3 + a^2 r + 2Ma^2}. \tag{12.59}$$

In the limit of large negative specific angular momenta $L < 0$, $E < 0$ provided $r\sqrt{\Delta} < 2aM$, or

$$r < 2M. \tag{12.60}$$

More generally, we can arrange for one particle to enter a trajectory into the black hole with the property that $E' < 0$ inside the ergosphere (12.13), whereby $E'' > E$ escapes with enhanced energy to infinity.

The two-particle split into positive and negative energy particles must be relativistic, with a relative velocity greater than $c/2$[33]. *This suggests that the Penrose process is to be considered for waves in terms of positive and negative frequencies.*

Exercises

1. Derive the frame-dragging angular velocity at a distance r from the Earth's center, given by

$$\omega_E(r) = 2\left(\frac{i_E}{r^2}\right)\left(\frac{R_g}{r}\right)\Omega_E, \qquad (12.61)$$

where R_E denotes the radius of the Earth, $R_g = GM_E/c^2$ the Schwarzschild radius of the Earth's mass M_E, i_E its specific moment of intertia, and Ω_E its angular velocity.

2. Consider two gyroscopes in space with antiparallel spin, aligned parallel to the axis of rotation of the Earth. Show that they drift apart, producing a relative displacement d due to spin–spin coupling with the Earth in a low-altitude orbit given by

$$\frac{d}{R_E} = 3\phi_E\phi_g\left(\frac{i_g}{R_E^2}\right). \qquad (12.62)$$

Here, i_g denotes the specific moment of inerta of the gyroscopes with angular velocites $\pm\Omega_g$. Over the integration time t, the accumulated phases are $\phi_E = \omega_E t$ and $\phi_g = \Omega_g t$. Calculate d for a 1-year integration time for a cm-sized gyroscope rotating at 1 kHz.

3. Bardeen[30] derived the evolution equation for accretion from the ISCO on to the black hole. This increases the black hole mass and spin according to

$$zM^2 = \text{const.} \qquad (12.63)$$

generally causing spin-up towards an extremal state of the black hole. Derive this integral.

4. Estimate the lifetime of a Schwarzschild black hole in response to Hawking radiation in the approximation of black body radiation.
5. Calculate the mass of a pair of Hawking radiating Schwarzschild black holes to form a short-lived binary, due to balance of radiation pressure against gravitational attraction. [Hint: this calculation closely follows the derivation of the Eddington luminosity.] What could be a relic signature of a cluster of such particles in the early universe?
6. Recall that radiation by a massless field $\Phi = e^{-i\omega t}e^{im\phi}$ has energy ω and angular momentum m, associated with the Killing vectors ∂_t and ∂_ϕ. Show that black holes may spontaneously radiate consistent with the first law of black hole thermodynamics and $TdS_H \geq 0$, satisfying

$$0 < \omega < m\Omega_H. \tag{12.64}$$

This is the general condition of superradiant scattering of bosonic fields by black holes, wherein scattered radiation is more intense than incoming radiation. Explain why superradiant scattering fails for fermionic fields.
7. Show that inside the ergosphere (12.13) particles with physical trajectories, defined by timelike trajectories inside the local light cone, may nevertheless possess negative energies. That is, show the existence of 4-momenta p^b with the property that $p^2 < 0$ and $p_t > 0$ inside the ergosphere. Show that this implies negative angular momentum, $p_\phi < 0$.
8. Does the no-hair property of topological equivalence of black hole event horizons hold true, when black holes are properly represented as quantum mechanical objects? What implication does this have for the spectrum of radiation?
9. The notion that the surface area of a black hole is discretized can be motivated by analogous expressions for the quantization of magnetic flux, $\int_0^{2\pi}\int_0^{\pi/2} F_{\theta\phi}d\theta d\phi = nh/2e$ for the flux through a hemisphere. The quantity $\Phi_{\theta\phi} = \int_0^{2\pi}\int_0^{\pi} R_{\theta\phi\theta\phi}d\theta d\phi$. Show that $\Phi_{\theta\phi} = 4\pi M^2 = (1/2)(A_H/2)$, where $A_H = 16\pi M^2$ denotes the surface area of a Schwarzschild black hole.
10. Using X-ray spectroscopy, tentative evidence for black-hole spin has been found for a supermassive black-hole candidate in MCG-6-30-15 in a "deep minimum state" discovered in ASCA observations by Y. Tanaka *et al.*[515] and K. Iwasama *et al.*[281] as shown in Figure (12.E.1) (confirmed in recent XMM observations[602, 174]), and similarly for stellar mass black-hole candidates in galactic sources XTE J1650-500[374, 377] and GX339-4[373]. G. Minuitti, A. C. Fabian and J. M. Miller[377] argue for evidence of rotating black holes on the basis of broadening of X-ray iron emission-lines, redshifted (from the rest frame energy of 6.3 keV) down to much lower energies (to below 4 keV). (These X-ray lines are modeled as fluorescence lines, excited

by corona emissions from the disk in a manner that is qualitatively similar to the X-ray emission coming off solar flares. A notable feature in their model is that fluctuations in flaring height at one side of the disk can affect the line-emissions at the other side by strong gravitational lensing in the gravitational field of the black hole.)

To see this connection to black hole spin, sketch the effect on X-ray line emissions (observed count rate as a function of energy in the 1.5–200 keV

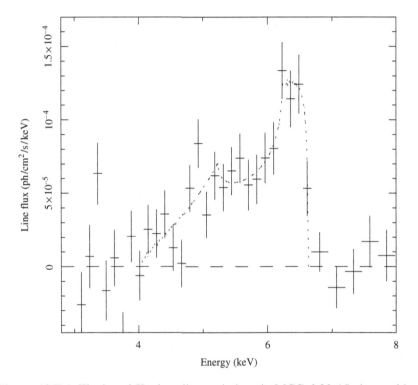

Figure 12.E.1 The broad $K\alpha$ iron line-emissions in MCG-6-30-15 observed by ASCA reveal relativistic orbital velocities with pronounced asymmetry about the restmass energy of 6.35 keV. This asymmetry is in quantitative agreement with the combined effect of redshift and Doppler shifts close to a central black hole, when seen nearly face-on at 30° inclination angle, as shown in the dotted line[177]. (Reproduced with permission from[515]). The $K\alpha$ emissions are time-variable. Shown are further the ASCA observations of a deep minimum state displaying an anomalously large red tail, reaching far below the limit of about 4 keV corresponding to the observed energy at the ISCO around a Schwarzchild black hole. This extension is in quantitative agreement with allowing the ISCO to move close to the black hole in accord with the Kerr metric[319] and a steep radial emissivity profile[281]. (Reproduced with permission from[281].)

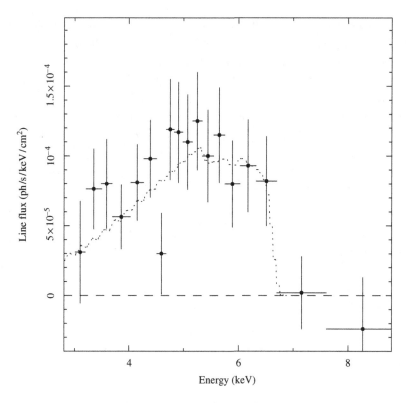

Figure 12.E.1 (cont.)

range) due to (a) Newtonian Doppler effects, (b) relativistic beaming, (c) gravitational redshift due to the potential well of the black hole, and (d) rotation of the black hole as it affects the inner radius of the accretion disk. Explain how X-ray line emissions can be used to distinguish rotating from non-rotating black holes.

11. Given that synchrotron emission per unit volume in an optically thin fluid scales with the magnetic field-energy density, estimate the predicted enhancement in brightness in the nozzle N in Fig. 9.2. Calculate the ratio of the lifetimes of optical-to-radio synchrotron emitting electrons.

13

Luminous black holes

"Inequality is the cause of all local movements".

Leonardo da Vinci (1452–1519).

With the second law of thermodynamics $dS \geq 0$, specific angular momentum increases with radiation:

$$a_p \equiv \frac{-\delta J_H}{-\delta M} \geq \Omega_H^{-1} \geq 2M > M \geq a, \qquad (13.1)$$

based on the Kerr solution which has $\Omega_H \leq 1/(2M)$. Consistent with the Rayleigh criterion, rotating black holes couple to radiation as a channel to lower the total energy (of black hole plus radiation). In isolation, this coupling is exponentially small, due to angular momentum barriers[542, 254, 522, 441, 523]: for all practical purposes, *isolated* stellar mass black holes rotate forever.

Black holes may become luminous in environments that successfully circumvent or suppress the angular momentum barriers. Broadly, this poses the questions: What astrophysical nuclei harbor active black holes? What is the lifetime and luminosity of a rotating black hole?

In this chapter, we discuss these questions at varying degrees of depth in core collapse supernovae, by considering black holes surrounded by a uniformly magnetized torus. Specifically, we shall identify a powerful spin-connection between the black hole and torus based on equivalence to pulsars when viewed in poloidal topology, and a spin–orbit coupling as a mechanism for linear acceleration of high specific angular momentum in open ergotubes. Both are in accord with the Rayleigh criterion.

13.1 Black holes surrounded by a torus

The topology of corecollapse of a uniformly magnetized progenitor star or in the tidal break-up of a magnetized neutron star around a black hole shows the

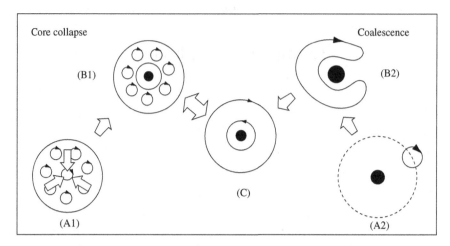

Figure 13.1 A uniformly magnetized torus around a black hole (C) is represented by two counteroriented current rings in the equatorial plane. It forms a
common end point of both core collapse (A1,B1,C) and black hole–neutron star
coalescence (A2,B2,C). Core collapse (A1–B1) in a magnetized star results in
a uniformly magnetized, equatorial annulus (C); tidal break-up (A2–B2) wraps
the current ring representing the magnetic moment of a neutron star around the
black hole which, following a reconnection, leaves the same (C). (Reprinted
from[568]. ©2003 The American Astronomical Society.)

formation of a uniformly magnetized torus (Figure 13.1). This assumes that the
progenitor star, the massive progenitor of a Type Ib/c supernova or the progenitor
neutron star, respectively, is magnetized. A magnetized star can be represented to
leading order by a single current loop or equivalently a density of magnetic dipole
moments. Nucleating a black hole in core collapse is a highly dissipative process,
which removes the central magnetization of the star and leaves a magnetized
annulus consisting of two counteroriented current rings as the projection of the
remaining stellar matter in the equatorial plane. Tidal break-up of a neutron star
around an existing black hole causes the winding of a current loop in the equatorial
plane. Following a single reconnection event, this leaves likewise a magnetized
annulus consisting of two counteroriented current loops in the equatorial plane.
Core collapse supernovae and binary black hole–neutron star coalescence both
give rise to the same outcome: a black hole surrounded by a magnetized torus,
represented by two counteroriented current loops or, equivalently, a uniform
density of magnetic moments.

A configuration consisting of a black hole surrounded by a torus magnetosphere raises several questions, on the state of the black hole, the torus, the
torus magnetosphere and any spin-connection between the black hole and the
torus.

13.2 Horizon flux of a Kerr black hole

Of some theoretical interest is the exact vacuum solution of an asymptotically uniform magnetic field surrounding a Kerr black hole, described by R. M. Wald[577]. The Killing fields of the Kerr metric are solutions to the vacuum vector potential of the electromagnetic field, and so is any linear superposition,

$$A_a = c_1 \chi_a + c_2 \xi_a. \tag{13.2}$$

This defines an axisymmetric magnetic field of asymptotically constant magnetic field, whose surfaces of constant magnetic flux $A_\phi =$const are

$$\Phi = 2\pi A_\phi \simeq 2\pi c_1 \chi_\phi \simeq 2\pi c_1 r^2 \sin^2 \theta \tag{13.3}$$

asymptotically in the limit as $r \to \infty$. Thus, we identify

$$c_1 = \frac{1}{2} B \tag{13.4}$$

in case of an asymptotic field strength B.

Consider the electric charge q on the black hole given by the flux integral

$$4\pi q = \int_S *F. \tag{13.5}$$

In view of Stokes' theorem, and the fact that the electromagnetic field tensor F_{ab} satisfies the vacuum Maxwell's equations $d * F = 0$, the integral in (13.5) is the same for all 2-surfaces S outside the black hole. We are at liberty to evaluate (13.5) for a 2-sphere S in the limit of arbitrarily large radius. Using (12.17), we therefore have

$$4\pi q = 2\pi \lim_{r \to \infty} \int_S *F = 16\pi J_H c_1 - 8\pi M c_2 = 8\pi (J_H B - M c_2). \tag{13.6}$$

It follows that

$$c_2 = -\frac{q}{2M} + aB, \tag{13.7}$$

and, hence, we have the general expression

$$A_a = \frac{1}{2} B \chi_a + \left(aB - \frac{q}{2M}\right) \xi_a \tag{13.8}$$

for the vector potential of an asymptotically uniform magnetic field around a black hole with charge q.

The horizon flux is given by $2\pi A_\phi$ evaluated at $\theta = \pi/2$,

$$\Phi_H = 4\pi M^2 \left[B + \left(\frac{q}{M} - 2aB\right) \Omega_H\right]. \tag{13.9}$$

An uncharged black hole satisfies

$$\Phi_H = 4\pi BM^2 \cos(\lambda). \tag{13.10}$$

This shows that the magnetic flux is expelled in response to rotation, and it approaches zero in the limit of an extreme Kerr black hole ($a = \pm M$, $\lambda = \pm\pi/2$). A finite charge q creates a magnetic moment $\mu = r_H A_\phi$ by corotation with the event horizon[104]

$$\mu = \frac{q J_H}{M} \tag{13.11}$$

and, hence, contributes a magnetic flux

$$\Phi'_H = 4\pi q M \Omega_H \tag{13.12}$$

through the horizon. This shows a gyromagnetic ratio $g = 2$[576, 129].

13.2.1 Maximal flux in the lowest energy state

The null generators of the horizon have tangent velocity 4-vectors $z^b = \xi^b + \Omega_H \chi^b$. The tangential electric field on the event horizon is purely poloidal. The horizon surface is in electrostatic equilibrium when its tangential electric field vanishes:

$$0 = F_{\theta a} z^a = c_2 \left(-\xi_{t,\theta} + \Omega_H \chi_{\phi,\theta}\right), \tag{13.13}$$

or

$$c_2 \left(-g_{tt} + \Omega_H g_{\phi t}\right)_\theta = 0. \tag{13.14}$$

Because the term between brackets is always nonzero, it must be that c_2 vanishes, and so

$$q = 2 B J_H. \tag{13.15}$$

This Wald charge (13.15) gives electrostatic equilibrium between the north pole and infinity[576, 232, 171, 172, 162]. The contribution of magnetic flux produced by the equilibrium charge (13.12) is

$$\Phi'_H = 8\pi BM^2 \sin^2(\lambda/2). \tag{13.16}$$

For gravitationally negligible magnetic field strengths, the electric charge (and its associated electric field) does not affect the gravitational field. Accordingly, the equilibrium condition in (13.10)–(13.16) preserves maximal horizon flux at all rotation rates:

$$\Phi_H = 4\pi BM^2. \tag{13.17}$$

A lowest energy state of the black hole can be estimated without using a vacuum solution by considering the total energy[560]

$$\mathcal{E} = \frac{1}{2}Cq^2 - \mu_H B,$$ (13.18)

where $C \simeq 1/r_H$ denotes the electrostatic capacitance of the black hole. The energy $\mathcal{E}$ is quadratic in the electric charge with a minimum at

$$q = BJ_H(r_H/M).$$ (13.19)

While this argument is approximate, ignoring the detailed topology of the magnetic field outside the black hole, it is robust and suffices to show that a nonzero electric charge, i.e. a nonzero equilibrium magnetic moment develops which maintains essentially maximal horizon flux. (At slow rotation rates, the exact Wald value $q = 2BJ_H$ is recovered.) This preserved a strong connection between black hole and torus at arbitrary rotation rates.

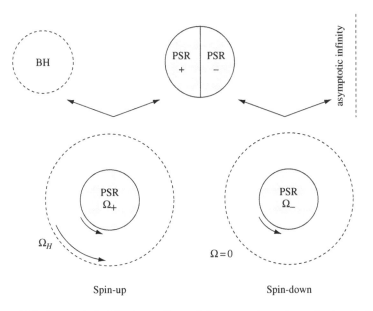

Figure 13.2 *Lower left*: The inner face of the torus (angular velocity Ω_+) and the black hole (angular velocity Ω_H) is equivalent to a pulsar surrounded by infinity with relative angular velocity $\Omega_H - \Omega_+$, in accord with Mach's principle. By equivalence in poloidal topology to pulsars, the inner face receives energy and angular momentum from the black hole as a causal process, whenever $\Omega_H - \Omega_+ > 0$. *Lower right*: The outer face of the torus (angular velocity Ω_-) is equivalent to a pulsar with angular velocity Ω_-, and always loses energy and angular momentum, by the same equivalence. (Reprinted from[568]. ©2003 The American Astronomical Society.)

13.3 Active black holes

Energy extraction mechanisms by scattering of positive energy waves onto rotating black holes – superradiant scattering of Ya. B. Zel'dovich[614], W. H. Press and S. A. Teukolsky[441], A. A. Starobinsky[507] J. M. Bardeen[33] – is a continuous wave analogue to the Penrose process. Its astrophysical applications are probably that of introducing instabilities in magnetized environments[557]. In the weak magnetic field-limit, R. Ruffini and J. R. Wilson[466] point out that horizon Maxwell stresses extract energy from a rotating black hole, already in the zero-frequency limit. This was suitably generalized by R. D. Blandford and R. L. Znajek[64], who identified a direct-current Poynting flux emanating from the black hole in force-free magnetospheres[64].

The poloidal topology of the inner torus magnetosphere shown in Figures (13.3)–(13.4) is insensitive to the detailed structure of spacetime. This can be seen, for example, by comparison with calculations in Schwarzschild spacetime[491, 545] (in part, on the basis of[49, 48, 50, 51]). The spin-connection provided by the inner torus magnetosphere, therefore, is robust, possibly

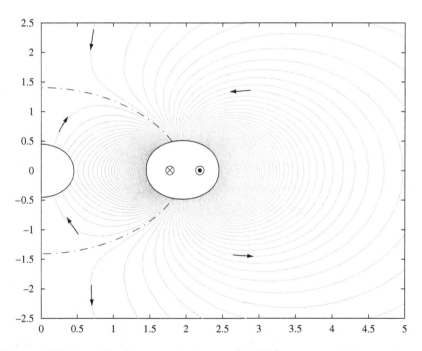

Figure 13.3 A uniformly magnetized torus (middle) represented by two counteroriented current rings, equivalent to a distribution of magnetic dipole moments, creates an inner and an outer torus magnetosphere around a black hole (left), delineated by a separatrix (dashed curve). (Reprinted from[568]. ©2003 The American Astronomical Society.)

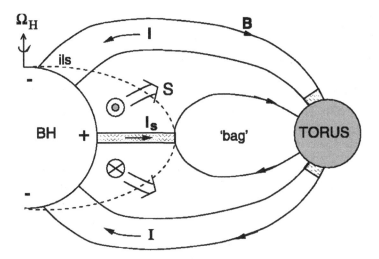

Figure 13.4 Viewed in poloidal cross-section, the magnetosphere of a uniformly magnetized torus is topologically equivalent to that of a pulsar. This equivalence holds for the inner face, facing the black hole (shown), and the outer face, facing asymptotic infinity (not shown). By this spin-connection, most of the black hole luminosity is incident onto the inner face of the torus (closed model). Notice the topology of magnetic field-lines as defined by their boundary conditions: magnetic field-lines connect the black hole and the inner face of the torus and closed magnetic field-lines make up an inner toroidal bag. The bag reaches down to the inner light-surface (dashed lines). The same holds true for the open magnetic field-lines connecting the outer face of the torus to infinity and a bag of closed magnetic field-lines reaching out to the outer light-cylinder (not shown). The spin-connection mediates most of the black hole luminosity (S) to the torus for reprocessing in various radiation channels in a state of suspended accretion. (Reprinted from[557]. ©1999 American Association for the Advancement of Science.)

modulated by time variability around rotating black holes, e.g. screw-instabilities on short timescales[241] or instabilities due to superradiant scattering on intermediate timescales[557] (see also[6, 103]). In contrast, the topology of the separatrix is generally subject to change regardless of black hole spin in response to outflows from the disk corona as discussed in Section 13.3.2.

Extracting energy from a rotating black hole introduces the "loading problem:" where does the energy go, how is the black hole luminosity distributed, and what are the observable radiation channels? We shall find that this is determined by the poloidal topology of the surrounding magnetosphere.

13.3.1 A spin-connection by equivalence to pulsars

The horizon of a black hole represents a compact null surface. It has in common with asymptotic infinity radiative boundary conditions: ingoing into the black hole and outgoing to infinity[64, 534, 399]. The horizon surface is generally different from asymptotic infinity by its angular velocity, which can readily exceed that of a surrounding torus. Open magnetic field-lines on the inner face of the torus may extend to the event horizon of the black hole. These *open* magnetic field-lines mediate angular momentum transport by Alfvén waves created by the inner face of the torus[557]. The spin-connection of the black hole to the torus is hereby equivalent to the spin-connection between the torus and infinity as shown in Figure (13.2).

By equivalence pulsars when viewed in poloidal topology, these spin-connections mediate angular momentum transport between the black hole and infinity via the torus, similar to those in pulsar winds. Flux surfaces outside the separatrix and those inside it are topologically equivalent, upon identifying the compact horizon surface with (noncompact) infinity. Equivalently to pulsars, the inner face of the torus can emit negative angular momentum Alfvén waves into the event horizon, while the outer face can emit positive angular momentum Alfvén waves to infinity. Both emissions satisfy causality.

In the approximation of flat spacetime, the magnetic flux surfaces produced by a superposition of current rings can be calculated analytically on the basis of the vector potential[282]

$$A_\phi = \frac{4IR}{\sqrt{R^2 + r^2 + 2rR\sin\theta}} \left[\frac{(2-k^2)K(k) - 2E(k)}{k^2} \right] \tag{13.20}$$

in terms of the complete elliptic integrals K and E, as a function of the argument

$$k^2 = \frac{4rR\sin\theta}{R^2 + r^2 + rR\sin\theta} \tag{13.21}$$

for a given ring current I. Magnetic flux surfaces are defined by $r\sin\theta A_\phi =$ const. Figure (13.3) shows the topology of the resulting inner and outer torus flux surfaces in vacuum by superposition of two concentric ring-current solutions of equal magnitude and opposite sign. They are separated by a separatrix (dashed line), which defines *ab initio* a bifurcation of the magnetic field on the rotation axis of the black hole. Figure (13.4) shows the poloidal topology of the inner torus magnetosphere in the force-free limit, which takes into account the presence of the inner light surface of R. L. Znajek[617]. The inner light surface delineates the inner most closed orbits of particles in corotation with the torus. The inner torus magnetosphere establishes a spin-connection, whereby most of the black hole luminosity is incident on the inner face of the torus[572, 557, 567, 53]

$$T \simeq -\dot{J}_H, \tag{13.22}$$

where T denotes the spin-up torque on the torus and $\dot{J}_H$ represents the angular momentum loss in the black hole. Energy extraction in the spin-connection between the black hole and the torus may be further augmented by supperradiant scattering. Low-frequency fast magnetosonic waves hereby are amplified, which scatter between the inner face of the torus and an angular momentum potential barrier closer to the rotating black hole. This process renders the inner torus magnetosphere unstable on an intermediate timescale of 0.1–1 s, which might account for sub-bursts seen in many GRB lightcurves[557, 6, 103]. See[103] for a recent discussion. In particular, the suspended accretion state can produce a "magnetic bomb:" a burst of X-ray emission upon prompt disconnection of the inner torus magnetosphere from the black hole by subsequent dissipation of magnetic field-energy[170], consistent in X-ray energies and time-scales with Type B events in GRS 1915 + 105. A spin-connection between the black hole and surrounding matter may be intermittent with relevance to the microquasar GRS1915 + 105 by sudden disconnection events in the inner torus magneto-sphere[170].

13.3.2 Spin–orbit coupling in ergotubes

Torus winds and ejection of matter from the hot torus corona may disrupt this structure by moving the separatrix between the inner and the outer torus magneto-sphere to infinity as schematically indicated in Figure (13.5). At large distances, the torus winds generally cross an Alfvén point, and become nearly radial at

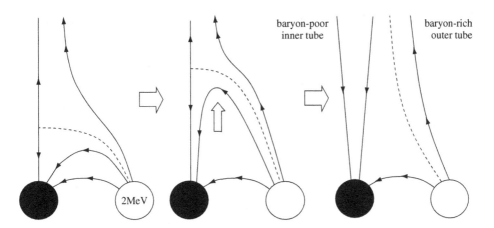

Figure 13.5 Topology of creating an open flux tube out of the inner torus magnetosphere, by moving the separatrix between the inner and the outer torus magnetosphere to infinity. The tube has slip–slip boundary conditions at infinity and on the horizon, and is surrounded by an outer flux tube supported by the inner face of the torus. The horizon half-opening angle of the inner *ergotube* is determined by poloidal curvature $\theta_H \simeq M_H/R$ of the inner torus magnetosphere. (Reprinted from[568]. ©2003 The American Astronomical Society.)

several scale heights. At smaller scale heights, large pressure gradients in the corona tend to push matter along some of the magnetic field-lines in the outer layers of the inner torus magnetosphere. Buoyancy and centrifugal forces may subsequently twist these field lines, ultimately leading to a fold and stretch, thereby forming a region of oppositely directed magnetic field lines. Stretched to infinity, field-lines thus created near the axis constitute an *open flux tube*; those anchored to the torus now extend to infinity and constitute the accompanying *outer flux tube*. They are separated by a charge and current sheet, whenever the outer tube carries a (super-Alfvénic) wind to infinity. Reconnection may occur in the boundary layer, which is of interest in the rearrangement of the magnetosphere near the axis. The resulting poloidal topology is schematically shown in Figure (13.5). This mechanism for a change in topology by a hot, pressure-driven flow might be aided by an axially focussed electromagnetic disk-wind. The latter was originally developed for AGN[61, 62, 63], which might find confirmation with the advance of subparsec scale observations on these sources on, for example, M87[286] (e.g.,[370, 302, 233] for numerical simulations). Tentative identification by Wardle and collaborators[581] of a baryon-poor component in the outflows in the quasar 3C 279 is particularly striking in this respect, which suggests that the same change in topology of the separatrix into an open ergotube might be taking place in AGN. The result would be a two-component beam-wind outflow[423, 502].

The angular momentum j_e of particles with charge e in a magnetic flux-tube symmetric around the axis of rotation of a black hole satisfies[558]

$$J_e = \frac{e\Phi}{2\pi},$$

(13.23)

and is proportional to the enclosed magnetic flux Φ. This angular momentum is in the electromagnetic field, since the canonical angular momentum is zero with respect to the symmetry axis. The interaction potential (relative to infinity) induced by black hole-spin satisfies the spin–orbit coupling[558]

$$E = \omega J_e,$$

(13.24)

were $\omega = \Omega_H$ on the horizon, similar to the spin–spin coupling (12.24). It equals

$$E = \Omega_B s$$

(13.25)

in terms of the cyclotron frequency and "induced spin"

$$\Omega_B = \frac{eB}{m_e c}, \quad s = m_e \frac{\omega A}{2\pi},$$

(13.26)

where A denotes the surface area of the flux tube at hand. This resembles the magnetic interaction energy $U = -\mu \cdot \mathbf{B}$ with the electron's magnetic moment $\mu = -(e/m_e c)\mathbf{s}$.

The spin–angular momentum potential (13.24) establishes a first-principle mechanism for driving charged outflows to infinity, in light of the fact that the specific angular momentum $j_e = J_e/m_e$ is effectively infinite given the mass m_e of the electron, whereby the gravitational Coulomb attraction between the black hole and the electron–positrons can be neglected. Equivalently, the induced electric forces far exceed the attractive gravitational forces. In dimensionful units, the potential energy on the horizon of an extremal black hole $(a = M)$ satisfies[573]

$$E_\nu = 1.5 \times 10^{22} \left(\frac{M_H}{7M_\odot} \right) \left(\frac{B}{10^{16}\text{G}} \right) \left(\frac{\theta_\nu}{0.1} \right)^2 \text{eV}, \qquad (13.27)$$

where θ_ν denotes the half-opening angle of the flux surface on the event horizon of the black hole. An open magnetic flux-tube along the axis of rotation of a black hole hereby represents an *ergotube*, werein black hole spin performs on charged outflows. This gravitational spin–angular momentum interaction potential E is different from the Penrose process, which is restricted to the ergosphere.

The force $-\partial E/\partial r$ shows the role of differential frame-dragging in creating an equivalent Faraday-induced electric potential on the electric charge. Notice that this induces an electric field *along* the magnetic field. Charged particles hereby accelerate, and carry along a Poynting flux to larger distances.

These observations show two equivalent pictures of first-principle interactions in the ergotube:

1. Gravitational spin–orbit coupling induces a potential energy of charged particles proportional to their angular momentum. This gravitational interaction produces a repelling force between aligned spin and angular momentum.
2. Black hole-spin creates electric fields *along* a flux-tube by Faraday-induction due to differential frame-dragging on magnetic flux-surfaces. This gravitational interaction creates poloidal electric currents towards electrostatic equilibration by charge separation.

The first was pointed out in[558]. *By spin–orbit coupling, the open ergotube along the rotation axis of the black hole is a linear accelerator of particles with large specific angular momentum.* In response to such outflows, the black hole evolves by conservation of energy, angular momentum, and electric charge. The second was pointed out in[576] and further elaborated on in[560]. The first is an integral of the second, where the second may further be viewed as a variant of Lorentz forces[573].

The potential (13.27) is large, and sufficiently so that generic pair creation processes are effective, e.g. via curvature radiation processes[64] or direct vacuum breakdown[558]. The ergotube rapidly fills with electric charges. As a result, charge separation sets in which counteracts any gradient in E or, equivalently, any electric field along the tube.

Charge separation in black hole magnetospheres was first described by R. D. Blandford and R. L. Znajek[64]. In the ideal limit, the ergotube becomes *force-free*. In this event, surfaces of constant magnetic flux become surfaces of constant electric potential and assume rigid rotation. The latter corresponds to vanishing Faraday-induced electric fields along flux surfaces. In this limit, flux surfaces are ideal conductors with zero electric dissipation, while mediating Poynting-flux dominated flows.

With or without charge-separation, (13.27) represents the "open loop" potential in the "black hole plus ergotube" configuration.

The force-free limit is described by a local equilibrium charge density (see van Putten & Levinson[568] and references therein)

$$\rho = -\frac{(\Omega_F - \omega)B}{2\pi} = \begin{cases} -\Omega_{F-}B/2\pi & \text{at infinity} \\ (\Omega_H - \Omega_{F+})/2\pi & \text{on H} \end{cases} \qquad (13.28)$$

where Ω_F denotes the local angular velocity of the flux surface at hand, reaching Ω_{F-} at infinity and Ω_{F+} on the event horizon. The density (13.28) is expressed in the orthogonal frame associated with the zero-angular momentum observers, whose angular velocity is ω. This generalizes the Goldreich–Julian charge density in pulsar magnetospheres[234], by taking into account frame-dragging.

13.3.3 Ejecting blobs by frame-dragging

A magnetized blob of perfectly conducting fluid (see Chapter 9) is rich in specific angular momentum carried by charged particles. About the axis of rotation, their lowest energy state has zero canonical angular momentum, whereby they carry an angular momentum $J = eA_\phi$. Consider a magnetized blob about the axis of rotation of the black hole, as sketched in Figure (13.6). The blob represents a section of an open magnetic flux tube, subtended by a finite half-opening angle on the event horizon of the black hole. The blob subtends a certain amount of magnetic flux $(2/\pi_A)/\phi$ of this open magnetic flux tube. The number $N(s)$ of particles per unit height s of the blob, therefore, satisfies

$$N(s) = (\Omega_b - \Omega) A_\phi/e, \qquad (13.29)$$

where e denotes the elementary charge.

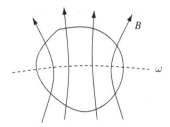

Figure 13.6 Schematic illustration of the ejection of a magnetized blob. In a perfectly conducting state, the blob assumes a well-defined angular velocity Ω_b and angular momentum $J = e(\Omega_b - \omega)A_\phi$ per charged particle, where ω denotes the local frame-dragging angular velocity. Gravitational spin–orbit coupling induces a potential energy $E = \omega J$. Blobs with $\Omega_b > \omega$ are ejected, while blobs with $\Omega_b < \omega$ are absorbed by the black hole. The blobs move along an open magnetic flux-tube (not shown).

Blobs of scale height h hereby receives an energy

$$E_{blob} = \omega JNh = \omega(\Omega_b - \omega)A_\phi^2 h. \tag{13.30}$$

This represents the energy of blobs ejected ballistically with conservation of angular momentum J, from a radius where the frame-dragging angular velocity equals ω. Moving ahead a little in notation, we write (13.30) as

$$E_{blob} = 4.6 \times 10^{-6} \left(\frac{\Omega_b}{\Omega_H}\right)^2 \left(\frac{\omega}{\Omega_b}\right)\left(1 - \frac{\omega}{\Omega_b}\right)\left(\frac{\eta}{10}\right)^{10/3}\left(\frac{h}{M}\right)\left(\frac{15\mathcal{E}_B}{\mathcal{E}_k}\right)\mathcal{E}_k, \tag{13.31}$$

in two-sided ejections by a maximally spinning black hole. Here, $\eta = \Omega_T/\Omega_H \simeq 2(M/R)^{3/2}$ denotes the ratio of the angular velocity of the torus to that of the black hole and $A_\phi = BM^2\theta_j^2$ with θ_j as in (13.39). $\mathcal{E}_B$ refers to poloidal magnetic field-energy in the torus magnetosphere and

$$\mathcal{E}_k = 6.8 \times 10^{-2} \left(\frac{\eta}{10}\right)^{2/3} M_T \tag{13.32}$$

refers to the kinetic energy in the torus of mass M_T, as discussed in Chapter 14. For a torus mass of $0.1M_\odot$, a single blob of size $h \sim M$ will have an energy of about 1.6×10^{46} erg for $\omega \simeq \Omega_H/2$ and $\Omega_b \simeq \Omega_H$. Note that causality in the spin–orbit ejection of magnetic blobs is self-evident. The ejection of a pair of blobs with energy (Eqn (13.32)) takes place in essentially a light-crossing time of about 0.3 ms for a stellar mass hole, corresponding to an instantaneous luminosity of the order of 3×10^{50} erg s^{-1}.

In Figure (13.5), the open magnetic flux tube thus acts much like the barrel of a gun in ejecting energetic, rotating blobs of particles according to $E = \Omega J$ and $J = eA_\phi$. The barrel is created upon moving the separatrix surrounding the inner torus magnetosphere to infinity. It will be appreciated that ejection refers to the charged particles streaming along the open field-lives, not the magnetic field (the barrel) itself.

13.3.4 Launching a magnetized jet

The equilibrium charge density ρ assumes opposite signs on the event horizon of the black hole and at infinity, whenever $\Omega_{F-} > 0$ and $\Omega_H > \Omega_{F+}$. This is commensurate with opposite signs in angular momentum of the electromagnetic field. Charge-separation fills the ergotube with angular momentum, which is positive towards infinity and negative towards the black hole: *the ergotube is polarized in high specific angular momentum.*

If the ergotube assumes a largely force-free state as envisioned in[64], the angular velocity and angular momentum of ejecta is constant[64, 568]. The resulting luminosity is similar but not the same as (13.30). As the flow becomes supercritical going into the black hole and out to infinity, applying the force-free limit gives the current continuity condition

$$I = (\Omega_{F+} - \Omega_H)A_\phi = \Omega_{F-}A_\phi \tag{13.33}$$

in the small-angle approximation. If the flux tube is force-free everywhere, it rotates uniformly with

$$\Omega_{F+} = \Omega_{F-} = \frac{1}{2}\Omega_H. \tag{13.34}$$

This establishes a maximal black hole luminosity

$$\mathcal{L}_{max} = \frac{1}{4}\Omega_H^2 A_\phi^2 \tag{13.35}$$

through the flux tube, as envisioned but not shown in[64]. If the force-free limit does not hold everywhere along the flux tube, then a finite voltage drop occurs due to a finite difference in angular velocities. In this event, Poynting flux "leaks" out of the ergotube. There is no unique recipe known for current closure in this

case. In one example with current at infinity closing over an outer flux tube
supported by the surrounding torus,

$$\Omega_{F+} - \Omega_{F-} = \Omega_H - 2\Omega_T, \tag{13.36}$$

giving

$$\mathcal{L} = \Omega_T(\Omega_H - 2\Omega_T)A_\phi^2. \tag{13.37}$$

The ergotube is now subluminous, and carries a finite amount of particle flux due
to dissipation in differential rotating sections.

By canonical pair-creation and charge-separation processes, ergotubes are
"loaded with a sea of high specific angular momentum" in electromagnetic
form. The black hole launches a jet in the ergotube carrying away this angular
momentum according to the following two descriptions, based on the previous
section:

1. *Rayleigh picture.* Gravitational spin–orbit coupling (12.24) and (13.24) causes the
   black hole to eject a "sea" of high positive-specific angular momentum associated with
   charged particles in the ergotube. Carried along is an electric current, which closes
   over the event horizon of the black hole and infinity. Moderated by a finite surface
   impedance of 4π[534], the open loop potential energy E (13.27) hereby produces a
   finite luminosity. By the polarized distribution of angular momentum in the ergotube,
   an equal amount of negative angular momentum is absorbed by the black hole, causing
   it to spin down by conservation of angular momentum.
2. *Faraday picture.* Black hole spin induces an open electric potential E/e (13.27) which
   drives a poloidal current. Particle flow through open boundary conditions into the black
   hole and out to infinity creates a continuous current. The "black hole plus ergotube"
   is hereby *not* in equilibrium, even when the ergotube assumes a force-free state. With
   closure over the event horizon of the black hole[534] and infinity, these open electric
   currents carry along angular momentum to infinity. Moderated by a finite surface
   impedance of 4π[534], the ergotube has a finite luminosity. Poloidal current closure
   over the black hole introduces Maxwell stresses on the event horizon, and causes it to
   spin down.

The link between these two descriptions is by fast magnetosonic waves, which
are excited by the linear acceleration process due to spin–orbit coupling. Notice
that the Rayleigh picture is general and explicitly causal. It is the driving agency
in the creation of black hole outflows. It does not rely on a particular electrostatic
state, such as the force-free limit.

The force-free limit of Blandford–Znajek[64] corresponds to a maximum in
pure Poynting-flux outflows[568]. In the force-free limit, causality is not self-
evident. It has been shown in an elegant analyis by A. Levinson[333], who
considers the limit of small-amplitude fast magnetosonic waves (and accompanying Alfvén waves) in response to perturbations of black hole spin.

The luminosity of the (two-sided) ergotube is limited by the half-opening angle θ_H on the event horizon. For an extremal black hole, we have[573]

$$\frac{E_{ergotube}}{E_{rot}} \simeq \frac{1}{2}\theta_H^4 \tag{13.38}$$

asymptotically in the small-angle approximation and for small η. This generally represents a small fracton of the rotational energy of the black hole. We expect that θ_H is determined by the poloidal curvature in the inner torus magnetosphere, i.e.

$$\theta_H \simeq \frac{M_H}{R}. \tag{13.39}$$

This chapter is summarized in Figure (13.7).

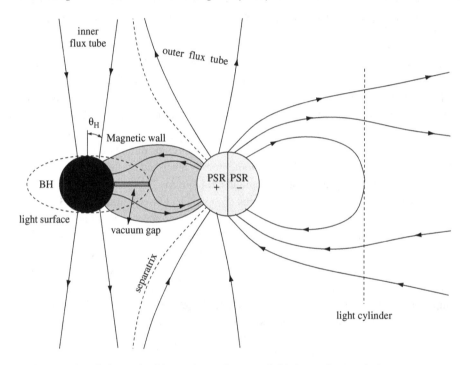

Figure 13.7 Schematic illustration of the poloidal topology of the magnetosphere of an active black hole surrounded by a uniformly magnetized torus. Most of the spin energy is dissipated in the event horizon, which defines the lifetime of rapid spin of the black hole. Most of the black hole luminosity is incident on the torus, while a minor forms a jet in a baryon-poor ergotube. (Reprinted from[568]. ©The American Astronomical Society.)

Exercises

1. Show that (13.5) is independent of S, given the same boundary δS.
2. Derive the equation of the inner light surface of[617], given an angular velocity Ω_T of the torus.
3. Establish Ferraro's law on the basis of zero dissipation and Faraday's law: a perfectly conductive flux surface is in rigid rotation.
4. Derive the equation for the rate of spin-down of a pulsar by Poynting-flux-dominated pulsar winds (see[234]).
5. Derive the modified equation for the Goldreich–Julian charge density along a flux surface with angular velocity Ω, close to the axis of rotation of a Kerr black hole (see[53]).
6. Show that the first law of thermodynamics implies that the efficiency of energy transport to the torus from the black hole is given the ratio of their angular velocities.
7. Show that the black hole spin-connection $m = 0$ is stable, and that self-gravity in the torus formed in core collapse is not important. Can self gravity in the torus be relevant in case of black hole–neutron star coalescence?, (see[189]).
8. What is the timescale for building up the magnetic field up to the critical stability value $E_B/E_k = 1/15$, assuming the energy is provided by the black hole?
9. Derive (13.23) on the basis of zero canonical angular momentum in the Landau states of the charged particles.
10. Derive the ratio of specific angular momentum $j_e = J_e/m_e$ to that of the black hole,

$$\frac{j_e}{a} = 8\frac{Gm_e M}{c\hbar}\left(\frac{B}{B_c}\right)\left(\frac{A}{A_H}\right)\cos^2(\lambda/2) \tag{13.40}$$

or

$$\frac{j_e}{a} = 4.0 \times 10^{15}\left(\frac{M}{M_\odot}\right)\left(\frac{B}{B_c}\right)\left(\frac{A}{A_H}\right)\cos^2(\lambda/2), \tag{13.41}$$

where A denotes the cross-sectional area of the flux tube at hand and $B_c = m_e^2 c^3/e\hbar = 4.4 \times 10^{13}$ G denotes the QED value of the magnetic field.

11. What is the *minimal* magnetic field strength for the spin–orbit coupling to eject a charged particle to infinity? Discuss (a) ejection by a black hole and (b) ejection from the Earth's surface.

12. Show that for an electron in a magnetic flux tube of cross-sectional area A on the surface of the Earth, we have the ratio of potential energies

$$\frac{E}{U_G} = \frac{2m_e \Omega_E A}{5\pi\hbar} \left(\frac{B}{B_c} \right), \tag{13.42}$$

where $U_G = Gm_e M/R$ and Ω_E denotes the angular velocity of the Earth. [*Hint*: Use the specific moment of inertia $i_E = (2/5)R_E^2$ for a mass $M_E = 5 \times 10^{27}$ g and radius $R_E = 6 \times 10^8$ cm.]

13. Derive (13.38), considering (13.39) and the force-free state of[64]. [*Hint*: in the force-free limit, current continuity implies a two-sided luminosity given by (13.35), where $A_\phi = \frac{1}{2}B\rho^2\theta_H^2$ with $\rho^2 \simeq 2M^2$ for small θ_H for an extremal black hole in Boyer–Lindquist coordinates (12.3). The rate of dissipation of black hole spin energy in the event horizon satisfies $D = (\Omega_H - \Omega_T)^2 A_\phi^2 \simeq \frac{1}{2}\Omega_H^2 A_\phi^2$ for small η, where $A_\phi = \frac{1}{2}\rho^2 B$ with $\rho^2 = M^2$.]

14. Sketch the equilibrium charge-distribution in the magnetosphere supported by a uniformly magnetized torus around a Schwarzschild black hole, a slowly rotating black hole whose angular velocity is less than that of the torus, and a rapidly rotating black hole[568].

15. Launching magnetized jets along open magnetic flux-tubes requires pair-production to sustain charged particle flows out to infinity and into the black hole. Generalize the discussion of ergotubes by including a dissipative, differentially rotating gap, for *in-situ* pair production.

14

A luminous torus in gravitational radiation

Alice laughed: "There's no use trying," she said; "one *ca'n't* believe impossible things." "I daresay you haven't had much practice," said the Queen. "When I was your age, I always did it for half-an-hour a day. Why, sometimes I've believed as many as six impossible things before breakfast."
Lewis Carroll, *Through the Looking-glass, and what Alice Found There*, Chapter 5.

A torus surrounding a luminous black hole receives black hole spin energy for reprocessing in various emission channels. A balance between spin energy received and energy radiated allows a torus to remain in place for the duration of rapid spin of the black hole – a *suspended accretion* state[569]. Amplification of this "seed" field to superstrong values requires a dynamo action in the torus. Conceivably, this dynamo is powered by black hole-spin energy in a long-lasting suspended accretion state.

In this chapter, we derive a bound on the magnetic field energy that a torus of given mass can support. It defines a black hole luminosity function in terms of the angular velocity and mass of the torus, both relative to the angular velocity and mass of the black hole. The torus is compact and lives around a stellar mass black hole. The competing torques of spin-up by the black hole and spin-down by radiation promote a slender shape. This raises the questions: What is the lifetime of rapid spin of the black hole and its luminosity? What are the radiation energies emitted by the torus?

We consider these questions by deriving a magnetic stability criterion for tori of finite mass, and by solving for the equations of balance, between input received by the black hole and radiative output by the torus in the approximation of viscosity dominated by turbulent magnetohydrodynamical stresses[558, 256]. The seed magnetic field of the torus is assumed to be provided by the progenitor star[75].

215

14.1 Suspended accretion

A state of suspended accretion[569] arises as schematically indicated in Figure (14.1), when the flux of energy and angular momentum into surrounding matter is balanced by radiative losses of the same. This is based on the spin-connection between the black hole and the resulting torus of Chapter 13.

The angular momentum transport between the black hole and infinity is governed by the angular velocity of the torus. Losses from the outer face of the torus generally stimulate accretion, whereas gain by the spin-connection to the black hole provides a spin-up torque (13.22) – as when a pulsar is being wrapped around by infinity in Figure (13.2). (We focus on angular velocities, not angular momentum. The inner face of the torus will still have positive angular momentum, when it is sufficiently wide. It would be incorrect to consider equivalence to a pulsar with negative spin, and hence negative angular momentum.) Infinity now receives negative angular momentum from the Alfvén waves created by the pulsar, i.e. the inner face of the torus spins

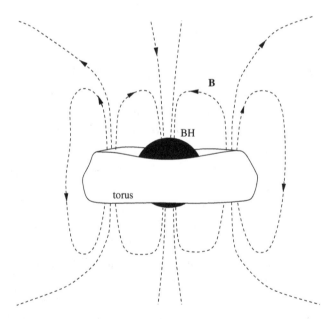

Figure 14.1 A Kerr black hole surrounded by a uniformaly magnetized torus, receiving most of the black hole luminosity. The torque $T \simeq -\dot{J}_H$ from the black hole puts the torus in a state of suspended accretion, balanced against losses in heat and its various radiation channels, mostly gravitational radiation. A small fraction of black hole spin energy is launched by linear acceleration of baryon-poor outflows of high specific angular momentum along an open ergotube in accord with the Rayleigh criterion. (Reprinted from[558]. ©2001 The American Physical Society.)

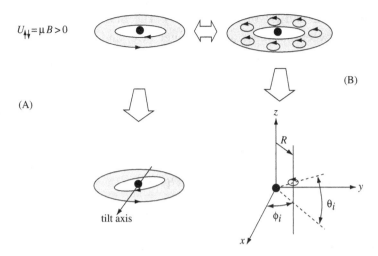

$U_{\uparrow\uparrow}=\mu B>0$

(A)

(B)

tilt axis

Figure 14.2 A uniformly magnetized torus is in its highest magnetic energy state. Magnetic instabilities are stabilized by tidal forces, provided that magnetic field energy is below a critical value. The two alternative leading-order partitions of the current distribution (*top*) have unstable poloidal modes, described by a relative tilt between the two current rings towards alignment (A) or towards buckling (B), characterized by perturbations out of the equatorial plane with poloidal angles $\theta_i \neq \theta_j$. We may consider the stability for poloidal motion along a cylinder of radius R. (Reprinted from[568]. ©2003 The American Astronomical Society.)

up, in response to which the black hole spins down by conservation of angular momentum.

On balance, between losses to infinity and gain from the black hole, a suspended accretion state results. In regards to angular momentum transport as a function of the mean radius R of the torus, we note that the spin-connection (13.22) to the black hole is governed by the horizon magnetic flux $\propto (M/R)^2$, while the spin-connection to infinity satisfies $\propto \Omega_T \simeq M/R^{3/2}$. A suspended accretion state is hereby stable in the mean radius of the torus. In describing this, we first consider a bound on the magnetic field energy.

14.2 Magnetic stability of the torus

Fluid motion *in* the equatorial plane around a black hole is generally stabilized by poloidal magnetic pressure. In contrast, poloidal motion *out* of the equatorial plane is generally destabilized by magnetic moment–magnetic moment self-interaction in the torus. This destabilizing effect on the poloidal motion can be modeled by partitioning the torus in a finite number of fluid elements with current loops, representing local magnetic moments. The two leading-order partitions are shown in configurations C and B1 of Figure (13.1). The first partitioning is subject to magnetic tilt between the inner and the outer face, and the second is subject to magnetic buckling of the torus.

14.2.1 A magnetic tilt instability

Following C in Figure (13.1), consider the magnetic interaction energy of a pair of concentric current rings, given by

$$U_\mu(\theta) = -\mu B \cos\theta. \tag{14.1}$$

Here, μ is the magnetic dipole moment of the inner ring, B is the magnetic field produced by the outer ring, and θ denotes the angle between μ and B. Note that $U_\mu(\theta)$ has a period 2π, and is maximal (minimal) when μ and B are antiparallel (parallel, as in Figure (14.3)). Consider tilting a fluid element of a ring out of the equatorial plane to a height z subject to motion approximately on a cylinder of constant radius R. (This is different from tilting a rigid ring, whose elements move on a sphere of constant radius.) A tilt hereby changes the distance to central black hole to $\rho = \sqrt{R^2 + z^2} \simeq R(1 + z^2/2R^2)$. In the approximation of equal mass

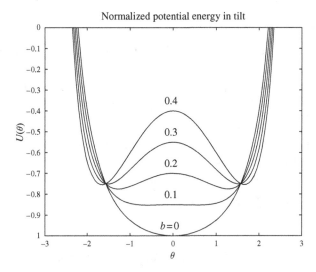

Figure 14.3 The potential energy associated with a poloidal tilt angle θ between the inner and the outer rings is the sum of a magnetic moment–magnetic moment interaction plus a tidal interaction with the central potential well of the black hole. It is shown for various normalized magnetic field energies $b = \mathcal{E}_B/\mathcal{E}_k$. The equilibrium θ becomes unstable when $d^2 U(\theta)/d\theta^2 < 0$, corresponding to a bifurcation into two stable branches of nonzero angles beyond $b > 1/12$. (Reprinted from[568]. ©2003 The American Astronomical Society.)

in the inner and outer face of the torus, simultaneous tilt of one ring upwards and the other ring downwards is associated with the potential energy

$$U_g(\theta) \simeq -\frac{M_T M_H}{R}\left(1 - \frac{1}{4}\tan^2(\theta/2)\right), \qquad (14.2)$$

with $\tan(\theta/2) = z/R$, where we averaged over all segments of a ring. Note that $U_g(\theta)$ has period π and is minimal when $\theta = 0$. Stability is accomplished provided that the total potential energy $U(\theta) = U_\mu(\theta) + U_g(\theta)$ satisfies

$$\frac{d^2 U}{d\theta^2} > 0. \qquad (14.3)$$

The potential $U(\theta)$ is shown in Figure (14.3), which shows the bifurcation in stability at $b = 1/12$, from a stable into an unstable equilibrium at $\theta = 0$. This stability exchange is accompanied by the appearance of two neighboring stable equilibria at nonzero angles, whereby it is second-order. Nevertheless, the torus may become nonlinearly unstable at large angles ($b >> b^*$). We therefore consider below the physical parameters at this bifurcation point.

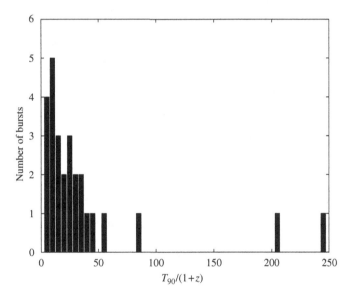

Figure 14.4 Shown is the histogram of redshift-corrected durations of 27 long bursts with individually determined redshifts from their afterglow emissions. It shows redshift-corrected durations $T_{90}/(1+z)$ of about one-half minute, a mean value $<z> = 1.25$ of redshifts and a redshift-correction factor $< T_{90}/(1+z) >$ $/ < T_{90} > = 0.45$. The mean of the observed and redshift-corrected durations is 83 s and 38 s, respectively (53 s and 23 s without the two long bursts). (Reprinted from[561]. ©2002 The American Astronomical Society.)

For two rings of radii R_+ with $(R_+ - R_-)/(R_+ + R_-) = O(1)$, we have $U_\mu \simeq \frac{1}{2} B^2 R^3 \cos\theta$, so that (14.3) gives $B_c^2 M_H^2 = (1/4)(M_H/R)^4 (M_T/M_H)$, or

$$B_c \simeq 10^{16}\,\mathrm{G}\left(\frac{7 M_\odot}{M_H}\right)\left(\frac{6 M_H}{R}\right)^2\left(\frac{M_T}{0.03 M_H}\right)^{1/2}. \qquad (14.4)$$

The critical value of the ratio of poloidal magnetic energy $(\mathcal{E}_B = f_B B^2 R^3/6)$ to kinetic energy $(M_T M_H/2R)$ in the torus becomes

$$\left.\frac{\mathcal{E}_B}{\mathcal{E}_k}\right|_c = \frac{f_B}{12}, \qquad (14.5)$$

where f_B denotes a factor of order unity, representing the volume of the inner torus magnetosphere as a fraction of $4\pi R^3/3$.

14.2.2 A magnetic buckling instability

We partition the magnetization of the torus into N equidistant fluid elements with dipole moments, $\mu_i = \mu/N = (1/2)BR^3/N$. We consider the vertical degree of freedom of fluid elements which move to a height z above the equatorial plane. By conservation of angular momentum, this motion is restricted to a cylinder of constant radius. Their position vectors in and off the equatorial plane will be denoted by

$$\mathbf{r}_i^e = (R\cos\phi_i, R\sin\phi_i, 0), \quad \mathbf{r}_i = (R\cos\phi_i, R\sin\phi_i, z_i), \quad \phi_i = 2\pi i/N. \quad (14.6)$$

A fluid element i assumes an energy which consists of magnetic moment–magnetic moment interactions and the tidal interaction with the central potential well. The total potential energy of the ith fluid element is given by

$$U_i = -\frac{\mu_i B'}{N}\Sigma_{j\neq i}\frac{|\mathbf{r}_i^e - \mathbf{r}_j^e|^3}{|\mathbf{r}_i - \mathbf{r}_j|^3}\cos\theta_{ij} + U_g(\theta_i), \qquad (14.7)$$

where $B' = B/N^*$ denotes the magnetic field strength of a magnetic dipole at distance $d = 2\pi R/N$, θ_{ij} denotes the angle between the ith magnetic moment and the local magnetic field of the jth magnetic moment, and $U_{gi} = -(M_T M_H/RN)(1 - z_i^2/2R^2)$ the tidal interaction of the ith fluid element with the black hole. Here, N^* is a factor of order N which satisfies the normalization condition

$$\Sigma_i U_i = -\mu B \qquad (14.8)$$

(in equilibrium). Upon neglecting azimuthal curvature in the interaction of neighboring magnetic moments, we have a magnetic moment–magnetic moment interaction

$$\mu_i B'\frac{|\mathbf{r}_i^e - \mathbf{r}_j^e|^3}{|\mathbf{r}_i - \mathbf{r}_j|^3}\cos\theta_{ij} \simeq \frac{\mu_i B'}{|i - j|^3}\left(1 - \left[1 + \frac{3}{2|i-j|^2}\right]\alpha_{ij}^2\right), \qquad (14.9)$$

where $\alpha_{ij} = (z_i - z_j)/d$,

$$\cos\theta_{ij} = -\sqrt{(1-\alpha_{ij}^2)/(1+\alpha_{ij}^2)} \simeq -(1-\alpha_{ij}^2), \qquad (14.10)$$

and $|\mathbf{r}_i - \mathbf{r}_j| \simeq |i-j|d\left(1+\alpha_{ij}^2/2|i-j|^2\right)$.

We shall use a small amplitude approximation, whereby $z_i/R = \tan\theta_i \simeq \theta_i$. We study the stability of this configuration, to derive an upper limit for the magnetic field strength. An upper limit obtains by taking into account only interactions between neighboring magnetic moments. (The sharpest limit obtains by taking into account interactions between one magnetic moment and all its neighbors.) Thus, we have $N^* = 2N$ and consider the total potential energy

$$U_i = \frac{\mu_i B'}{N}\Sigma_{|i-j|=1}\left(1-\frac{5}{2}\alpha_{ij}^2\right)+U_g(\theta_i), \qquad (14.11)$$

where $\alpha_{ij} = N(\theta_i - \theta_j)/2\pi$. The Euler–Lagrange equations of motion are

$$\frac{M_T R}{N}\ddot{\theta}_i + \frac{\partial U_i}{R\partial\theta_i} = 0. \qquad (14.12)$$

This defines the system of equations for the vector $\mathbf{x} = (\theta_1, \theta_2, \cdots, \theta_N)$ given by

$$\frac{M_T R}{N}\ddot{\mathbf{x}} + \frac{M_T M_H}{NR^2}\mathbf{x} = \frac{5\mu B}{2N}\begin{pmatrix} 2 & -1 & 0 & \cdots & 0 & -1 \\ -1 & 2 & -1 & \cdots & 0 & 0 \\ & & & \cdots & & \\ -1 & 0 & \cdots & 0 & -1 & 2 \end{pmatrix}\mathbf{x}. \qquad (14.13)$$

The least stable eigenvector is $\mathbf{x} = (1, -1, 1, \cdots, -1)$ (for N even), for which the critical value of the magnetic field is

$$B_c^2 M_H^2 = \frac{1}{5}\left(\frac{M_T}{M_H}\right)\left(\frac{M_H}{R}\right)^4. \qquad (14.14)$$

This condition is very similar to (14.4), and gives a commensurable estimate

$$\left.\frac{\mathcal{E}_B}{\mathcal{E}_k}\right|_c = \frac{1}{15} \qquad (14.15)$$

and lifetime of rapid spin of the black hole.

A high-order approach can be envisioned, in which the inner and outer face of the torus are each partitioned by a ring of magnetized fluid elements. This is of potential interest in studying instabilities in response to shear, in view of the relative angular velocity $\Omega_+ - \Omega_- > 0$. Magnetic coupling between the two faces of the torus through the aforementioned tilt or buckling modes inevitably leads to transport of energy and angular momentum from the inner face to the outer face of the torus by the Rayleigh criterion.

14.3 Lifetime and luminosity of black holes

For black hole angular velocities much larger than that of the torus, the spin-connection (13.22) causes most of the spin-energy to be dissipated in the event horizon of the black hole.

The rate of dissipation of spin energy of the black hole in its horizon is determined by the angular velocity of the torus[557], given by[534, 563]

$$T_H \dot{S}_H = \Omega_H(\Omega_H - \Omega_+)f_H^2 A^2, \tag{14.16}$$

where T_H denotes the horizon temperature and S_H its entropy. Most of the black hole luminosity is hereby incident onto the surrounding torus. The lifetime of rapid spin is therefore given by

$$T_s \simeq \frac{E_{rot}}{T\dot{S}_H}. \tag{14.17}$$

The aforementioned magnetic stability condition gives rise to a critical field-strength (14.4). Observational evidence for super-strong magnetic fields may be found in SGRs and AXPs (see, for example,[306, 524, 179, 275, 227]). We then have[568]

$$T_s \simeq 90\,\text{s} \left(\frac{M_H}{7M_\odot}\right)\left(\frac{\eta}{0.1}\right)^{-8/3}\left(\frac{\mu}{0.03}\right)^{-1}. \tag{14.18}$$

In the application to long GRBs, this estimate is consistent with durations of tens of seconds of long gamma-ray bursts[305].

The black hole luminosity is a fraction $\eta = \Omega_T/\Omega_H$ times the dissipation rate in the event horizon of the black hole, or

$$L_H = 4.4 \times 10^{51} \eta_{0.1}^{8/3} \mu_{0.03}\,\text{erg s}^{-1} \tag{14.19}$$

at the critical value of stability (10.8).

14.4 Radiation channels by the torus

The suspended accretion state lasts for the lifetime of rapid spin of the black hole. This is a secular timescale of tens of seconds, much larger than the millisecond period of the accretion disk. This reduces the problem of calculating the emission to algebraic equations of balance in energy and angular momentum flux, in the presence of various emission channels: gravitational radiation, MeV-neutrino emissions and magnetic winds. Asymptotic expressions for small ratios η of the angular velocity of the disk to that of the black hole, and small slenderness ratios δ of the torus give energy outputs which are $O(\eta)$ in gravitational radiation, $O(\delta\eta)$ in MeV-neutrinos, and $O(\eta^2)$ in magnetic winds. The latter dissipates into radiation, powering an accompanying supernovae in good agreement with observations. The results can be expressed in terms of fractions of black hole-spin energy, i.e.

$$E_{gw}/E_{rot}, \quad E_w/E_{rot}, \quad E_\nu/E_{rot}. \quad (14.20)$$

The gravitational wave-emissions are due to quadrupole emissions due to a mass-inhomogeneity δM_T according to Peters and Mathews[419]

$$L_{gw} = \frac{32}{5} (\omega \mathcal{M})^{10/3} F(e) \simeq \frac{32}{5} (M_H/R)^5 (\delta M_T/M_H)^2, \quad (14.21)$$

where $\omega \simeq M^{1/2}/R^{3/2}$ denotes the orbital frequency of the torus with major radius R, $\mathcal{M} = (\delta M_T M_H)^{3/5}/(\delta M_T + M_H)^{1/5} \simeq M_H (\delta M_T/M_H)^{5/3}$ denotes the chirp mass, and $F(e)$ denotes a geometric factor representing the ellipticity e of the oribital motion. Application of (14.21) to PSR1913+16 with ellipticity $e = 0.62$[271] provided the first evidence for gravitational radiation consistent with the linearized equations of general relativity to within 0.1%[518]. Here, we apply the right-hand side of (14.21) to a nonaxisymmetric torus around a black hole, whose mass quadrupole inhomogeneity δM_T is determined self-consistently in a state of suspended accretion for the lifetime of rapid spin of the black hole. A quadrupole mass-moment appears spontaneously due to nonaxisymmetric waves whenever the torus is sufficiently slender.

In the suspended accretion state, most of the black hole spin energy is dissipated in the event horizon for typical ratios $\eta \sim 0.1$ of the angular velocity of the torus to that of the black hole. Hence, the lifetime of rapid spin of the black hole is effectively determined by the rate of dissipation of black hole spin energy in the event horizon, itself bounded by a finite ratio $\mathcal{E}_B/\mathcal{E}_k < 1/15$ of the poloidal magnetic field energy-to-kinetic energy in the torus[568]. This gives rise to long durations of tens of seconds for the lifetime of rapid spin of the black hole. The resulting gravitational wave emissions should be limited in bandwidth, changing in frequency about 10% during the emission of the first 50% of its energy output. This change mirrors a decrease of 10% in the angular velocity of a maximally spinning black hole in converting 50% of its spin energy. Thus, gravitational radiation is connected to Kerr black holes, representing a connection between the linearized equations of general relativity and, respectively, fundamental objects predicted by the fully nonlinear equations of general relativity.

Gravitational radiation in collapsars and hyperaccretion flows onto a central black hole have been considered in a number of other studies[385, 352, 74, 149, 212, 376, 297], also in model-independent search strategies associated with GRBs[188, 384]. These studies focus on gravitational radiation produced by the release of gravitational binding energy during collapse and in accretion processes onto newly formed black holes (e.g.[215]). Accretion flows are believed to be strongly turbulent. Any radiation produced in the process will have a relatively broad spectrum. The aforementioned studies on gravitational radiation do not include the spin-energy of a newly formed black hole. The results appear to indicate an energy output which leaves a range of detectability by current ground-based detectors of up to about 10 Mpc. These events should therefore be considered in the context of core-collapse events independent of the GRB phenomenon, in light of our current estimates on the local GRB event rate of 1 per year within 100 Mpc. Currently published bounds on gravitational wave emissions from GRBs are provided by bar detectors[537, 24]. Quite generally, upper bound experiments are important in identifying various detection strategies.

14.5 Equations of suspended accretion

The suspended accretion state of the torus is described by balance of energy and angular momentum flux, received from the black hole and radiated to infinity by the torus.

Let $\tau_\pm$ denote the torques on the inner and outer face, each with mean angular velocity $\Omega_\pm$. These two competing torques promote azimuthal shear in the torus, leading to a super-Keplerian state of the inner face and a sub-Keplerian state of the outer face. The torus becomes geometrically thick and may reach a slenderness sufficient to excite nonaxisymmetric wave-modes $m = 1$ (minor-to-major radius less than 0.7506) or $m = 2$ (minor-to-major radius less than 0.3260). This resulting black hole–blob binary or a blob–blob binary bound to the black hole produces gravitational radiation at essentially twice the angular frequency of the torus.

Gravitational radiation exerts a torque τ_{gw} on the torus. Denoting the angular velocities of the inner and outer face by $\Omega_\pm$ and the mean angular velocity $\Omega_T = (\Omega_+ + \Omega_-)/2$, the equations of suspended accretion are[560]

$$\tau_+ = \tau_- + \tau_{gw}$$
$$\Omega_+ \tau_+ = \Omega_- \tau_- + \Omega_T \tau_{gw} + P_\nu, \tag{14.22}$$

where $L_{gw} = \Omega_T \tau_{gw}$ represents the luminosity in gravitational radiation and P_ν represents dissipation, which will be found to be primarily in MeV-neutrino emissions. These equations are closed by a constitutive relation for the dissipation process. In what follows, closure is set by attributing dissipative heating by magnetohydrodynamical stresses. Closure by attributing P_ν to magnetohydrodynamical stresses gives overall scaling with magnetic field energy E_B. The resulting total energy emissions, representing integrations of luminosities over the lifetime of rapid spin of the black hole, become thereby independent of E_B as fractions of E_{rot}.

Following the analysis on multipole mass-moments in the torus, consider

$$\Omega = \Omega_T \left(\frac{a}{r}\right)^q, \tag{14.23}$$

where a denotes the major radius of the torus and $3/2 < q < 2$ the rotation index. The rotation index is bounded below by Keplerian motion and bounded above by the Rayleigh stability criterion for $m = 0$, where m denotes the azimuthal wave number. The slenderness of the torus in terms of the minor-to-major radius gives $\Omega_\pm \simeq \Omega(1 \pm \delta)$, where

$$\delta = \frac{qb}{2a}, \tag{14.24}$$

so that $[\Omega] = \Omega_+ - \Omega_- \simeq \Omega_T b/a$.

By dimensional analysis, closure by magnetohydrodynamical stresses satisfies

$$P_\nu = \gamma A_r^2 [\Omega]^2, \tag{14.25}$$

where $[\Omega] = \Omega_+ - \Omega_-$, γ is a factor of order unity and $A_r = ah < B_r^2 >^{1/2}$ denotes the root-mean-square of radial magnetic flux $2\pi A_r$ averaged over the interface of radius a, scale height h, and contact area $2\pi ah$ between the inner and the outer face. The poloidal magnetic flux of open magnetic field-lines supported by the torus is denoted by $2\pi A$. These open field-lines connect either to the horizon of the black hole or to infinity in the form of magnetic winds. The effective viscosity per unit of poloidal magnetic flux can be expressed as

$$z = \gamma \left(\frac{b}{a}\right)\left(\frac{A_r}{A}\right)^2, \tag{14.26}$$

which satisfies

$$z \sim \text{const.} \tag{14.27}$$

asymptotically for small slenderness ratio $b/a \ll 1$. The asymptotic relation (14.27) corresponds to a flat infrared spectrum of magnetohydrodynamical flow up to the first geometrical break $m^* = a/b$ in the azimuthal wavenumber m.

The net poloidal flux $2\pi A$ of open field-lines supported by the torus – by its two counteroriented current rings or, equivalently, its distribution of magnetic dipole moments – partitions into fractions f_H and f_w which support winds into the horizon of the black hole and, respectively, to infinity. A remainder of magnetic field-lines forms an inner and outer toroidal 'bag' of closed magnetic field-lines up to, respectively, the inner light surface associated with the inner face and the outer light cylinder associated with the outer face. We thus have, by equivalence to pulsar magnetospheres when viewed in poloidal topology

$$\tau_+ = (\Omega_H - \Omega_+) f_H^2 A^2, \quad \tau_- = \Omega_- f_w^2 A^2 \tag{14.28}$$

for the angular momentum flux received by the inner face ($\Omega_H > \Omega_+$) and that lost by the outer face due to Maxwell stresses in magnetic winds. The associated wind luminosities are

$$L_\pm = \Omega_\pm \tau_\pm. \tag{14.29}$$

The inner face hereby receives a fraction Ω_+/Ω_H of the rotational energy E_{rot} of the black hole. Since Ω_+ is generally appreciably smaller than Ω_H, most of the rotational energy E_{rot} is dissipated in the event horizon of the black hole.

14.6 Energies emitted by the torus

The first (14.22) may be used to eliminate τ_+ in the second, giving $P_\nu = \frac{1}{2}[\Omega]\tau_{rad} + [\Omega]\tau_-$. With the constitute relation (14.25), it follows that $\tau_{rad} = 2A_r^2[\Omega] - 2\tau_-$. This defines a gravitational wave-luminosity

$$L_{gw} = \Omega_T \tau_{rad} = \Omega_T^2 A^2 \left[2\left(\frac{A_r}{A}\right)^2 \left(\frac{[\Omega]}{\Omega_T}\right) - 2f_w^2 \right] = \alpha \Omega_T^2 A^2, \qquad (14.30)$$

where

$$\alpha = 2\left(qz - f_w^2\right) > 3z - 2f_w^2 \simeq 3z - \frac{1}{2} \qquad (14.31)$$

in view of $[\Omega]/\Omega_T = qb/a$. Thus, we find that a suspended accretion state exists with positive gravitational wave-luminosity, whenever viscosity is sufficiently strong for slender tori. In case of a symmetric flux-distribution, a sufficient condition is $z > 1/6$, whereby the amplitude of the nonaxisymmetry in the torus is determined self-consistently with the steady-state gravitational wave-luminosity.

The frequency of quadrupole gravitational radiation is essentially twice the angular frequency of the torus for $m = 1$ and $m = 2$, since the phase velocities of these nonaxisymmetric waves are neglible as seen in the corotating frame.

Asymptotic expressions for the algebraic solutions to the equations of suspended accretion obtain in η and δ using $\Omega_\pm = \Omega_T \pm [\Omega]/2$ and substitution of (14.28) and (14.30) into the first (14.22). The result is

$$\eta = \frac{f_H^2}{f_H^2(1+\delta) + f_w^2(1-\delta) + \alpha} \sim \frac{1}{4\alpha}, \qquad (14.32)$$

where the right-hand side represents the asymptotic result for a symmetric flux-distribution in the limit of large α. Likewise, we find

$$\frac{L_{gw}}{L_H} = \frac{\alpha f_H^{-2} \Omega_T^2}{\Omega_+(\Omega_H - \Omega_+)} = \frac{\alpha f_H^{-2} \eta}{1 + \delta - \eta(1+\delta)^2}. \qquad (14.33)$$

The radiation energies emitted by the torus can be expressed as fractions of the rotational energies, assuming maximal rotation rates. This is convenient, and will serve as estimates for rapidly rotating black holes. Substitution of (14.32) into (14.33) gives the output gravitational radiation

$$\frac{E_{gw}}{E_{rot}} = \eta \frac{L_{gw}}{L_H} = \frac{\alpha\eta}{\alpha(1+\delta) + f_w^2(1-\delta^2)} \sim \eta. \qquad (14.34)$$

The result holds asymptotically in the limit of strong viscosity (large α) and small slenderness (small δ). Note that the energy output is effectively η, the efficiency of energizing the black hole–torus system by black hole spin energy[560, 569, 562]. This shows that most of the black hole luminosity is emitted in gravitational radiation. The energy in the remaining subdominant radiation channels follows likewise: winds satisfy $E_w = \alpha^{-1} f_w^2 (1-\delta)^2 E_{gw}$, whereby, in the same asymptotic limit,

$$\frac{E_w}{E_{rot}} = \frac{\eta f_w^2 (1-\delta)^2}{\alpha(1+\delta) + f_w^2(1-\delta^2)} \sim \eta^2 \qquad (14.35)$$

$$\frac{E_\nu}{E_{rot}} = \delta \frac{E_{gw}}{E_{rot}} + \frac{2\delta}{1-\delta} \frac{E_w}{E_{rot}} \sim \delta\eta. \qquad (14.36)$$

Here, the expressions simplify in case of a symmetric flux-distribution ($f_H = f_w = 1/2$), which will be valid in case of small η (wide tori).

According to the above, the primary output in gravitational radiation has energy and frequency

$$E_{gw} \simeq 0.2 M_\odot \left(\frac{\eta}{0.1}\right)\left(\frac{M_H}{7M_\odot}\right), \quad f_{gw} \simeq 500\,\text{Hz} \left(\frac{\eta}{0.1}\right)\left(\frac{7M_\odot}{M_H}\right), \qquad (14.37)$$

powered by the spin energy of an extreme Kerr black hole. Here, energies are in units of $M_\odot = 2 \times 10^{54}$ erg. This appreciation (14.37) of GRBs surpasses the true energy $E_\gamma \simeq 3 \times 10^{50}$ erg in gamma-rays[196] by several orders of magnitude. Subdominant emissions power an accompanying supernova and GRB in the core-collapse scenario of Type Ib/c supernovae.

These emissions represent of the order of 10% of the rotational energy of the black hole. The associated mass inhomogeneity $\delta M_T = \epsilon M_T$ in the torus is determined self-consistently with the gravitational wave-luminosity in suspended accretion. According to the quadrupole luminosity function for gravitational radiation by mass inhomegeneities δM_T with angular velocity $\Omega_T \simeq M_H^{1/2} a^{-3/2}$, we have

$$\delta M_T \simeq 0.5\% M_H (R/5M_H)^{7/4}, \qquad (14.38)$$

corresponding to a relative mass-inhomogeneity $\epsilon \simeq 20\%$ for a torus $M_T = 0.2M_\odot$ around a black hole of mass $M_H = 7M_\odot$.

The energy output in torus winds is a factor η less than that in gravitational radiation, or

$$E_w = 4 \times 10^{52}\ \text{erg} \left(\frac{\eta}{0.1}\right)^2 \left(\frac{M}{7M_\odot}\right). \qquad (14.39)$$

These winds provide a powerful agent towards collimation of the enclosed baryon-poor outflows from the black hole[335], as well as a source of neutrons for pick-up by the same[336]. The energy output in thermal and MeV-neutrino emissions is a factor δ less than that in gravitational radiation, or

$$E_\nu = 2 \times 10^{53} \, \text{erg} \left(\frac{\eta}{0.1}\right) \left(\frac{\delta}{0.30}\right) \left(\frac{M_H}{7M_\odot}\right). \tag{14.40}$$

At this dissipation rate, the torus develops a temperature of a few MeV and produces baryon-rich winds.

14.7 A compactness measure

Strong sources of gravitational radiation from astrophysical sources are relativistically compact, in the sense that their linear size R is a few times their Schwarzschild radius R_g. For gravitationally bound systems, this implies a simple scaling relationship between energy $E_{gw} = EM_\odot$ and frequency $f_{gw} = f$ Hz, given by

$$fE = 35 \left(\frac{\epsilon}{0.01}\right) \left(\frac{7R_g}{R}\right)^{3/2}, \tag{14.41}$$

where ϵ denotes the efficiency of converting mass-energy into gravitational radiation. Notable candidates for burst sources of gravitational radiation are binary coalescence of neutron stars and black holes, whose event rates were estimated early on by R. Narayan, T. Piran and P. Shemi[387] and E. S. Phinney[424], and theoretically by H. A. Bethe and G. E. Brown[85, 52, 44] and in subsequent work by V. Kalogera, *et al.*[287], and collaborators[287, 44], newborn neutron stars[182], and gamma-ray burst supernovae[569, 353, 568, 573]. Collectively, these astrophysical sources also make a contribution to the stochastic background in gravitational waves.

A compact relativistically compact nucleus tends to radiate predominantly in gravitational radiation, rather than electromagnetic radiation. Consider, therefore, the compactness parameter

$$\gamma = 2\pi \int_0^{E_{gw}} f_{gw} dE, \tag{14.42}$$

which expresses the amount of rotational energy relative to the linear size of the system. This is invariant under rescaling of the mass of a central black hole according the Kerr metric[293]. For spin-down of an extreme Kerr black hole, we have

$$\gamma = 0.0035 \left(\frac{\eta}{0.1}\right)^2 \tag{14.43}$$

using the trigonometric expressions in Table (12.1), where the right-hand side is in units of c^5/G. Values $\gamma > 0.005$ rigorously rule out radiation from a rapidly spinning neutron star, whose upper bound of 0.005 for their spin-down emissions in gravitational radiation obtains from a Newtonian derivation for a sphere with uniform mass-density.

Exercises

1. In the context of active galactic nuclei, consider a torus of about one solar mass around a supermassive black hole. Derive the scaling (14.18).

2. Verify that AGN, microquasars or soft X-ray transients have disk inhomogeneities that are too small to produce significant luminosities in gravitational radiation.

3. At the calculated dissipation rates, verify that the torus in the GRB supernova model reaches a temperature of about 2 MeV.

4. Derive (14.19).

5. Short GRBs are probably associated with the coalescence of two neutron stars or a neutron star with a black hole. While the waveform of binary inspiral is well understood (see[140]), the gravitational waveform in the final merger phase is highly uncertain. Estimate the energy emission in the final plunge of torus debris inside the ISCO, assuming that the debris follows geodesic motion.

6. Derive (14.41) and (14.43).

7. If the torus develops various multipole mass-moments ($m = 1, 2$ and higher), its instantaneous spectrum in gravitational radiation will comprise several lines. Does this affect the total luminosity, as follows from the equations of suspended accretion (14.22)?

8. The torus in suspended accretion is a *catalytic converter* of black hole spin energy. Illustrate this by comparing the energy emitted in gravitational radiation with its restmass energy for a torus of $0.1 M_\odot$.

9. Consider the open magnetic field-lines which define the spin-connection between the black hole and the torus, as described by (14.28). Show that in the limit of small η, the fraction of magnetic flux supported by the torus which provides this spin-connection is approximately constant during spin-down of the black hole. [*Hint*: For small η, most of the black hole spin energy is dissipated in the event horizon. The bag of closed magnetic field-lines attached

229

to the inner face of the torus extends to the inner light surface. Consider the initial and final position of the inner light surface, as the black hole evolves from an extremal state and to a synchronous state with $\Omega_H = \Omega_T$.]

10. The angular momentum vector of a torus is aligned with its spin axis. If its spin axis is misaligned with that of the black hole, the torus shows Lense–Thirring precession [332, 35, 18, 508] due to frame-dragging. Give an order of magnitude estimate for the change in gravitational radiation frequency [565]. Compare this with the expected change in frequency due to slow-down of black-hole spin.

15

GRB supernovae from rotating black holes

"It is not certain that everything is uncertain." Pascal (1623–1662),

Pensées.

GRB030329/SN2003dh[506, 265] confirmed the earlier indication of GRB980425/ SN1998bw[224] that Type Ib/c supernovae are the parent population of long GRBs. The branching ratio of Type Ib/c SNe to GRB-SNe can be calculated from the ratio $(1-2) \times 10^{-6}$ of observed GRBs-to-Type II supernovae[439], a beaming factor of 450[570] to 500[196] and a rate of about 0.2 of Type Ib/c-to-Type II supernovae[540], giving

$$\mathcal{R}[\text{Ib/c} \rightarrow \text{GRB}] = \frac{N(\text{GRB-SNe})}{N(\text{Type Ib/c})} \simeq (2-4) \times 10^{-3}. \qquad (15.1)$$

This ratio is remarkably small, suggesting a higher-order down-selection process. It can be attributed to various factors in the process of creating GRBs in Type Ib/c supernovae[437], e.g. not all baryon-poor jets successfully punch through the remnant stellar envelope[358], and not all massive progenitors making Type Ib/c supernovae nucleate rapidly rotating black holes. It is unlikely that either one of these down-selection processes by itself accounts for the smallness of $\mathcal{R}$. Rather, a combination of these might effectively contribute to a small branching ratio. We favor an association with binaries[390, 539] based on the Type II/Ib event SN1993J[367] and the proposed association of GRB-supernovae remnants with soft X-ray transients[53].

In Chapter 11, we alluded to candidate inner engines to GRB/XRF-supernovae in terms of (M, J, K): the black hole mass M, angular momentum J and kick velocity K. Black holes nucleated in nonspherical collapse receive a kick by Bekenstein's gravitational radiation recoil mechanism[41], whenever core-collapse is aspherical. Systemic massmoments by tidal deformation and random multipole massmoments produce a distribution in kick velocities. Some black holes will leave the central high-density core prematurely, before completion of the stellar collapse process. These events are *decentered*[564]. Other black holes will remain *centered*. They surge into a high-mass object surrounded by a high-density accretion disk or torus, allowing them to become luminous in a state of suspended accretion. Figure (15.1) illustrates these two alternatives.

In this chapter, we shall identify[564]

1. Centered nucleation of black holes in Type Ib/c supernovae.
2. A small branching ratio $\mathcal{R}$ with the probability of low kick velocities.
3. (De-)centered events with (single) double bursts in gravitational waves.
4. Radiation-driven supernovae powered by black hole spin energy.
5. The true energy in gamma rays in ergotubes of finite opening angle.

A related but different mechanism for explaining the small branching ratio based on kick velocities in core collapse poses fragmentation into two or more objects[149]. In this scenario, GRBs are associated with the formation of a fireball in the merger of binaries possessing small kick velocities. It is motivated, in part, in the search for delay mechanisms in creating a GRB, after the onset of the supernova on the basis of X-ray line emissions in GRB011211.

However, X-ray line emissions produced in radiatively powered supernovae allow the same time of onset of the GRB and the supernova, obviating the need

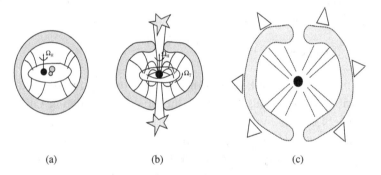

     (a)   (b)   (c)

Figure 15.1 Cartoon of decentered (a) and centered (b)–(c) nucleation of black holes (not to scale) corresponding, respectively, to high and low kick velocities. Decentered nucleation is typical by gravitational radiation recoil, whereby the black hole leaves the high-density center prematurely. It produces a short burst in gravitational radiation. Other transient compact objects may form before accreting onto the black hole[149]. An associated supernova can be powered by accretion[357]. In centered nucleation, a high-mass black hole forms surrounded by a high-density torus, producing a GRB by dissipation of kinetic energy in a baryon-poor outflow launched along an open ergotube (b). The MeV-torus catalyzes black hole-spin energy mostly into a long-duration burst in gravitational radiation and, to a lesser degree, into magnetic winds. Dissipation of these winds radiatively drives a supernova with late X-ray line-emissions when the remnant stellar envelope has expanded and become optically thin. (Adapted from[565]. ©2004 The American Physical Society.)

for any delay mechanism[573]. This is naturally accounted for by high-energy radiation from torus winds[563], closely related to the supernova mechanism of[53, 330].

15.1 Centered nucleation at low kick velocities

In core collapse of massive stars, rotating black holes nucleate by accumulation of mass and angular momentum from infalling matter. The Kerr solution describes the constraint (12.1). Quite generally, initial collapse of a rotating core produces a torus[453, 164], which initially satisfies

$$J_T > GM_T^2/c. \qquad (15.2)$$

Nucleation of black holes hereby takes place through a *first-order* phase-transition: a torus forms, whose mass increases with time by accumulation of matter, diluting its angular momentum until it satisfies (12.1) allowing collapses into an extremal black hole. The alternative of a second-order phase transition which initially forms a sub solar mass black hole, requires rapid shedding of excess angular momentum by gravitational radiation. However, limited mass densities in core collapse probably render this mechanism ineffective in competition with mixing on the free-fall timescale of the core. Nevertheless, gravitational radiation emitted from a nonaxisymmetric torus prior to the nucleation of the black hole is potentially interesting[453, 164].

Gravitational radiation in the formation of black holes through a first-order phase transition is important in nonspherical collapse, even when its energy emissions are small relative to the initial mass of the black hole. The Bekenstein gravitational radiation-recoil mechanism operates already in the presence of initial asphericities of about 10^{-3}. The recoil thus imparted is about $300\,\mathrm{km\ s^{-1}}$ or less. The radius of the accretion disk or torus around a newly formed stellar mass black hole is about $R_T \sim 10^7\,\mathrm{cm}$. A torus of a few tenths of a solar mass forms by accumulation of matter spiralling in, compressed by a factor of at least $(r/r_{ISCO})^4$ as it stalls against the angular momentum barrier outside the innermost stable circular orbit (ISCO) of the newly nucleated black hole. The time of collapse of stellar matter from a radius r is approximately the free-fall timescale,

$$t_{ff} \simeq 30\,\mathrm{s} \left(\frac{M_{He}}{10M_\odot}\right)^{-1/2} \left(\frac{r}{10^{10}\,\mathrm{cm}}\right)^{3/2}, \qquad (15.3)$$

where M_{He} denotes the mass of the progenitor He star. It follows that a newly formed low-mass black hole is typically kicked out of the central high-density region into surrounding lower-density regions *before* core collapse is completed. The black hole then continues to grow off-center by accretion of relatively low-density matter – a high-density accretion disk never forms. With low but nonzero probability, the black hole has a small recoil, allowing it to remain centered and surge into a high-mass black hole surrounded by a high-density torus.

After nucleation of the black hole, an accretion disk may form provided the specific angular momentum j_m of infalling matter exceeds that of the ISCO according to (12.48). The evolution of the newly nucleated black hole continues to be governed by angular momentum loss of the surrounding matter, until the inequality in (12.48) is reversed.

The black hole rapidly grows without bound when the inequality (12.48) is reversed:

$$j_m < l(a/M)GM/c. \tag{15.4}$$

We shall refer to this collapse phase as *surge*. Surge continues, until once again (12.48) holds. We solved numerically for equality in (12.48) in dimensionless form,

$$l\left(\frac{\beta j(s)}{m^2(s)}\right) = \frac{k_1 \beta s^2}{m(s)}, \tag{15.5}$$

where

$$\beta = k_2 \frac{\omega c s_0}{GR\rho_c} = 4.22 k_2 P_d^{-1} R_1^{-1} (M_{He}/10M_\odot)^{-1/3} \tag{15.6}$$

in terms of the dimensionless integrals

$$j(s) = 4\pi \int_0^s \hat{\rho} s^4 ds, \quad m(s) = 4\pi \int_0^s \hat{\rho} s^2 ds \tag{15.7}$$

of the normalized Lane–Emden density distribution with $\hat{\rho} = 1$ at the origin and the zero $\hat{\rho} = 0$ at $s_0 = 6.89685$[295]. Here, $(k_1, k_2) = (1, 1)$ in cylindrical geometry for which $j_m = \omega r^2$, and $(k_1, k_2) = (5/3, 2/3)$ in spherical geometry for which $j_m = (2/3)\omega r^2$; P_d denotes the binary period in days, R_1 denotes the radius in units of the solar radius 6.96×10^{10} cm[295], and M_{He} the mass of the progenitor He star.

Figure (15.2) shows the solution branches as a function of dimensionless period $1/\beta$. The upper branch shows that rapidly spinning black holes plus accretion disk form in small-period binaries[53, 330], following a surge for periods beyond the bifurcation points

$$\text{cylindrical geometry} (\beta = 6.461): \tag{15.8}$$

$$\frac{a}{M} = 0.9541, \quad \frac{E_{rot}}{E_{rot}^{max}} = 0.6624, \quad \frac{M}{M_{He}} = 0.4051. \tag{15.9}$$

$$\text{spherical geometry} (\beta = 5.157): \tag{15.10}$$

$$\frac{a}{M} = 0.7679, \quad \frac{E_{rot}}{E_{rot}^{max}} = 0.3220, \quad \frac{M}{M_{He}} = 0.3554. \tag{15.11}$$

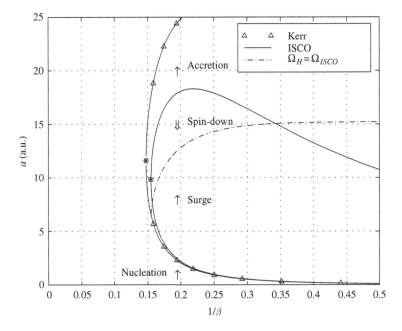

Figure 15.2 Centered nucleation of black holes in core collapse of a uniformly rotating massive star: accumulated specific angular momentum of the central object (arbitrary units) versus dimensionless orbital period $1/\beta$. Arrows indicate the evolution as a function of time. Kerr black holes exist *inside* the outer curve (diamonds). A black hole nucleates following the formation and collapse of a torus, producing a short burst in gravitational radiation. In centered nucleation, the black hole surges to a high-mass object by direct infall of matter with relatively low specific angular momentum, up to the inner continuous curve (ISCO). At this point, the black hole either spins up by continuing accretion or spins down radiatively against gravitational radiation emitted by a surrounding nonaxisymmetric torus. This state lasts until the angular velocity of the black hole equals that of the torus (dot–dashed line). These curves are computed for a Lane–Emden mass distribution with polytropic index $n = 3$ in the limit of conservative collapse, neglecting energy and angular momentum loss in radiation and winds. Shown are the results in cylindrical geometry. (Reprinted from[565]. ©2004 The American Astronomical Society.)

The resulting mass and energy fractions as a function of $1/\beta$ are shown in Figure (15.3). Given the tidal interaction between the two stars prior to collapse, these two geometries serve to bound the range of values in more detailed calculations, e.g. through multidimensional numerical simulations.

Bardeen's spin-up corresponds to continuing accretion beyond surge, wherein matter remaining in the remnant envelope forms an accretion disk outside the ISCO. At this point, magnetohydrodynamical stresses within the disk as well as disk winds may drive continuing accretion. Accretion from the ISCO onto the black hole further increases the black hole mass and spin according to (12.63),

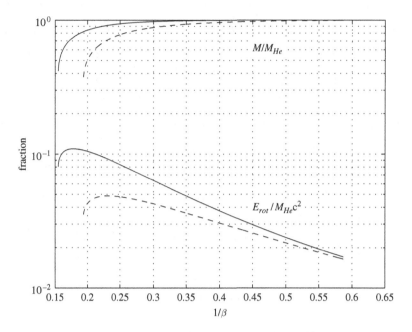

Figure 15.3 The black hole mass M and rotational energy E_{rot} are shown, formed after surge in centered nucleation. They are expressed relative to the mass M_{He} of the progenitor He star. The results are shown in cylindrical geometry (continuous) and spherical geometric (dashed). Note the broad distribution of high-mass black holes with large rotational energies of 5–10% (spherical to cylindrical) of $M_{He}c^2$. (Reprinted from[565]. ©2004 The American Astronomical Society.)

causing spin-up towards an extremal state of the black hole. In Figure (15.2) this is indicated by accretion *upwards* beyond the upper ISCO branch.

Radiative spin-down corresponds to a long-duration burst of gravitational radiation emitted by a nonaxisymmetric torus[557, 573], described by a frequency and energy

$$E_{gw} = (4 \times 10^{53} \text{ erg}) M_7 \eta_{0.1} \left(\frac{E_{rot}}{E_{rot}^{max}} \right), \quad f_{gw} = (500 \text{ Hz}) M_7 \eta_{0.1}, \qquad (15.12)$$

where $M_7 = M/7 M_\odot$ and $\eta = \Omega_T / \Omega_H$ denotes the relative angular velocity of the torus. This takes place if the torus is uniformly magnetized with the remnant magnetic field of the progenitor star. In Figure (15.1), this radiative spin-down is indicated by a transition *downwards* from the upper ISCO branch to the branch on which the angular velocities of the black hole and of matter at the ISCO match ($\Omega_H = \Omega_{ISCO}$ and $\eta = 1$). This radiative transition lasts for the lifetime of rapid spin of the black hole – a dissipative timescale of tens of seconds[563]. Additional matter accreted is either blown off the torus in its winds, or accumulates and accretes onto the black hole after spin-down.

15.2 Branching ratio by kick velocities

In what follows, we consider a two-dimensional Gaussian distribution of black hole kick velocities in the equatorial plane associated with the tidal deformation of the progenitor star by its companion, and assume a velocity dispersion $\sigma_{kick} \simeq$ $100\,\text{km s}^{-1}$ in Bekenstein's recoil mechanism.

The probability of centered nucleation during $t_{ff} \simeq 30\,\text{s}$ is that of a kick velocity $K < v^* = 10\,\text{km s}^{-1}$, i.e.

$$P_c = P(K < v^*) \simeq 0.5\% \left(\frac{v^*}{10\,\text{km/s}}\right)^2 \left(\frac{\sigma_{kick}}{100\,\text{km/s}}\right)^{-2}. \qquad (15.13)$$

While the numerical value has some uncertainties, the selection mechanism by gravitational radiation recoil effectively creates a small probability of centered nucleation. We identify the branching ratio of Type Ib/c SNe into GRBs with the probability of centered nucleation,

$$\mathcal{R}[\text{Ib/c} \rightarrow \text{GRB}] = P_c \simeq 0.5\%, \qquad (15.14)$$

effectively creating a small, higher-order branching ratio.

15.3 Single and double bursters

The proposed centered and decentered core collapse events predict a differentiation in gravitational wave signatures. These signatures are of interest to the newly commissioned gravitational wave detectors LIGO, Virgo and TAMA, both as burst sources and through their collective contributions to the stochastic background in gravitational radiation[563].

The black hole nucleation process is accompanied by a short burst in gravitational radiation, specifically in response to nonaxisymmetric toroidal structures and fragmentation[41, 453, 149, 438, 164]. Its gravitational radiation signature depends on details of the hydrodynamical collapse. Centered nucleation is followed by a long burst in gravitational radiation.

Single bursts in centered and decentered nucleation of black holes is hereby common to all Type Ib/c supernovae. This may apply to Type II events as well. Type II events are possibly associated with low spin rates and could represent delayed core collapse via an intermediate "nucleon" star (e.g. SN1987A[53]). Their gravitational wave emissions are thereby essentially limited to that produced by kick (if any) and collapse of this nucleon star. The gravitational radiation signature black hole nucleation depends on details of the hydrodynamical collapse. This remains largely unknown to date, except for indications on the formation of nonaxisymmetric tori before black hole nucleation[453, 164]. For a recent review of the short-duration ($\ll 1$ s) bursts of gravitational waves in core-bounce, more closely related to Type II supernovae, see[213] and references therein.

Double bursts in gravitational radiation are expected as short bursts are followed by long bursts in centered nucleation of black holes. The second burst takes place after a quiescent or subluminous[376] surge of the black hole into a high-mass object. On account of (15.9–15.11), rapidly rotating black holes are formed whose spin energy is about one-half the maximal spin energy of a Kerr black hole. In a suspended accretion state, these black holes spin-down in the process of emitting a fraction η into gravitational waves.

15.4 Radiatively driven supernovae

Following centered nucleation of a black hole in a collapsar, dissipated torus winds irradiate the remnant stellar envelope with high-energy continuum emissions. This provides a copious energy source for the excitation of X-ray lines and kinetic energy, whose impact will produce an aspherical supernova. Both of these processes are remarkably *inefficient*. Excitation of X-ray lines by continuum emissions has an estimated efficiency of less than 1%[229]. Deposition of kinetic energy by approximately luminal torus winds has an efficiency of $\beta/2$ where β represents the velocity of the ejecta relative to the velocity of light.

The result is a *radiatively* driven supernova by ejection of the remnant envelope[563]. When the remnant envelope has expanded sufficiently for its optical depth to this continuum emission to fall below unity, excited X-ray line emissions are observable such as those in GRB011211[454]. This supernova mechanism is novel in that the supernova energy derives *ab initio* from the spin energy of the black hole, and is otherwise similar but not identical to pulsar-driven supernova remnants by vacuum dipole radiation[400], and magnetorotational-driven Type II supernovae by Maxwell stresses[58, 327, 59, 596, 7] and associated heating[314]. This supernova mechanism is similar but not identical to that of[330, 53]. It posits that the time of onset of the supernova is the same as the GRB, which is distinct from the delayed GRB scenario in[149]. We predict that the intensity of line emissions and the kinetic energy in the ejecta are positively correlated.

The energy output (14.39) in torus winds is consistent with the lower bound of[229] on the energy in continuum emissions for the line emissions in GRB011211[563]. In our proposed mechanism for supernovae with X-ray line emissions, therefore, we envision efficient conversion of the energy output in torus winds into high-energy continuum emissions, possibly associated with strong shocks in the remnant envelope and dissipation of magnetic field energy into radiation. We note that the latter is a long-standing problem in the pulsars, blazars, and GRBs alike (see[337] and references therein). Conceivably, this process is aided by magnetoturbulence downstream[321, 92]. These supernovae will be largely nonspherical, as determined by the collimation radius of the magnetic torus winds, see, for example[97], and references therein.

Matter ejecta in both GRB 991216[434] and GRB 011211[454] show an expansion velocity of $\beta \simeq 0.1$. The efficiency of kinetic energy deposition of the torus wind onto this remnant matter is hereby $\beta/2 = 5\%$. With E_w as given in (14.39), this predicts a supernova remnant with

$$E_{SNR} \simeq \frac{1}{2}\beta E_w \simeq 2 \times 10^{51} \text{ erg}, \tag{15.15}$$

which is very similar to energies of non-GRB supernovae remnants. Ultimately, this connection is to be applied the other way around: obtain estimates for E_w from kinetic energies in a sample of supernova remnants around black hole binaries, assuming that $\beta \sim 0.1$ holds as a representative value for the initial ejection velocity obtained from E_w. This assumption may be eliminated by averaging over observed values of β in a sample of GRB supernova events with identified line-doppler shifts.

The asymptotic relations (14.34)–(14.36) indicate that the emissions by the torus in various channels are strongly correlated. The torus winds ultimately dissipate into radiation. Thus, *calorimetry on the supernovae associated with GRBs provides a method for predicting the frequency of the correlated emissions in gravitational radiation.* For quadrupole gravitational radiation, we have

$$f_{gw} \simeq 470 \text{ Hz} \left(\frac{\beta}{0.1}\right)^{-1} \left(\frac{E_{SN}}{4 \times 10^{51} \text{ erg}}\right)^{1/2} \left(\frac{7M_\odot}{M}\right)^{3/2}. \tag{15.16}$$

This provides a unique link between the gravitational wave-spectrum and the supernova explosion.

Efficient conversion of the energy output in torus winds into high-energy continuum emissions may take place in shocks in the remnant envelope and by dissipation of magnetic field-energy. The magnetic field-strength (14.4) indicates the existence of a transition radius beyond which the magnetic field strength becomes subcritical. While this transition may bring about a change in the spectrum of radiation accompanying the torus wind, it is unlikely to affect conversion of wind energy to high-energy emissions at larger distances. The reader is referred to[532] and[166] for radiative processes in superstrong magnetic fields.

15.5 SN1998bw and SN2002dh

In GRB supernovae from rotating black holes, *all* emissions are driven by the spin energy of the central black hole, and hence *all* ejecta are expected to be nonspherical.

The radiatively driven supernova mechanism produces aspherical explosions, whereby E_{SN} in (15.15) is distinct from, and generally smaller than, the observed isotropic equivalent kinetic energy $E_{k,iso}$ in the ejecta. The canonical value for E_{SN} agrees remarkably well with the estimated explosion energy of 2×10^{51} erg in SN1998bw[268], based on asphericity in the anomalous expansion velocities of the ejecta. This estimate is consistent with the partial explosion energy of about 10^{50} erg in ejecta with velocities in excess of $0.5c$, where c denotes the velocity of light[341]. Conversely, $E_{k,iso}$ assumes anomalously large values in excess of 10^{52} erg, depending on the degree of asphericity.

Explosion energies (15.15) represent normal SNe Ic values[268]. The term "hypernova"[404] applies only to the apparent energy $E_{k,iso} \simeq 2 - 3 \times 10^{52}$ erg in GRB980425[280, 612] upon assuming spherical geometry, not to the true kinetic energy E_{SN} in the actual aspherical explosion.

The GRB emissions are strongly anisotropic, produced by beamed baryon-poor jets along the rotational axis of the black hole. Based on consistency between the true GRB event rate, based on[196, 570], and GRB980425, these beamed emissions are possibly accompanied by extremely weak gamma-ray emissions over wide angles or perhaps over all directions. The beaming factor of the baryon-poor jet is 450–500[196, 570]. Evidently, the degree of anisotropy in the GRB emissions exceeds the axis ratio of $2:3$ in the associated supernova ejecta[268] by about two orders of magnitude. While viewing the source on-axis gives rise to the brightest GRB and the largest $E_{k,iso}$, a viewing of the source off-axis could give rise to an apparently dim GRB with nevertheless large $E_{k,iso}$. This may explain the apparent discrepancy between the dim GRB980425 in the presence of a large $E_{k,iso}$, yet normal E_{SN}([268]; (15.15) above), in SN1998bw.

The remarkable similarity between the optical light curve of SN2003dh associated with GRB030329[506] supports the notion that GRBs are driven by standard inner engines. GRB030329 was a bright event in view of its proximity, though appeared with a slightly subenergetic $E_{\gamma,iso}$. We attribute this to viewing strongly anisotropic GRB emissions slightly off the rotational axis of the black hole. Based on spectral data[291], note that the energy $E_{k,iso}$ of SN2003dh is probably between that of SNe1997ef (e.g.[394, 81]) and SN1998bw, although SN2003dh and SN1998bw feature similar initial expansion velocities. If SN2003dh allows a detailed aspherical model similar to that of SN1998bw, we predict that the true kinetic energy E_{SN} will attain a normal value.

The observational constraint $E_{SNR} \simeq 2 \times 10^{51}$ erg on SN1998bw[268] and consistency with the energy requirement in high-energy continuum emissions for the X-ray line emissions in GRB011211, therefore, suggest an expectation value of $f_{gw} \simeq 500$ Hz according to (15.16) and (15.15). It would be of interest to refine this estimate by calorimetry on a sample of SNRs which are remnants of GRBs. Given the true GRB event of about one per year within a distance of 100 Mpc, we anticipate about one GRB-SNR within 10 Mpc. These remnants will contain a black hole in a binary with an optical companion, possibly representing a soft X-ray transient.

15.6 True GRB afterglow energies

The true energy in gamma-rays and subsequent afterglow emissions is the total energy (12.13) times an efficiency factor.

For a canonical value $\epsilon \simeq 30\%$ of the efficiency of conversion of kinetic energy-to-gamma rays (for various estimates, see[298, 143, 408, 243]), we have according to (13.38),

$$\frac{E_\gamma}{E_{rot}} \simeq \frac{1}{2}\epsilon\theta_H^4 \simeq 2 \times 10^{50}\epsilon_{0.30}\eta_{0.1}^{8/3} \text{ erg.} \tag{15.17}$$

The baryon content and the loading mechanism of these jets (and essentially of GRB fireballs in any model) is as yet an open issue. In one scenario proposed recently by Levinson and Eichler[336] baryon loading is accomplished through pickup of neutrons diffusing into the initially baryon-free jet from the hot, baryon-rich winds from the MeV torus, to recombine with protons to form ^{4}He. In their estimate of the total number of picked-up neutrons, they arrive at an asymptotic bulk Lorentz factor of the jet of 10^2–10^3. A specific prediction from their model is that inelastic nuclear collisions inside the jet leads to very high-energy neutrinos ($\gg$1 TeV) with a very hard spectrum, providing a possible source for the upcoming km^3 neutrino detectors for sources up to about a redshift of 1.

Frame-dragging responds slowly as the black hole mass and angular momentum change; the radial electric field may change rapidly by Gauss' theorem, if current is not closed. If so, intermittency will result and the outflow becomes a $\gamma e^\pm$ outflow from a differentially rotating ergotube. Neither alternative is excluded on the basis of observations. In fact, GRB lightcurves are generally highly intermittent featuring submillisecond timescale variability. It may well be that the spin–orbit coupling launches "rotating blobs" (13.30) – Poynting flux-dominated and magnetized $\gamma e^\pm$ ejecta – in the process of intermittent behavior in the ergotube. Their baryonic content depends on the number of neutrons picked up in their escape. This may be contrasted with the "cannon-ball" model of Dar and de Rújula[148], which are assumed to be baryon-rich.

Concluding, GRB supernovae from rotating black holes are consistent with the observed durations and true energies in gamma-rays, the observed total kinetic energies in an associated aspherical supernova, and X-ray line emissions produced by underlying continuum emissions. On this basis, we predict band-limited gravitational wave-line emissions contemporaneous with the GRB according to the scaling relations (9.8) at an event rate of probably once a year within a distance of 100 Mpc. Figure (15.4) summarizes the associated calorimetry.

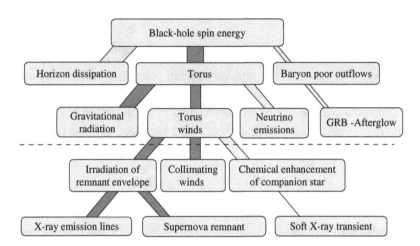

Figure 15.4 A radiation energy diagram for the dissipation and radiation of black hole spin energy catalyzed by a surrounding torus. Most of the spin energy is dissipated in the horizon – an unobservable sink of energy. Most of the spin energy released is incident on the inner face of the torus, while a minor fraction forms baryon-poor outflows through the inner flux-tube to infinity. We associate the latter with the input to the observed GRB afterglow emissions. The torus converts its input primarily into gravitational radiation and, to a lesser degree, into winds, thermal, and neutrino emissions. Direct measurement of the energy and frequency emitted in gravitational radiation by the upcoming gravitational wave experiments provides a calorimetric compactness test for Kerr black holes (dark connections). Channels for calorimetry on the torus winds are indicated below the dashed line, which are incomplete or unknown. They provide in principle a method for constraining the angular velocity of the torus and its frequency of gravitational radiation. This is exemplified by tracing back between torus winds and their remnants (dark connections). As the remnant envelope expands, it reaches optical depth of unity and releases the accumulated radiation from within. This continuum emission may account for the excitation of X-ray line emissions seen in GRB 011211, which indicates a torus wind energy of a few times 10^{52} erg. Matter ejecta ultimately leave remnants in the host molecular cloud, which remain to be identified. Torus winds may further deposit a fraction of their mass onto the companion star[87], thereby providing a chemical enrichment in a remnant soft X-ray transient. (Reprinted from[568]. ©2003 The American Astronomical Society.)

Exercises

1. Some 20% of the GRBs in the BATSE catalog show precursor emissions in the form of weak gamma-ray emissions prior to the main GRB event by some tens (up to hundreds) of seconds[324]. Upon associating these precursor emissions with the nucleation phase, discuss a possible correlation between the delay time to the main GRB event and its true energy in gamma-rays. (For a different explanation, see[447].)

2. Compare gravitational wave emissions due to black-hole kick velocities by the Bekenstein radiation-recoil mechanism and due to the formation of a nonaxisymmetric torus prior to black hole nucleation.

3. Does Bekenstein's radiation recoil mechanism apply as a mechanism for neutron star kicks?

4. The association of GRB supernovae and their parent population of Type Ib/c supernovae with centered and decentered, respectively, nucleation of black holes, suggests that GRB supernovae represent a narrow distribution of events in a much larger continuum of centered–decentered nucleation of black holes in core-collapse supernovae. Discuss qualitatively the characteristics of the latter, and their potential observational signatures.

5. The association of GRB-supernovae and their parent population of Type Ib/c supernovae with centered with small kick velocity and, respectively, decentered nucleation of black holes with large kick velocity, suggests that such events are not rare, but rather should be quite common relative to the true-but-unseen GRB events. Discuss intrinsically weak GRBs in the context of the continuum of centered–decentered nucleation of black holes in core collapse supernovae.

6. The origin of the largely unbeamed, single pulse-shaped gamma-ray emissions in weak GRBs may come from dissipation of torus winds into high-energy photons, impacting the remnant envelope from within. Estimate the time-delay between the onset of the core-collapse supernova and the observed gamma-rays in this case. Compare your results with observations of GRB021101.

7. The late-time light curve of supernova associated with GRBs shows evidence for energetic input from radiative decay of ^{56}Ni. At the estimated MeV temperature, the torus is expected to have a mass loss rate of about $10^{30}\,\mathrm{erg\,s^{-1}}$[568]. Estimate the fraction of this wind that converts into ^{56}Ni on the basis of[443].

8. The "open model" of [64] predicts that most of the black-hole luminosity L_{BZ} is channeled into a jet (e.g., the cartoon Fig. 6 of [560] for open and closed models). For rapidly spinning black holes, derive L_{BZ}/L_d of jet-to-disk luminosity for a common value of the poloidal magnetic field-strength penetrating the horizon and the inner boundary of the disk (cf. [346]). Compare the results with the kinetic energy 2e51 erg in SN1998bw and the true GRB-energies of 3e50 erg, assuming a disk-powered supernova mechanism.

16

Observational opportunities for LIGO and Virgo

"Measure what is measurable, and make measurable what is not so."
Galileo Galilei (1564–1642), in H. Weyl,
Mathematics and the Laws of Nature.

"Wir müssen wissen. Wir werden wissen."
David Hilbert (1862–1943),
engraved on his tombstone in Göttingen.

Gravitational wave detectors LIGO[2, 34], Virgo[78, 4, 503] shown in Figure 16.1, GEO[147, 601] and TAMA[15] are broad band detectors, sensitive in a frequency range of about 20–2000 Hz. The *laser interferometric* detectors are based on Michelson interferometry, and have a characteristic right angle between their two arms for optimal sensitivity for spin-2 waves[476]. At low frequencies (approximately less than 50 Hz), observation is limited by unfiltered seismic noise. In a middle band of up to about 150 Hz, it is limited by thermal noise and, at high frequencies above a few hundred Hz, by shot noise[495]. The design bandwidth of these detectors is chosen largely by the expected gravitational wave frequencies emitted in the final stages of binary neutron star coalescence, i.e. frequencies up to a few hundred Hz produced by compact stellar mass objects. At these frequencies, the detectors operate in the short wavelength limit, wherein the signal increases linearly with the length of the arms. It is therefore advantageous to build detectors with arm lengths as long as is practically feasible, given that many noise sources are independent of the arm length.

The first-generation gravitational-wave detectors are the narrow-band bar detectors pioneered by J. Weber[584, 538]. For an instructive overview of bar detectors, the reader is referred to[382, 398]. While bar detectors can reach sensitivities of astrophysical interest, they are limited by practical system noise, and those currently in use are narrow-band detectors. Binary coalescence produces a broad-band chirp, which requires a dynamical range in frequency sensitivity by

Figure 16.1 Aerial view of the Hanford (WA) detector site (*top*), showing the characteristic 90° angle between two 4 km arms (*top*, *right*) for laser interferometry on the quadrupolar gravitational waves. The Hanford site houses two interferometric detectors LH1 and LH2, while the sister site in Livingston (LA) houses a single interferometer (LL1). The French-Italian experiment Virgo (*bottom*) is located in Cascina near Pisa, Italy, using 3 km arms. In the shot-noise region above a few hundred Hz, the performance of these interferometric detectors is largely determined by laser power, from initially a few watts to greater than 1 kW in advanced detectors. (Courtesy of LIGO and Virgo.)

a factor of at least a few. This and other considerations have led to the design of laserinterferometric detectors. The first ideas on interferometric detectors are described by M. E. Gertsenshtein and V. I. Pustovoit[228] and, independently, by J. Weber in his laboratory notebooks. The first thorough study is due to R. Weiss[590]. A worldwide effort to develop the technology ensued with prototypes in Cambridge and Pasadena (USA), Munich (Germany) and Glasgow (UK). Weiss at MIT pushed forward the idea of a US national facility and, along with colleagues K. S. Thorne[531] and R. Drever, proposed to the National Science Foundation for the creation of a gravitational-wave observatory. Its present incarnation is the LIGO Laboratory, consisting of a pair of 4 km detectors in Washington and Louisiana. On the European side, the French-Italian Virgo Project initiated by A. Brillet and A. Giazotto develops at a very similar stage with a 3 km detector at Cascina (near Pisa), Italy. The LIGO and Virgo detectors overlap considerably in design, choice of hardware and sensitivity. Similar, somewhat shorter, detectors are the German-UK GEO 600 m detector in Germany and the TAMA 300 m detector in Japan, as well as an 80 m test facility in Australia[285].

These broad-band detectors are configurated for anticipated gravitational wave frequencies produced in binary coalescence of stellar mass compact objects – neutron stars and black holes – as well as potential gravitational wave bursts from supernovae and rapidly spinning neutron stars. While the early stages of these gravitational wave experiments are aimed at a first detection from *some* source, known or serendipitous, ultimately the aim is to develop a new tool for *gravitational wave-astronomy*. For example, what is the gravitational wave luminosity of a galaxy or a nearby cluster of galaxies such as Virgo? Is there a detectable contribution from the early universe to the stochastic background in gravitational waves? The latter is perhaps more amenable to the low-frequency regime of 0.1–100 mHz to be probed by the European-US Laser Interferometric Space Antenna (LISA).

First detections by these detectors are probably determined by sources which have an optimal combination of strength and event rate, where the latter statistically determines the anticipated distance. The only exception is a continuous source, such as rapidly spinning neutron star produced in a recent supernova, or spun-up by accretion[54].

For a broad review of various sources, see, for example, Cutler and Thorne[140]. For post-Newtonian waveform analysis of the initial chirp in compact binaries[139], see[60]. For black hole–black hole coalescence, the transition from chirp to a common type of horizon envelope state should be smooth, maybe very luminous[532, 533, 486] and more frequent[440] than neutron star–black hole coalescence[424, 387, 258]. In neutron star–black hole coalescence, on the other hand, a short-lived intermediate black hole torus state is possible if the black hole spins rapidly[403].

In this section, we describe the method of estimating signal-to-noise ratios for interferometric gravitational wave detectors. We apply this to GRB supernovae from rotating black holes, both as nearby point sources and in their contribution to the stochastic background in gravitational radiation. The reader is referred to the specialized literature for discussions on solid state bar detectors.

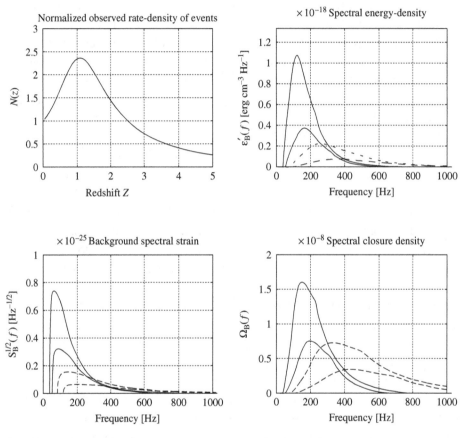

Figure 16.2 The stochastic background in gravitational radiation from GRB supernovae, locked to the star formation rate according to $N(z)$ of Chapter 7. Shown is the spectral energy density $\epsilon'_B(f)$, the strain amplitude $S_B^{1/2}(f)$, and the spectral closure density $\Omega_B(f)$. The results are calculated for a uniform mass distribution $M_H = 4 - 14 \times M_\odot$ (top curves) and $M_H = 5 - 8 \times M_\odot$ (lower curves). The results are shown for $\eta = 0.1$ (solid curves) and $\eta = 0.2$ (dashed curves). The extremal value of $\Omega_B(f)$ is in the neighborhood of maximal sensitivity of LIGO and Virgo. (Corrected and reprinted from[565]. ©2004 The American Physical Society.)

16.1 Signal-to-noise ratios

The sensitivity of a detector for a given source is commonly expressed in terms of a signal-to-noise ratio. Optimal sensitivity obtains using matched filtering. While not all sources are amenable to this procedure, matched filtering provides an important theoretical upper limit for the signal-to-noise ratio. Because detector noise is strongly frequency-dependent, the signal-to-noise ratio is commonly expressed in the Fourier domain. The following discussion and notation is based on the exposition of E. Flanagan[191] and M. Maggiore[360].

We define the spectral energy density $S_h(f)$ of the detector noise $n(t)$ by the one-sided integral

$$< n(t) >^2 = \int_0^\infty S_h(f) \, df, \tag{16.1}$$

and the Fourier transform $\tilde{h}(f)$ of the signal $h(t)$ at the detector by

$$\tilde{h}(f) = \int_{-\infty}^\infty e^{2\pi i f t} h(t) \, dt, \quad h(t) = \int_{-\infty}^\infty e^{-2\pi i f t} \tilde{h}(f) \, df. \tag{16.2}$$

At the detector, a gravitational wave signal is essentially planar. The detector sensitivity is a function of the relative orientation (θ, ϕ) between the wave vector and the normal to the plane spanned by the two detector arms. Let $F_{+\times}$ denote the angular detector response functions to the two polarizations $h_{+\times}$ of the gravitational wave, as a function of (θ, ϕ), as the polarization angle ψ of the wave. In this notation[191]

$$h(t) = F_+ h_+ + F_\times h_\times, \quad h_{+,\times} = H_{+,\times}/r, \tag{16.3}$$

where r denotes the distance to the source. The square of the signal-to-noise is the ratio of the spectral energy densities:

$$\rho^2 = 4 \int_0^\infty \frac{|F_+ \tilde{H}_+(f) + F_\times \tilde{H}_\times(f)|^2}{S_h(|f|)} df. \tag{16.4}$$

The functions $F_{+,\times}^2$ have angular averages 1/5[531] and are orthogonal. Averaging over all relative orientations between the detector and the wave vector gives the expectation value

$$< \rho^2 > = \frac{4}{5r^2} \int_0^\infty \frac{|\tilde{H}_+(f)|^2 + |\tilde{H}_\times(f)|^2}{S_h(|f|)} df. \tag{16.5}$$

Parseval's theorem,

$$\int_{-\infty}^\infty h^2(t) dt = \int_{-\infty}^\infty |\tilde{h}(f)|^2 df, \tag{16.6}$$

allow us to expresses the total energy equivalently in terms of a distribution in the time domain and a distribution in the frequency domain. We interpret

$dE/df = |\tilde{h}(f)|^2$ as the spectral energy density. By (6.36), the energy flux in the time domain and frequency domain satisfies

$$\frac{dE}{dtdA} = \frac{1}{16\pi} \left(\dot{h}_+(t)^2 + \dot{h}_\times(t)^2 \right), \tag{16.7}$$

$$\frac{dE}{df} = \frac{4\pi^2 f^2}{2} \left(|\tilde{H}_+(f)|^2 + |\tilde{H}_\times(f)|^2 \right) \tag{16.8}$$

following the convention[191] to express results in one-sided frequency-distributions and $dA = r^2 d\Omega$. The expectation value of the signal-to-noise becomes

$$< \rho^2 > = \frac{2}{5\pi^2 r^2} \int_0^\infty \frac{1}{f^2 S_h(|f|)} \frac{dE}{df} df. \tag{16.9}$$

Orientation averaging of the angular detector response functions $F_{+\times}$ takes the expectation value (15.15) a factor of 5 below that for optimal orientation and polarization. Flanagan[191] proposes to rewrite the results therefore in terms of the quantities

$$h_{char}(f)^2 = \frac{2}{\pi^2 r^2} \frac{dE}{df}, \quad h_{rms}(f) = \sqrt{fS_h(f)}, \quad h_n(f) = \sqrt{5} h_{rms}, \tag{16.10}$$

where we dropped the tildes to denote the Fourier transform. Thus, (16.9) becomes

$$< \rho^2 > = \int \frac{h_{char}(f)^2}{h_n(f)^2} \frac{df}{f}, \tag{16.11}$$

which differs from the expression for optimal orientation only using the "enhanced" detector noise h_n for the true detector noise h_{rms}. Based on (7.20), (16.10) generalizes to

$$h_{char}(f)^2 = \frac{2(1+z)^2}{\pi^2 d_L(z)^2} \left(\frac{dE}{df} [(1+z)f] \right)_e, \tag{16.12}$$

where e refers to evaluation in the comoving frame of the source. The reader is referred to[191] for further discussions.

16.2 Dimensionless strain amplitudes

The strain amplitude for a band-limited signal is commonly expressed in terms of the dimensionless characteristic strain amplitude of its Fourier transform. For a signal with small relative bandwidth $B \ll 1$, we have (adapted from[191])

$$h_{char} = \frac{1+z}{\pi d_L(z)} \left(\frac{2E_{gw}}{f_{gw,s} B} \right)^{1/2}, \tag{16.13}$$

which may be re-expressed as

$$h_{char} = 6.55 \times 10^{-21} \left(\frac{M}{7M_\odot}\right) \left(\frac{100\,\text{Mpc}}{d_L}\right) \left(\frac{0.1}{B}\right)^{1/2}, \qquad (16.14)$$

upon ignoring dependence on redshift z. Note that h_{char} is independent of η. The signal-to-noise ratio as an expectation value over random orientation of the source is

$$\left(\frac{S}{N}\right)^2 = \int \left(\frac{h_{char}}{h_n}\right)^2 d\ln f \simeq \left(\frac{h_{char}}{h_{rms}}\right)^2 \frac{B}{5}, \qquad (16.15)$$

where $h_n = h_{rms}/\sqrt{5}$, and $h_{rms} = \sqrt{fS_h(f)}$ in terms of the spectral noise energy density $S_h(f)$ of the detector. The factor $1/5$ refers to averaging over all orientations of the source[191]. In light of the band-limited signal at hand, we shall consider a plot of

$$h_{char}\sqrt{B/5} \qquad (16.16)$$

versus $f_{gw,s}$ according to the dependence on black hole mass given in (16.18), using a canonical value $\eta = 0.1$. The instantaneous spectral strain amplitude h follows by dividing h_{char} by the square root of the number of 2π-wave periods $N \simeq f_{gw,s}T_{90}$ according to (14.18). It follows that

$$h = 3 \times 10^{-23} \left(\frac{0.1}{B}\right)^{1/2} \left(\frac{\eta}{0.1}\right)^{4/3} \left(\frac{\mu}{0.03}\right)^{1/4} \left(\frac{M}{7M_\odot}\right) \left(\frac{100\,\text{Mpc}}{d_L}\right). \qquad (16.17)$$

16.3 Background radiation from GRB-SNe

The cosmological contribution (7.24) can be evaluated semi-analytically for band limited signals $B = \Delta f/f_e$ of the order of 10% around (9.8), where f_e denotes the average gravitational wave frequency in the comoving frame. In what follows, we will use the scaling relations

$$E_{gw} = E_0 \left(\frac{M}{M_0}\right), \quad f_e = f_0 \left(\frac{M_0}{M}\right) \qquad (16.18)$$

where $M_0 = 7M_\odot$,

$$E_0 = 0.203 M_\odot (\eta/0.1), \quad f_0 = 455\,\text{Hz}(\eta/0.1). \qquad (16.19)$$

These scaling relations assume extreme Kerr black holes. For non-extremal black holes, as calculated in Chapter 14, an additional factor E_{rot}/E_{rot}^{max} is to be included in the energy relation. This factor carries through to the final results, whence it is not taken into account explicitly.

By (16.18), we have

$$\epsilon_B'(f) \simeq N(0) \left(\frac{E_0}{f}\right) \left(\frac{M}{M_0}\right) D(z), \tag{16.20}$$

where $D(z) = R_{SF}(z; 0)/R_{SF}(0; 0)(1+z)^{5/2}$ and $1+z = f_e/f = f_0 M_0/fM$. The average over a uniform mass distribution $[M_1, M_2]$ $(\Delta M = M_2 - M_1)$ satisfies

$$< \epsilon_B'(f) > \; \simeq N(0) \left(\frac{E_0}{f}\right) \left(\frac{M_0}{\Delta M}\right) \int_{M_1}^{M_2} \left(\frac{f_0}{f} u^{-1}\right) D(u) \left(\frac{f_0}{f} du^{-1}\right) \tag{16.21}$$

i.e.

$$< \epsilon_B'(xf_0) > = N(0) \left(\frac{E_0}{f_0}\right) \frac{M_0}{\Delta M} f_B(x), \tag{16.22}$$

where

$$f_B(x) = x^{-3} \int_{M_0/M_2 x}^{M_0/M_1 x} u^{-3} D(u) du, \quad x = f/f_0. \tag{16.23}$$

The function $f_B(x) = f_B(x, M_1, M_2)$ displays a maximum of order unity, reflecting the cosmological distribution $z \simeq 0 - 1$, preceded by a steep rise, reflecting the cosmological distribution at high redshift, and followed by a tail x^{-2}, reflecting a broad distribution of mass at $z \simeq 0$[138]. Because $E_{gw}' \propto M_H^2$, these peaks are dominated by high-mass sources, and, for the spectral strain amplitude, at about one-fourth the characteristic frequency of f_0.

We may apply (16.22) to a uniform mass distribution $[M_1, M_2] = [4, 14] M_\odot$, assuming that the black hole mass and the angular velocity ratio η of the torus to that of the black hole are uncorrelated. Using (16.18), we have, in dimensionful units,

$$< \epsilon_B'(f) > = 1.08 \times 10^{-18} \hat{f}_B(x) \, \text{erg cm}^{-3} \text{Hz}^{-1}, \tag{16.24}$$

where $\hat{f}_B(x) = f_B(x)/\max f_B(\cdot)$. The associated dimensionless strain amplitude $\sqrt{S_B(f)} = (2G/\pi c^3)^{1/2} f^{-1} F_B^{1/2}(f)$, where $F_B = c\epsilon_B'$ and G denotes Newton's constant, satisfies

$$\sqrt{S_B(f)} = 7.41 \times 10^{-26} \left(\frac{\eta}{0.1}\right)^{-1} \hat{f}_S^{1/2}(x) \text{Hz}^{-1/2}, \tag{16.25}$$

where $\hat{f}_S(x) = f_S(x)/\max f_S(\cdot)$, $f_S(x) = f_B(x)/x^2$. Likewise, we have for the spectral closure density $\tilde{\Omega}_B(f) = f\tilde{F}_B(f)/\rho_c c^3$ relative to the closure density $\rho_c = 3H_0^2/8\pi G$

$$\tilde{\Omega}_B(f) = 1.60 \times 10^{-8} \left(\frac{\eta}{0.1}\right) \hat{f}_\Omega(x), \tag{16.26}$$

where $\hat{f}_\Omega = f_\Omega(x)/\max f_\Omega(\cdot)$, $f_\Omega(x) = x f_B(x)$. This shows a simple scaling relation for the extremal value of the spectral closure density in its dependency on

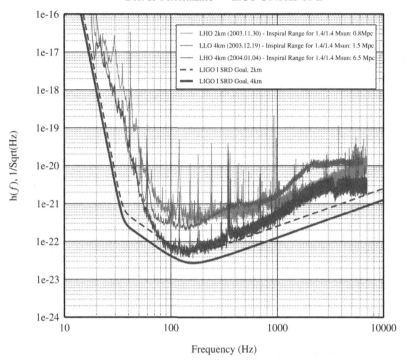

Figure 16.3 LIGO spectral noise amplitude: measured in the third Science Run, compared with the planned LIGO I noise curve (solid). Sensitivity ranges for binary neutron star inspiral provide a low-frequency performance parameter. (Courtesy of LIGO.)

the model parameter η. The location of the maximum scales inversely with f_0, in view of $x = f/f_0$. The spectral closure density hereby becomes completely determined by the SFR, the fractional GRB rate thereof, η, and the black hole mass distribution. Figure (16.1) shows the various distributions. The extremal value of $\Omega_B(f)$ is in the neighborhood of the location of maximal sensitivity of LIGO and Virgo shown in Figure (16.3).

16.4 LIGO and Virgo detectors

LIGO consists of two 4 km detectors, located at Livingston, Louisiana, and Hanford, Washington (Figure 16.1). The Livingston site houses a single laser beam interferometer (LL1), while the Hanford site houses two laser beam interferometers (LH1 and LH2). Virgo consists of a single 3 km detector in Cascina (near Pisa), Italy. TAMA in Mitaka, Tokyo, and GEO at Hanover are detectors of comparable design, except their arm lengths are 300 and 600 m, respectively. Gingin at Perth is an 80 m test facility, operated by the University of Western Australia in collaboration with LIGO and GEO.

The sensitivity of these detectors is limited by their noise. There are several, largely independent noise sources. The following highlights mostly qualitatively some of the relevant noise sources. This discussion is based on an overview by M. Punturo[445] on noise sources in the Virgo configuration. For a closely related detector description of LIGO and GEO, see[495, 1].

1. **Seismic noise** is apparent in the low-frequency regime with spectral amplitude $\tilde{x}_S(f) \propto f^{-2}$ in most environments. Seismic noise is a function of weather conditions, winds, and ocean waves. Low-frequency perturbations also derive locally, e.g. a nearby train or wood logging at the Livingston site, or an occasional rock concert in nearby Tokyo[185]. Good seismic isolation is essential in continuous operations of the detectors, and important in reaching the desired sensitivity in the low-frequency regime. At present, LIGO, Virgo, and Gingin all have different seismic suspension systems. It will be interesting to compare their performances in the near future.

   The $1/f^2$ seismic disturbances are attenuated by high-order seismic isolation systems and suspensions, whose spectral transfer functions are approximately of the form $H(f) = (f_0/f)^n$ ($6 \leq n \leq 12, 0.1 \leq f_0 \leq 10$ Hz). This effectively renders seismic noise subdominant above 10–50 Hz, depending on the design. The km-sized arms further introduce a finite angle between the Earth's gravitational acceleration and the wavefront of the laser beam, due to a finite curvature of the Earth. Horizontal separation between mirrors is hereby coupled to their vertical motions. This poses the challenge of attenuating vertical noise due to seismic disturbances and thermal noise in the vertical degrees of freedom in the mirror suspension system. (Hanging mirrors orthogonal to the laser beam by attaching additional magnets to the mirrors[528] is of potential interest, with the inherent technical challenge of controlling additional noise sources.)

2. **Gravity gradient noise**. The static gravitational field is modulated by seismic disturbances $\tilde{x}_S(f)$. There are some model-dependent predictions[39, 474] which differ by about 1 order of magnitude. This noise source defines a low-frequency limit for ground-based gravitational wave detectors. This has motivated future plans for underground detectors to be built in tunnels, potentially extending the available band from 10 down to 1 Hz.

3. **Magnetic noise**. The present seismic suspension towers – in different forms – all use static magnets at room temperature to damp oscillations. In the case of Virgo, this gives a magnetic noise in response to $\tilde{x}_S(f)$ consisting of diamagnetic Marionetta-tower coupling and eddy currents on tower walls. Finite temperature Marionetta fluctuations due to eddy currents are dissipated, which produces a noise source independent of $\tilde{x}_S(f)$. The sum of these three contributions is commonly referred to as "magnetic noise."

4. **Shot noise** represents counting noise. It is determined by the finite number of photons involved at a given power level of the beam.

$$h_{shot}(f) = \frac{1}{8L_{arm}F} \left(\frac{4\pi\hbar\lambda c}{\eta C P_{beam}} \right)^{1/2} \left(1 + \left(\frac{f}{f_{FP}} \right)^2 \right)^{1/2}, \qquad (16.27)$$

where $\eta = 0.93$ denotes the photodiode efficiency, $C = 50$ the recycling factor, $F = 50$ the finesse, and $\lambda = 1.064\ \mu$m the wavelength of the laser (currently in use), and

$$f_{FP} = \frac{c}{4L_{arm}F} \qquad (16.28)$$

the Fabry–Perot cutoff frequency. Thus, shot noise satisfies the scaling $h_{shot}(f) \propto P_{beam}^{-1/2} L_{arm}^{-1} f$ in the high-frequency limit $f/f_{FP} \gg 1$. Laser power is hereby the defining factor in the performance of the detector at frequencies above a few hundred Hz.

5. **Radiation pressure noise.** Closely related to the shot noise is radiation pressure noise $h_{rad}(f)$, in response to the reflection of the laser beam by the freely suspended mirrors. Again, this noise contribution is essentially counting noise, now increasing with $P^{1/2}$. It has a distribution $1/f^2$ in frequency.

6. **Quantum limit.** Increasing laser power reduces shot noise, and increases radiation pressure noise. The quantum limit corresponds to the noise at the point where $h_{shot} = h_{rad}$, i.e.

$$h_{QL}(f) = \frac{1}{2\pi f L_{arm}} \sqrt{\frac{\hbar}{m_c}} \sqrt{\frac{2}{\eta} \left[1 + \left(\frac{f}{f_{FP}} \right)^2 \right]}, \qquad (16.29)$$

where m_c denotes the mirror mass. The relation (16.29) holds for the configuration used in initial Virgo and LIGO instruments. It is modified and can be manipulated to advantage in signal-recycled interferometers, planned for second-generation instruments and currently used in GEO-600.

7. **Thermal noise.** The suspension system, although made from low-loss materials, includes damping and hence dissipation of energy[345]. The associated creation of noise from dissipation is a function of temperature according to the fluctuation-dissipation theorem[475]. A large number of individual thermal noise sources have been identified, e.g. the excitation of pendulum, violin, tilt, and rotational modes in the mirror and its suspension, as well as the coupling of its vertical modes to the output of the detector via the vertical-to-horizontal coupling angle $\theta_0 = L_{arm}/2R_E$, where R_E is the radius of the Earth. Additionally, the test mass which contains the mirror has internal modes, which are excited to some degree at a finite temperature.

8. **Thermodynamic noise in the mirrors.** Braginsky *et al.*[79] show that thermal noise in the mirror couples to the reflective mirror surface by thermal expansion of the bulk, and to the coating through a finite temperature dependence in the refractive index. By their nature, these are low-frequency noise sources estimated to be $8 \times 10^{-24}(10\ \text{Hz}/f)(3\ \text{km}/L_{arm})\ \text{Hz}^{-1/2}$ and, respectively, $2 \times 10^{-24}(10\ \text{Hz}/f)^{1/4}(3\ \text{km}/L_{arm})\text{Hz}^{-1/2}$.

9. **Mechanical shot noise (creep)** appears due to inelastic stretch of suspension wires, causing fluctuations in the form of shot noise parametrized by a typical rate and strength of creep events. The product of these two is determined experimentally.

10. **Residual gas pressure** produces viscous damping in the pendulum mode. This, dissipation of eddy currents in mirror magnets, and residual gas in the interferometer arms, causes a finite Q value of the pendulum mode. Residual gas also introduces accoustic coupling to external walls and, hence their disturbances. Fluctuations in the residual gas pressure also cause variations in the beam phase, due to coupling of pressure to the refractive index in the arms. This introduces stringent requirements of ultra-high vacuum in the detector arms, a significant factor in the cost and implementation of realistic instruments. The refractive index of the dielectric coating of the mirror itself are also subject to temperature fluctuations.

Finally, noise is also introduced by laser power heating and distortion of the reflective surface of the mirrors ("distortion by laser heating"), as well as laser power fluctuations in the presence of absorption asymmetries in the two Fabry–Perot cavities ("nonlinear opto-thermal coupling").

Collectively, the instrumental noise is shown in the spectral domain in Figures (16.3) and (16.4).

16.5 Signal-to-noise ratios for GRB-SNe

Gamma-ray bursts from rotating black holes produce emissions in the shot noise region of LIGO and Virgo, where the noise strain energy density satisfies $S_h^{1/2}(f) \propto f$. We will discuss the signal-to-noise ratios in various techniques. We discuss matched filtering as a theoretical upper bound on the achievable signal-to-noise ratios. We discuss the signal-to-noise ratios in correlating two detectors both for searches for burst sources and for searches for the stochastic background in gravitational radiation.

The signal-to-noise-ratio of detections using matched filtering with accurate waveform templates is given by the ratio of strain amplitudes of the signal to that of the detector noise. Including averaging over all orientations of the source, we have[191, 140]

$$\left(\frac{S}{N}\right)_{mf} = \frac{(1+z)\sqrt{2E_{gw}}}{\pi d_L(z) f^{1/2} h_n}. \tag{16.30}$$

Here, we may neglect the redshift for distances of the order of 100 Mpc. Consequently, for matched filtering this gives

$$\left(\frac{S}{N}\right)_{mf} \simeq 8 \left(\frac{S_h^{1/2}(500 \text{ Hz})}{5.7 \times 10^{-24} \text{ Hz}^{-1/2}}\right)^{-1} \left(\frac{\eta}{0.1}\right)^{-3/2} \left(\frac{M}{7M_\odot}\right)^{5/2} \left(\frac{d}{100 \text{ Mpc}}\right)^{-1}. \tag{16.31}$$

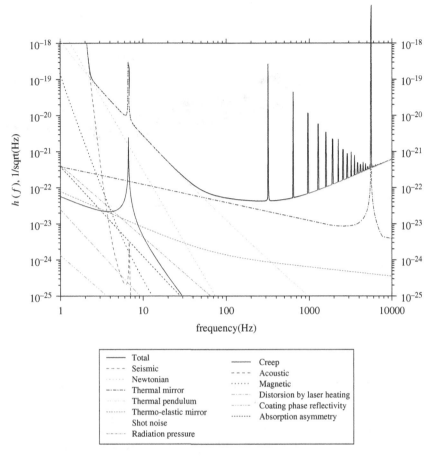

Figure 16.4 Virgo dimensionless spectral noise amplitude, modeled according to various noise sources. (Courtesy of Virgo.)

The expression (16.31) shows a strong dependence on black hole mass. For a uniformly distributed mass distribution, we have the expectation value $\overline{S/N} = 18$ for an average over the black hole mass distribution $M_H = 4 - 14 \times M_{\odot}$ as observed in galactic soft X-ray transients; we have $\overline{S/N} = 7$ for a narrower mass distribution $M_H = 5 - 8 \times M_{\odot}$. The cumulative event rate for the resulting strain-limited sample satisfies $\dot{N}(S/N > s) \propto s^{-3}$.

The signal-to-noise ratio (16.31) in matched filtering is of great theoretical significance, in defining an upper bound in single-detector operations. Figure (16.5) shows the characteristic strain-amplitude of the gravitational wave-signals produced by GRBs from rotating black holes, for a range $M = 4 - 14 \times M_{\odot}$ of black hole masses and a range $\eta = 0.1 - 0.15$ in the ratio of the angular velocities of the torus to the black hole. The ratio of the characteristic strain-amplitude

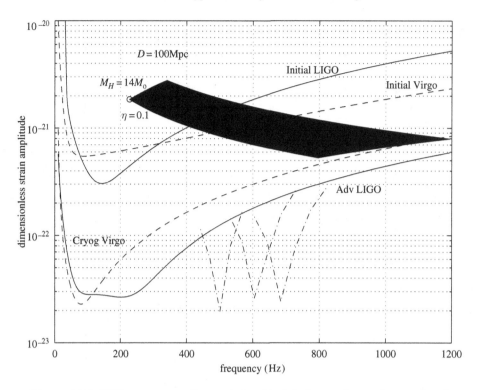

Figure 16.5 GRB supernovae from rotating black holes produce a few tenths of $M_\odot$ in long duration bursts of gravitational radiation, parametrized by black hole mass $M = 4 - 14M_\odot$ and the ratio $\eta \sim 0.1 - 0.15$ of the angular velocity of the torus to that of the black hole. The signal is band-limited with relative bandwidth $B \simeq 10\%$. The dark region shows $h_{char}B^{1/2}/\sqrt{5}$ of the orientation-averaged characteristic dimensionless spectral strain-amplitude h_{char}. The source distance is $D = 100$ Mpc, corresponding to an event rate of once per year. The dimensionless strain-noise amplitudes $h_{rms}(f) = \sqrt{fS_h(f)}$ of Initial/Advanced LIGO (lines), Initial/Cryogenic Virgo (dashed;[445]) are shown with lines removed, including various narrow-band modes of Advanced LIGO (dot-dashed), where $S_h(f)$ is the spectral energy density of the dimensionless strain noise of the detector. Short GRBs from binary black hole neutron star coalescence may produce similar energies distributed over a broad bandwidth, ranging from low frequencies during inspiral up to 1 kHz during the merger phase. (Reprinted from[565]. © 2004 The American Physical Society.)

of a particular event to the strain-noise amplitude of the detector (at the same frequency) represents the signal-to-noise ratio in matched filtering. We have included the design sensitivity curves of initial LIGO and Virgo, and Advanced LIGO and a potential Virgo upgrade using cryogenics to reduce thermal noise sources. The Virgo sensitivity curve is a current evaluation, to be validated in the coming months, during the commissioning phase of Virgo.

Evidently, matched filtering requires detailed knowledge of the waveform through accurate source modeling. The magnetohydrodynamical evolution of the torus in the suspended accretion state has some uncertainties, e.g. the accompanying accretion flow onto the torus from an extended disk. These uncertainties may become apparent in the gravitational wave spectrum over long durations. (Similar uncertainties apply to models for gravitational radiation in accretion flows.) For this reason, it becomes of interest to consider methods that circumvent the need for exact waveforms. In the following, we shall consider detection methods based on the correlation of two detectors, e.g. the collocated pair in Hanford, or correlation between two of the three LIGO and Virgo sites.

As mentioned in Section 16.1, the gravitational wave-spectrum is expected to be band-limited to within 10% of (9.8), corresponding to spin-down of a rapidly black hole during conversion of 50% of its spin energy. We may exploit this by correlating two detectors in narrow-band mode – a model-independent procedure that circumvents the need for creating wave templates in matched filtering. An optimal choice of the central frequency in narrow-band mode is given by the expectation value of (9.8) in the ensemble of GRBs from rotating black holes.

This optimal choice corresponds to the most likely value of M_H and η in our model. As indicated, present estimates indicate an optimal frequency within 0.5 to 1 kHz. (A good expectation value awaits calorimetry on GRB-associated supernova remnants.) A single burst produces a spectral closure density Ω_s, satisfying $T_{90}\Omega_s = 2E'_{gw}f_{gw}/3H_0^2 d^2$ in geometrical units. The signal-to-noise ratio obtained in correlating two detector signals over an integration period T satisfies[10]

$$\left(\frac{S}{N}\right)^2 = \frac{9H_0^4}{50\pi^4}T\int_0^\infty \frac{\Omega_s^2(f)df}{f^6 S_{n1}(f)S_{n2}(f)}. \tag{16.32}$$

This may be integrated over the bandwidth $\Delta f_{gw} \ll f_{gw}$, whereby

$$\left(\frac{S}{N}\right) \simeq \frac{1}{\sqrt{2}}\left(\frac{1}{BN}\right)^{1/2}\left(\frac{S}{N}\right)^2_{mf} \tag{16.33}$$

where $1/BN < 1$ by the frequency–time uncertainty relation. The number of periods N of frequency f_{gw} during the burst of duration T_{90} satisfies $N \simeq 2T_{90}/P \simeq 4\times10^4\eta_{0.1}^{-8/3}\mu_{0.03}^{-1/2}$. Hence, we have $1/BN \sim 10^{-3}$. Following (16.31) and (16.32), we find

$$\left(\frac{S}{N}\right) \simeq 12 f_4^{D1}f_4^{D2}\left(\frac{S_h^{1/2}(500\,\mathrm{Hz})}{5.7\times10^{-24}\,\mathrm{Hz}^{-1/2}}\right)^{-1}_{D1}$$

$$\left(\frac{S_h^{1/2}(500\,\mathrm{Hz})}{5.7\times10^{-24}\,\mathrm{Hz}^{-1/2}}\right)^{-1}_{D2}\eta_{0.1}^{-5/3}M_7^5 d_8^{-2}B_{0.1}^{-1/2}\mu_{0.03}^{1/4}, \tag{16.34}$$

where $\eta_{0.1} = \eta/0.1$, $M_7 = M_H/7M_\odot$, $d_8 = d/100\,\text{Mpc}$, $B_{0.1} = B/0.1$ and $\mu_{0.03} = \mu/0.03$, and the factors $f_4^{Di} = f^{Di}/4$ refer to enhancement in sensitivity in narrow-band mode, relative to broad-band mode. The cumulative event rate for the resulting flux-limited sample satisfies $\dot{N}(S/N > s) \propto s^{-3/2}$.

Given the proximity of the extremal value of $\Omega_B(f)$ in (16.26) and the location of maximal sensitivity of LIGO and Virgo, we consider correlating two collocated detectors for searches for the contribution of GRB supernovae to the stochastic background in gravitational waves. According to (16.32) and (16.26) for a uniform mass-distribution $M_H = M_\odot[4, 14]$, correlation of the two advanced detectors at LIGO Hanford gives

$$\left(\frac{S}{N}\right)_B \simeq 20 \left(\frac{S_h^{1/2}(500\,\text{Hz})}{5.7 \times 10^{-24}\,\text{Hz}^{-1/2}}\right)_{H1}^{-1}$$

$$\left(\frac{S_h^{1/2}(500\,\text{Hz})}{5.7 \times 10^{-24}\,\text{Hz}^{-1/2}}\right)_{H2}^{-1} \eta_{0.1}^{-7/2} T_{1\,\text{yr}}^{1/2}. \tag{16.35}$$

Here, the coefficient reduces to 9 for a mass distribution $M_H/M_\odot = [5, 8]$, and less for nonextremal black holes. The estimate (16.35) reveals an appreciable dependence on η.

16.6 A time-frequency detection algorithm

Gravitational wave emissions produced by GRB supernovae from rotating black holes have emission lines that evolve slowly in time. These time-varying frequencies may be searched for by time-frequency methods, or by identifying curves in the $\dot{f}(f)$-diagram[571].

The orbital period T_o of millisecond serves as a short timescale, and the lifetime T_s of rapid spin of the black hole of tens of seconds serves as a long timescale. We consider Fourier transforms on an intermediate timescale during which the spectrum is approximately monochromatic, using the the output of the two colocated detectors – with output

$$s_i(t) = h(t) + n_i(t) \quad (i = 1, 2), \tag{16.36}$$

where $h(t)$ denotes the strain amplitude of the source at the detector and $n_i(t)$ the strain-noise amplitude of H1 and H2.

We can search for these trajectories by performing Fourier transforms over time-windows of intermediate size, during which the signal is approximately monochromatic. The simulations show a partitioning in $N = 128$ subwindows of $M = 256$ data points, in the presence of noise with an instantaneous signal-to-noise ratio of 0.15. The left two windows show the absolute values of the Fourier coefficients, obtained from two simulated detectors with uncorrelated noise. The trajectory of a simulated slowly evolving emission line becomes apparent in the correlation between these two spectra (right window). The frequency scales with Fourier index i according to $f = (i-1)/\tau$ $(i = 1, \cdots M/2+1)$, where τ denotes the time period of the subwindow. Evaluating the spectrum over the intermediate timescale τ,

$$T_o \ll \tau \ll T_s, \tag{16.37}$$

we choose τ as follows. Consider the phase $\Phi(t) = \omega t + (1/2)\epsilon\omega t^2$ of a line of slowly varying frequency $\dot{\Phi}(t) = \omega(1 + (1/2)\epsilon t)$, where $B = \epsilon T_s \simeq 0.1$ denotes the change in frequency over the duration T_s of the burst. For a duration τ, the phase evolution is essentially stationary, provided that $(1/2)\omega\epsilon\tau^2 \ll 2\pi$, or

$$\tau/T_s \ll \sqrt{2/BN} \simeq 1/30. \tag{16.38}$$

For example, a typical burst duration of 1 min. may be divided into $N = 120$ subwindows of 0.5 s, each representing about 250 wave periods at a frequency of 500 Hz as used in the simulation shown in Figure 16.6.

Consider the discrete evolution of the spectrum of the signal over N subwindows $I_n = [(n-1)\tau, n\tau]$, by taking successive Fourier transforms of the $s_i(t)$ over each I_n. The two spectra $\tilde{s}_i(m, n)$, where m denotes the mth Fourier coefficient, can be correlated according to

$$c(m, n) = \tilde{s}_1(m, n)\tilde{s}_2^*(m, n) + \tilde{s}_1^*(m, n)\tilde{s}_2(m, n). \tag{16.39}$$

The signal $h(t)$ contributes to a correlation between the $s_i(t)$, and hence to non-negative values c_{mn}. In general, the presence of noise introduces values of c_{mn} which are both positive and negative. Negative values of c_{mn} only appear in

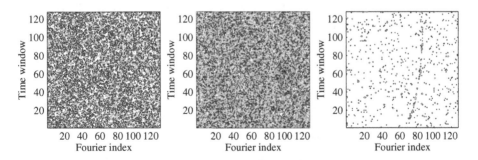

Figure 16.6 Simulated slowly evolving lines in gravitational radiation produced by GRB-SNe from rotating black holes, corresponding to the timescale of spin-down of the black hole. This produces trajectories in the temporal evolution of the spectrum of the signal. (Reprinted from[565]. ©2004 The American Physical Society.)

response to (uncorrelated) noise. A plot of positive values c_{mn}, therefore, displays the evolution of the spectrum of the signal. For example, we may plot all values of c_{mn} which are greater than a certain positive number, e.g. those for which $c_{mn} > 0.3 \times \max_{mn} c_{mn}$. Results of a simulation are shown in Figure (16.6).

The TFT algorithm may be applied to two independent detectors, or one single detector, i.e. the two collocated detectors at LIGO Hanford or, respectively, the LIGO detector at Livingston or the Virgo detector at Pisa. The latter applies, provided that the intermediate timescale (16.38) is much larger than the autocorrelation time in the detectors. LIGO and Virgo detectors have sample frequencies of 16 and 20 kHz respectively. This provides the opportunity for down-sampling a detector signal $s(t)$ into two separate and interlaced sequences $s_1(t_i)$ and $s_2(t_i')$ $(t_i' = t_i + \Delta t)$ that sample $f_{gw} \simeq 500$ Hz, while remaining sufficiently separated for the noise between them to be uncorrelated. The coefficients (16.38) would then be formed out of the Fourier coefficients $s_1(m, n)$ and $e^{im\Delta t}s_2(m, n)$.

The TFT algorithm is of intermediate order, partly first-order in light of the Fourier transform, and partly second-order in light of the correlation between the Fourier coefficients of the two detector signals. Consequently, its detection sensitivity is between matched filtering and direct correlation in the time domain. The gain in signal-to-noise ratio obtained in taking Fourier transforms over subwindows may circumvent the need for narrow-band operation.

Application of the TFT algorithm to searches for the contribution of GRBs to the stochastic background radiation could be pursued by taking the sum of the coefficients (16.39) over successive windows of the typical burst duration, in light of the GRB duty cycle of about 1[138]. The contributions of the signals from a distant event add linearly, but are distributed over a broad range of frequencies around 250 Hz. A further summation over all subwindows of 0.5 s would result in a net sum of over 10^6 coefficients during a 1 year observational period. The result should be an anomalous broad bump in the noise around 250 Hz with a signal-to-noise ratio of order unity, assuming advanced detector sensitivity.

16.7 Conclusions

There is an advantageous coincidence in the frequency range of long bursts of gravitational waves from GRB supernovae and the LIGO and Virgo detectors. The active nucleus in GRB supernovae is expected to emit frequencies of a few hundred Hz, which falls in the shot-noise of the detectors as shown in Figure (16.3). Here, detector improvements will take place with the installation of high powered lasers.

Gamma-ray burst supernovae occur about once per year within a distance of 100 Mpc. Their associated signatures in the electromagnetic spectrum, through

Table 16.1 *Model predictions^a versus observations on GRB supernovae.*

QUANTITY	UNITS	EXPRESSION	OBSERVATION
E_{gw}	erg	$4 \times 10^{53} \, \eta_{0.1} M_{H,7}$	
f_{gw}	Hz	$500 \, \eta_{0.1} M_7^{-1}$	
Ω_B	1	6×10^{-9} @250 Hz	
E_{SN}	erg	$2 \times 10^{51} \, \beta_{0.1} \eta_{0.1}^2 M_{H,7}$	2×10^{51} erg^b
E_γ	erg	$2 \times 10^{50} \, \epsilon_{0.30} \eta_{0.1}^{8/3} M_{H,7}$	3×10^{50} erg^c
$E^d_{\gamma \to X}$	erg	$4 \times 10^{52} \, \bar{\epsilon} \eta_{0.1}^2 M_{H,7}$	$> 4.4 \times 10^{51}$ erg^e
T_s	s	$90 \, \eta_{0.1}^{-8/3} M_{H,7} \mu_{0.03}^{-1}$	T_{90} of tens of s^f
Event rate	yr^{-1}		1 within $D = 100$ Mpc^g
$\mathcal{R}$[Ib/c $\to$ GRB]	1	$0.5\% \left(\frac{K}{10\,\mathrm{km\,s^{-1}}}\right)^2 \left(\frac{\sigma_{kick}}{100\,\mathrm{km\,s^{-1}}}\right)^{-2}$	$(2-4) \times 10^{-3}$

^{a} Based on a critical ratio $\mathcal{E}_B/\mathcal{E}_k \simeq 1/15$ of poloidal magnetic field energy-to-kinetic energy in a nonaxisymmetric torus surrounding an extremal black hole. Energies and durations T_{90} are correspondingly lower by a factor E_{rot}/E_{rot}^{max} for nonextremal black holes.
^{b} SN1998bw with aspherical geometry.
^{c} True energy in gamma rays produced along open magnetic ergotubes.
^{d} Continuum gamma ray emission produced by torus winds with undetermined efficiency $\bar{\epsilon}$ as energy input to X-ray line emissions.
^{e} lower bound.
^{f} broad distribution of durations.
^{g} Local estimate.

the supernova and radio afterglow emissions, enable coincident detections in the gravitational wave and electromagnetic spectrum.

Detection of both a long-duration gravitational wave burst and a Type Ib/c supernova enables the determination of the emitted energy in gravitational waves. This provides an estimate for the compactness parameter $2\pi E_{gw} f_{gw}$ which can be used to compare the emissions from rapidly rotating black holes by those from rapidly rotating neutron stars.

If GRB emissions are not conical, but represent strongly anisotropic emissions accompanied by weak radiation over arbitrary angles such as, perhaps, in GRB980425/SN1998bw[169, 570], we may search for coincidences of gravitational wave bursts with such apparently weak GRBs. Independently, upcoming all-sky surveys such as Pan-STARRS[309] may be used to trigger searches around the time of onset for all Type Ib/c supernovae, a fraction of less than 1% of which are candidates for GRB supernovae.

Up to days after the event, these may appear as a radio supernova representing the ejection of the remnant stellar envelope by the magnetic torus winds. Months thereafter, wide-angle radio afterglows may appear[338, 405]. Ultimately, the remnant is a black hole in a binary with an optical companion[568], which may appear as a soft X-ray transient[87] (see further[340, 514]). Thus, long GRBs provide a unique opportunity for integrating LIGO and Virgo detections with current astronomical observations.

Collectively, GRB supernovae occur about $0.5 \times 10^6 \, \mathrm{yr}^{-1}$, and contribute about 10^{-8} in spectral closure density around $250 \, \mathrm{Hz}$ to the stochastic background in gravitational waves, as shown in (16.2). This coincides with the location of minimal noise in the LIGO and Virgo detectors, where it may be detectable in a 1-year integration time.

The more common Type Ib/c supernovae which do not produce a GRB are less likely to produce a long duration burst in gravitational radiation. Nevertheless, their possible decentered nucleation of black holes combined with their higher event rate by some 2 orders of magnitude may produce an interesting contribution to the stochastic background at high frequencies ($> 1 \, \mathrm{kHz}$).

Detection of the anticipated energy E_{gw} in gravitational radiation provides a method for identifying Kerr black holes in GRB supernovae on the basis of calorimetry. We hope this theory of GRB supernovae and the suggested TFT method in Figure (16.6) provides some guidelines to this experiment.

Exercises

1. Consider a narrow-band search with enhanced sensitivity, assuming the frequency to be well chosen with regard to the expected distribution of frequencies. This is illustrated by the dot-dashed wedges in Figure (16.3). Show that the detection rate increases if $B/h_n = $ const., where B denotes the bandwidth and h_n the strain amplitude noise of the detector. (These are "pencil" searches in frequency space.)

2. Assuming the branching ratio $\mathcal{R}[\text{Ib/c} \rightarrow \text{GRB}]$ to be independent of redshift, estimate on the basis of Figure (16.2) the contribution of short-duration bursts of gravitational radiation by nucleation of black holes in Ib/c supernovae. Assume the emission to be due to (a) the kick velocity of the black hole by the Bekenstein recoil mechanism, whereby $E_{gw} = (\sigma_k/c)M_H c^2$ and (b) by a nonaxisymmetric torus prior to the formation of a black hole. Express the results as a function of the energy E_{short} as a fraction of $1M_\odot$ and frequency f_{short} as measured in the comoving frame. Evaluate the prospect of detecting this high-frequency contributions by Type Ib/c supernovae.

3. On the basis of Figure (16.4), derive a high-frequency performance parameter by deriving the sensitivity range for GRB supernovae assuming (a) random orientations and (b) beamed towards the detector. In (b), use the fact that quadrupole gravitational wave emissions are slightly anisotropic, whose amplitude is larger by a factor of about 1.58 relative to the orientation averaged value, e.g.[140].

4. Devise an algorithm for assigning a signal-to-noise ratio to time frequency trajectories shown in Figure (16.6).

5. Calculate the probability of detecting first the stochastic background radiation from GRB supernovae, while not detecting a nearby burst event – and vice versa.

6. Calculate canonical values for the shot noise and the quantum limit due to radiation pressure according to (16.27) and (16.29).

7. Write a proposal on "First light in gravitational waves" by (a) model-independent searches, and (b) model-dependent searches.

265

17

Epilogue: GRB/XRF singlets, doublets? Triplets!

"Physics is not a finished logical system. Rather, at any moment it
spans a great confusion of ideas, some that survive like folk epics from
the heroic periods of the past, and others that arise like utopian novels
from our dim premonitions of a future grand synthesis." (1972).

Stephen Weinberg, in *Gravitation and Cosmology*

Gamma-ray bursters are serendipitously discovered transients of nonthermal emissions of cosmological origin. They come in two varieties: (a) short bursts with durations of a few tenths of a second, and (b) long bursts with durations of a few tens of seconds. The latter are now observed in association with supernovae, while no such association is observed for the former. The parent population of Type Ib/c supernovae may well represent the outcome of binary evolution of massive stars, such as SN1993J. In light of these observations, a complete theory is to explain GRBs as a rare kind of supernovae. Long-duration GRB-supernovae require a baryon-poor inner engine operating for similar durations, for which the most promising candidate is a rapidly rotating Kerr black hole. Formed in core collapse of a massive star, the black hole is parametrized by its mass, angular momentum, and kick velocity (M, J_H, K).

At low kick velocity K, core-collapse produces a high-mass and rapidly rotating black hole. The Kerr solution predicts a large energy reservoir in angular momentum. Per unit of mass, this far surpasses the energy stored in any baryonic object, including a rapidly rotating neutron star. By its energetic interaction with a magnetosphere supported by a surrounding high-density torus, the black hole becomes an active nucleus inside the remnant envelope of the massive progenitor for the duration of its rapid spin. We have, where possible, analyzed this active nucleus in case of a torus magnetized at superstrong magnetic fields following the closed model in which the torus radiates off most of the black-hole output. A number

266

of interesting questions on microphysics are left for future developments, such as the nature of the dynamo action in the torus and gaps in ergotubes.

GRB-supernovae from rotating black holes may turn out to be an ideal laboratory for studying general relativity in the nonlinear regime, just as PSR 1913 + 16 and PSR 0737 − 3039 are the ideal laboratory for studying general relativity in the linearized regime. By frame-dragging, black-hole spin interacts with angular momentum in an open ergotube along its axis of rotation as well as with a surrounding torus. This is described by and energy and torque

$$E = \omega j, T \simeq -\dot{j}_H, \qquad (17.1)$$

where $J = eA_\phi$ denotes the angular momentum of a charged particle in the ergotube. Thus, black holes become luminous, ejecting baryon-poor blobs (intermittent) or jets (continuous), while delivering most of their energetic output to the surrounding magnetized matter. The latter catalytically converts black-hole spin energy in various radiation channels, powering a long-duration burst in gravitational radiation and megaelectronvolt neutrinos, and an aspherical supernova by dissipation of its magnetic winds against the remnant stellar envelope from within.

Multiple bursters are known explosive endpoints of − some massive stars, as discovered with the detection of a burst in neutrino emissions *and* a supernova (SN1987A), as well as the observation of a GRB *and* a supernova in GRB980425/SN1998bw and GRB030329/SN2003dh. While in case of SN1987A, the neutrino emissions provided first-principle evidence of matter in a state of high density and high temperature, representing a nucleon star or a rapidly rotating neutrino-torus, possibly in transition to collapse into a black hole, the GRB − supernovae association promises a first step towards observational evidence of a luminous black hole as their inner engines. This significance of observational evidence of luminous black holes goes much further than GRB supernovae, as it is believed to extend to extragalactic quasars and galactic microquasars. Perhaps the singular difference between GRB supernovae and these two other classes of astrophysical transients, is that the former is luminous in gravitational radiation and megaelectronvolt neutrino emissions, whereas the latter is not, all else being qualitatively the same.

From query to quest, our model predicts a large burst in gravitational radiation from the inner engines of GRB supernovae which surpasses the current calorimetric estimates in electromagnetic radiation by orders of magnitude. It flashes the endpoint of a massive star, which shifts our view on gamma-ray bursters (singlets) from GRB-SNe (doublets) to GRB-SN-GWB (triplets). Triplets might also exist as XRF-SN-GWBs, provided that the XRF-SN association is confirmed by future observations.

Presently, burst sources of gravitational radiation from GRB supernovae are pursued by model-independent LIGO searches for short-duration bursts of around 100 ms or less, including bursts associated with the long-duration burst GRB030329/SN2003dh. This approach does not represent "best use of data," as it misses the opportunity to integrate the detector signal against the duration of the GRB. Our theory points towards long-duration bursts of gravitational waves in GRB supernovae, as nearby point sources or through their collective contribution to the stochastic background radiation in gravitational waves. The output in gravitational radiation is predicted to be contemporaneous with the GRB and the onset of the supernova, satisfying

$$E_{gw} \simeq 4 \times 10^{53} \text{ erg } M_{H,7}\eta_{0.1} \quad f_{gw} \simeq 500\text{Hz } M_{H,7}^{-1}\eta_{0.1}. \qquad (17.2)$$

We propose to perform targeted searches by LIGO, Virgo, TAMA and GEO triggered by gamma-ray bursts and supernovae, selected as Type Ib/c events of, for example, RAPTOR (Los Alamos), Super-LOTIS (Livermore), KAIT (Berkeley), and Pann-Starrs (Hawaii). Where these observations, theory, and experiment shall meet, the explosive endpoint of massive stars will ultimately be understood by direct measurements.

Appendix A. Landau's derivation of a maximal mass

Chandrasekhar derived a maximal mass $\simeq 1.4 M \odot$ of a white dwarf. A white dwarf consists of degenerate electrons, i.e. Fermionic gas at low temperature described by a polytropic equation of state with polytropic index $\gamma = 4/3$ in the relativistic regime and with polytropic index $\gamma = 5/3$ in the non-relativistic regime.

The Chandrasekhar mass limit of a white dwarf is based on the maximal pressure provided by a degenerate Fermionic fluid against self gravity. The same principle applies to degenerate neutrons, i.e. to neutron stars. Landau[316] gives the following argument for a maximal mass; see, for example, Shapiro and Teukolsky[490].

Consider a star of radius R, consisting of N fermions at constant density $n = 3N/4\pi R^3$. At relativistic pressures the Pauli exclusion principle gives rise to momentum $2 \times (\hbar/2) n^{1/3}$ by the Heisenberg uncertainty principle, applied to both spin orientations of the particles. The associated Fermi energy of the particles is hereby

$$E_F \simeq \hbar c (3/4\pi)^{1/3} N^{1/3}/R. \tag{A.1}$$

at relativistic pressures. This may be compared with non-relativistic pressures for which $E_F \simeq p_F^2/2m_B$ with $p_F \simeq \hbar/R$. Fermions have an average gravitational energy $E_g \simeq -3GMm_B/5R$, where $M = Nm_B$. The total energy is

$$E = R^{-1} \left[\left(\frac{3}{4\pi} \right)^{1/3} N^{1/3} \hbar c - \frac{3}{5} GNm_B^2 \right] \tag{A.2}$$

at relativistic pressures; the first term on the right-hand side reduces to $\sim R^{-1}$ in the non-relativistic regime. Therefore, instability sets in only when $E < 0$. This gives rise to a critical particle number

$$N_* = \frac{5\sqrt{5}}{6\sqrt{\pi}} \left(\frac{\hbar c}{Gm_B^2} \right)^{3/2} \simeq 2 \times 10^{57}, \tag{A.3}$$

and a critical mass

$$M_* = N^* m_B \simeq 1.5 M_\odot. \tag{A.4}$$

The relativistic pressures set in for $E_F \geq mc^2$, where m denotes the mass $m_e = 9.1 \times 10^{-28}$ g of the electron or $m_n = 1.67 \times 10^{-24}$ g of the neutron. The associated radii are

$$R \leq \left(\frac{3}{4\pi}\right)^{1/3} \left(\frac{\hbar}{mc}\right) N_*^{1/3} \simeq 3 \times 10^8 \, \text{cm} \tag{A.5}$$

for a white dwarf and

$$R \leq \left(\frac{3}{4\pi}\right)^{1/3} \left(\frac{\hbar}{mc}\right) N_*^{1/3} \simeq 2 \times 10^5 \, \text{cm} \tag{A.6}$$

for a neutron star.

Sirius B, discovered by W. S. Adams[5] in 1914, is a white dwarf now known to have a mass $M \simeq 1.05 M_\odot$ and a radius of $R \simeq 5150$ km. Recently, XMM-Newton observations by Cottam, Paerels and Mendez[136] captured for the first time the gravitational redshift from the surface of a neutron star EXO 0748-676 (in a binary), determining the mass-to-radius ratio to be 0.152 $M_\odot \, \text{km}^{-1}$. Given the known mass of about $1.45 M_\odot$, the radius is hereby determined to be about 16 km.

Appendix B. Thermodynamics of luminous black holes

Axisymmetric state transitions of a Kerr–Newman black hole immersed in a charge-free magnetosphere connected to a distant (nonrotating) source satisfy the first law of black hole thermodynamics[105, 145]

$$\delta M = \Omega_H \delta(J_H + J_{em}) + T_H \delta S_H + [V]_\infty^H \delta q, \tag{A.7}$$

where M is the energy as measured at infinity, and J_H, J_{em} are the angular momentum in the black hole and the electromagnetic field. Here, q denotes the horizon charge and $[V]_\infty^H = (-\xi^a A_a)_H - (-\xi^a A_a)_\infty$ denotes the electric horizon potential relative to that of the distant source; $\xi^a = \eta^a - \beta k^a$ denotes the redshift corrected velocity four-vector of zero angular momentum observers (ZAMOs), in the presence of frame dragging β and azimuthal and asymptotically timelike Killing vectors, respectively, $\eta^a = (\partial_t)^a$ and $k^a = (\partial_\phi)^a$. The interaction is described by the surface integrals $A_\phi(\theta) = \int_0^\theta F_{\theta\phi} d\theta'$, $\Sigma_H(\theta) = (1/2) \int_0^\theta *F_{\theta\phi} d\theta'$ and $I(\theta) = 2\pi \int_0^\theta \sqrt{-g} j^r d\theta'$ of, respectively, $B_n/2\pi$, $\sigma_H = E_n/4\pi$ and the radial current density j^r over a polar cap with half-angle θ. Here, $\Sigma'_H = (1/2) * F_{\theta\phi} = \sigma_H dS$ and $A'_\phi = F_{\theta\phi} = B_n dS/2\pi$ in terms of the surface element $dS = 2\pi\tilde{\omega}\rho d\theta$. By conservation of electric charge, $\dot{\Sigma}_H + I + I_H$, where $I_H = I_H(\theta)$ is the poloidal surface current. In the quasistatic limit with no electromagnetic waves to infinity, the black hole magnetosphere evolves along stationary states with constant total energy and angular momentum, giving

$$\left\{ \begin{array}{l} \dot{J}_H + \dot{J}_{em} = \int_H I A_\phi dS, \\[2mm] T_H \dot{S}_H = \dfrac{1}{2}\|\dot{A}_\phi\|^2 + 2\|I_H\|^2, \end{array} \right. \tag{A.8}$$

where[105, 145]

$$J_{em} = \int_H \sigma_H A_\phi dS \tag{A.9}$$

and with the norm $\|f\|$ for horizon functions $f = f(\theta)$ defined below.

The result follows from integration of Maxwell's equations and regularity of the electromagnetic field on the horizon as seen by freely falling observers (FFOs)[534]. In axisymmetry, we can integrate Maxwell's equations $(\sqrt{-g}F^{ab})_{,a} = -4\pi\sqrt{-g}j^b$ and $(\sqrt{-g}*F^{ab})_{,a} = 0$ with respect to θ, using the identities $\sqrt{-g}F^{tr} = *F_{\theta\phi}$ and $\sqrt{-g}*F^{tr} = -F_{\theta\phi}$, to obtain $\sqrt{-g}F^{r\theta} = -2I_H$, $\sqrt{-g}*F^{r\theta} = -\dot{A}_\phi$. Freely falling observers are described by a velocity four-vector u^b; their motion conserves not only energy and angular momentum, but also $C = K_{ab}u^a u^b$, where $K_{ab} = 2\rho^2 l_{(a}n_{b)} + r^2 g_{ab}$ is a Killing tensor in terms of the principle null vectors $l^a \sim \Delta^{-1}(r^2 + a^2)[\eta^a + \Omega_H k^a - \Delta(\partial_r)^a]$ and $n^a \sim (2\rho^2)^{-1}(r^2 + a^2)[\eta^a + \Omega_H k^a + \Delta(\partial_r)^a]$ in the limit as one approaches the horizon[579, 104]. Here, $\Delta = r^2 + a^2 - 2Mr$ and $\rho^2 = r^2 + a^2\cos^2\theta$. Freely falling observers, therefore, satisfy $u_t + \Omega_H u_\phi - \Delta u_r/(r^2 + a^2) = o(1)$ upon approaching the horizon and, hence,

$$w := (-u^r/u^t)g_{\theta\theta}g_{rr}/\sqrt{-g} \sim (r_H^2 + a^2\cos^2\theta)/[(r_H^2 + a^2)\sin\theta] \qquad (A.10)$$

in this limit. Notice that $u_t = o(1)$ upon approaching the horizon corresponds to $u^t\alpha^2$[423, 513, 444, 333].

The angular momentum J_H of the black hole evolves, upon neglecting radiative losses, according to

$$\dot{J}_H = -2\pi\int_0^\pi \sqrt{-g}T_\phi^r d\theta = -\frac{1}{2}\int_0^\pi \sqrt{-g}F^{rc}F_{\phi c}d\theta, \qquad (A.11)$$

where $4\pi T_{ab} = F_a^{\cdot c}F_{ac} - g_{ab}F^{cd}F_{cd}/4$ is the energy momentum tensor of the electromagnetic field. We expand $F^{rc}F_{\phi c} = F^{r\theta}F_{\phi\theta} + F^{rt}F_{\phi t} = F^{r\theta}F_{\phi\theta} + *F^{r\theta}*F_{\phi\theta}$. By the θ-integral form of Maxwell's equations given above, the first equation in (A.8) follows with the surface integral (A.9) interpreted as the angular momentum in the surrounding electromagnetic field. To evaluate the rate of change of black hole mass

$$\dot{M}_H = 2\pi\int_0^\pi \sqrt{-g}T_t^r d\theta = \frac{1}{2}\int_0^\pi \sqrt{-g}F^{rc}F_{tc}d\theta, \qquad (A.12)$$

we expand $F^{rc}F_{tc} = F^{r\theta}F_{t\theta} + F^{r\phi}F_{t\phi} = F^{r\theta}F_{t\theta} + *F^{r\theta}*F_{\theta t}$. The components $F_{t\theta}$ and $*F_{t\theta}$ in the right-hand side must be expressed in the surface quantities at hand. The poloidal electric and magnetic field seen by FFOs must be finite: $F_{\theta b}u^b \sim u^t\left(F_{\theta t} + \Omega_H F_{\theta\phi} - w\sqrt{-g}F^{\theta r}\right) = O(1)$ and, hence, $F_{t\theta} \sim -\Omega_H F_{\theta\phi} - w\sqrt{-g}F^{r\theta}$ upon approaching the horizon; likewise, for $*F_{\theta t}$. This gives $\dot{M}_H = 2\|\dot{\Sigma}_H + I\|^2 + \frac{1}{2}\|\dot{A}_\phi\| + \Omega_H\dot{J}_H$ and, combined with (A.7), the second equation in (A.8) in the norm

$$\|f\|^2 = \int_0^\pi f^2(\theta)w(\theta)d\theta; \qquad (A.13)$$

$\dot{M} - \dot{M}_H = [V]_\infty^H \dot{q}$ defines the chemical potential of a charge δq on the black hole.

Appendix C. Spin–orbit coupling in the ergotube

In what follows, the metric is used with signature $(-, +, +, +)$ and expressed in geometrical units with $G = c = 1$ (hence M [cm] and time [cm]), while natural units are used for all other quantities (m_e[1 cm^{-1}], $e = \{4\pi\alpha\}^{1/2}$[1], and B[cm^{-2}]). Hence, $B_c = m_e^2 c^3/e\hbar = 4.414 \times 10^{13}$ G or $m_e^2/e = 2.21 \times 10^{21}$ cm^{-2} with numerical conversion factor $\{4\pi\hbar c\}^{1/2}$. The conversion factor for power [cm^{-2}] to power [erg s^{-1}] is $\hbar c^2 \sim 0.945 \times 10^{-6}$. For a general account of field theory, see[278, 56, 270].

We consider a black hole in an axisymmetric magnetic field B parallel to the axis of rotation, equilibrated to its lowest energy state by accumulation of a Wald charge (Chapter 13). The wave functions of charged particles can be expanded locally in coordinates (ρ, ϕ, s, t) as

$$e^{-i\omega t} e^{i\nu\phi} e^{ip_s s} \psi(\rho), \tag{A.14}$$

where s denotes arclength along the magnetic field. Comparison with the theory of plane-wave solutions[278] gives a localization on the νth flux surface at which

$$g_{\phi\phi}^{1/2} = \sqrt{2\nu/eB} \tag{A.15}$$

with Landau levels $E_{n\alpha} = \{m_e^2 + p_s^2 + |eB|(2n+1-\alpha)\}^{1/2}$, where m_e is the electron mass and $\alpha = \pm 1$ refers to spin orientation along B. These states enclose a flux

$$A_\phi = \frac{1}{2}Bk^2 = \nu/e. \tag{A.16}$$

Here, the angular momentum ν refers to the azimuthal phase velocity of the charged particles. It will be appreciated that these Landau states have zero canonical angular momentum ($\pi_\phi = 0$). This corresponds to the lowest energy state on orbits enclosing a fixed magnetic flux, as can be seen by explicitly solving the full Dirac question[278] in cylindrical coordinates. Note further that these orbital

Landau states have effective cross-sections $\Sigma_\nu = 2\pi/|eB|$. The gauge-covariant frequency of the Landau states near the horizon follows from

$$-\xi^a(i^{-1}\partial_a + eA_a)\psi = (\omega - \nu\Omega_H)\psi. \tag{A.17}$$

The jump

$$V_F = [-\xi^a(i^{-1}\partial_a + eA_a)]_\infty^H \psi = \nu\Omega_H \tag{A.18}$$

between the horizon and infinity defines the Fermi level of the particles at the horizon. In contrast, the Wald field about an uncharged black hole has $V_F = \nu\Omega_H - eaB_0$, which shows that it is out of electrostatic equilibrium. Note that the canonical angular momentum of the Landau states vanishes: $k^a\hat{\pi}_a\psi = (i^{-1}\partial_\phi - eA_\phi)\psi = 0$. (This corresponds to the lowest energy state on orbits enclosing a fixed magnetic flux, as can be seen by solving the full Dirac equation[278] in cylindrical coordinates. These orbital Landau states have effective cross-sectional areas $\Sigma_\nu = 2\pi/|eB|$.) The Fermi level (A.18) combines the spin coupling of the black hole to the vector potential A_a and the particle wave function ψ. The equilibrium state in the sense of $\partial_t q \sim 0$, or at most $q/\partial_t q \sim a/\partial_t a$, derives from this complete V_F. For this reason, we shall study the state of electrostatic equilibrium as an initial condition, to infer aspects of the late time evolution.

The strength of the spin–orbit coupling which drives a Schwinger-type process on the surfaces of constant flux may be compared with the spin coupling to the vector potential A_a. The latter can be expressed in terms of the EMF_ν over a loop which closes at infinity and extends over the axis of rotation, the horizon and the νth flux surface with flux Ψ_ν. Thus, we have $EMF_\nu = \Omega_H\Psi_\nu/2\pi[64, 534]$, which gives rise to the new identity

$$eEMF_\nu = \nu\Omega_H. \tag{A.19}$$

It should be mentioned that (A.19) continues to hold away from electrostatic equilibrium (i.e. $q \neq 2BJ$), since $\xi^a A_a = 0$ and, hence, $\nu - eA_\phi = 0$ on the horizon. Since the latter is a conserved quantity, it, in fact, continues to hold everywhere in the Wald field approximation.

In the assumed electrostatic equilibrium state, $\xi^a A_a = 0$, and the generalization of (A.19) to points (s, ν) away from the horizon is

$$[-\xi^a(i^{-1}\partial_a + eA_a)]_\infty^{(s,\nu)}\psi = -\nu\frac{g_{t\phi}}{g_{\phi\phi}}(s,\nu) = -\frac{1}{2}eBg_{t\phi}(s,\nu) = -eA_t(s,\nu) \tag{A.20}$$

for particles of charge $-e$. Thus, (A.20) localizes (A.19) by expressing the coupling of the black hole spin to the wave functions in terms of the electrostatic potential $V = A_t$ in Boyer–Linquist coordinates. Note that the zero angular momentum observers move along trajectories of zero electric potential.

Appendix D. Pair creation in a Wald field

The action of a gravitational field is perhaps most dramatic in the case of pair creation. Pair creation results in response to large gradients in a potential energy.

A formal calculation scheme for pair creation in curved spacetime is based on wavefront analysis. This is well-defined between asymptotically flat in- and out-vacua in terms of their Hilbert spaces of radiative states. Any jump in the zero energy levels of these two Hilbert spaces becomes apparent by studying the propagation of wavefronts between the in- and out-vacuum[153, 56]. It is perhaps best-known from the Schwinger process[388, 158, 144, 157] and in dynamical spacetimes in cosmological scenarios[56]. The energy spectrum of the particles is ordinarily nonthermal, with the notable exception of the thermal spectrum in Hawking radiation from a horizon surface formed in gravitational collapse to a black hole[254].

There are natural choices of the asymptotic vacua in asymptotically flat Minkowski spacetimes, where a timelike Killing vector can be used to select a preferred set of observers. This leaves the in- and out-vacua determined up to Lorentz transformations on the observers and gauge transformations on the wavefunction of interest. These ambiguities can be circumvented by making reference to Hilbert spaces on null trajectories – the past and future null infinities $\mathcal{J}^{\pm}$ in Hawking's proposal – and by working with gauge-covariant frequencies. The latter received some mention in Hawking's original treatise[254], and is briefly as follows.

Hawking radiation derives from tracing wavefronts from J^+ to J^-, past any potential barrier and through the collapsing matter, with subsequent Bogolubov projections on the Hilbert space of radiative states on J^-. This procedure assumes gauge covariance, by tracing wavefronts associated with gauge-covariant frequencies in the presence of a background vector potential A_a. The generalization to a rotating black hole obtains by taking these frequencies relative to real zero-angular momentum observers (ZAMOs), whose worldlines are orthogonal to the azimuthal Killing vector as given by $\xi^a \partial_a = \partial_t - (g_{t\phi}/g_{\phi\phi})\partial_\phi$. Then $\xi^a \sim \partial_t$ at infinity and $\xi^a \partial_a$ assumes corotation upon approaching the horizon, where g_{ab} denotes the Kerr metric. This obtains consistent particle–antiparticle conjugation by complex conjugation among all observers, except for the interpretation of a particle or an antiparticle. Consequently, Hawking emission from the horizon of a rotating black hole gives rise to a flux to infinity

$$\frac{d^2 n}{d\omega dt} = \frac{1}{2\pi} \frac{\Gamma}{e^{2\pi(\omega - V_F)/k} + 1}, \tag{A.21}$$

for a particle of energy ω at infinity. Here, $k = 1/4M$ and Ω_H are the surface gravity and angular velocity of the black hole of mass M, Γ is the relevant absorption factor.

The Fermi level V_F derives from the (normalized) gauge-covariant frequency as observed by a ZAMO close to the horizon, namely, $\omega - V_F = \omega_{ZAMO} + eV = \omega - \nu\Omega_H + eV$ for a particle of charge $-e$ and azimuthal quantum number ν, where V is the potential of the horizon relative to infinity. The results for antiparticles (as seen at infinity) follow with a change of sign in the charge, which may be seen to be equivalent to the usual transformation rule $\omega \rightarrow -\omega$ and $\nu \rightarrow -\nu$.

In case of $V = 0$, Hawking radiation is symmetric under particle–antiparticle conjugation, whereby Schwarzschild or Kerr black holes in-vacuo show equal emission in particles and antiparticles. For a Schwarzschild black hole, then, the resulting luminosity of (A.21) is thermal with Hawking temperature $T \sim 10^{-7}(M_\odot/M)K$, which is negligible for black holes of astrophysical size[407, 517]. The charged case forms an interesting exception, where the Fermi level $-eV$ gives rise to spontaneous emission by which the black hole equilibrates on a dynamical timescale[231, 521, 144]. In contrast, the Fermi level $\nu\Omega_H$ of a rotating black hole acting on neutrinos is extremely inefficient in producing spontaneous emission at infinity[542]. This is due to an exponential cutoff due to a surrounding angular momentum barrier, which acts universally on neutrinos independent of the sign of their orbital angular momentum. This illustrates that (A.21) should be viewed with two different processes in mind: (a) nonthermal spontaneous emission in response to a nonzero Fermi-level and (b) thermal radiation beyond[254].

Upon exposing a rotating black hole to an external magnetic field, this radiation picture is expected to change, particularly in regard to V_F and the absorption coefficient Γ. The radiative states are now characterized by conservation of magnetic flux rather than conservation of particle angular momentum, which has some interesting consequences.

The particle outflow derives from the distribution function (A.21) by calculation of the transmission coefficient through a barrier in the so-called level-crossing picture[150]. The WKB approximation (e.g. as derived by ZAMOs) gives the inhomogeneous dispersion relation

$$(\omega - V_F)^2 = m_e^2 + |eB|(2n+1-\alpha) + p_s^2, \tag{A.22}$$

where $V_F = V_F(s, \nu)$ is the s-dependent Fermi level on the νth flux surface. The classical limit of (A.22) is illustrative, noting that the energy ϵ of the particle is always the same relative to the local ZAMOs that it passes. Indeed, since $w^a(ma_a - eA_a)$ is conserved when w^a is a Killing vector[577], $\eta^a(mu_a - eA_a) = \pi_t$ and $k^a(mu_a - eA_a) = \pi_\phi$ are constants of motion, where u^a is the four-velocity of the guiding center of the particle, and $\pi_t = E_{n\alpha}$, $\pi_\phi = 0$ in a Landau state. With $\xi^a A_a = 0$, $\epsilon = -\xi^a mu_a = -\xi^a(mu_a - eA_a) = -\eta^a(mu_a - eA_a) = \pi_t$. This conservation law circumvents discussions on the role of $E \cdot B$ (generally nonzero in a Wald field). The energy of the particle relative to infinity is ω. This relates to the energy ϵ as measured by the ZAMOs following a shift $V_F(s, \nu)$ due to their angular velocity. Thus, (A.22) pertains to observations in ZAMO frames, but is expressed in terms of the energy at infinity ω. It follows that particle–antiparticle pair creation (as in pair creation of neutrinos[542]) is set by

$$\eta = |\partial V_F/\partial_s| \sim \left\| \partial_r \left(\frac{1}{2} eB g_{t\phi} \right) \right\| = |\partial_r(eA_t)| = eBaM \frac{r^2 - a^2 \cos^2 \theta}{(r^2 + a^2 \cos^2 \theta)^2} \sin^2 \theta, \tag{A.23}$$

using $\partial_s \sim \partial_r$. Radiation states at infinity are separated from those near the horizon by a barrier where $p_s^2 < 0$ about $V_F(s_0) = \omega$. The WKB approximation gives the transmission coefficient

$$|T_{n\alpha}|^2 = e^{-\pi[m_e^2 + |eB|(2n+1-\alpha)]/\eta}. \tag{A.24}$$

Since the Wald field B is approximately uniform, any additional magnetic mirror effects can be neglected. Also, $\eta \leq \frac{1}{8} eB(M/a) \tan^2 \theta \leq \frac{1}{4} eB$ and $|eB|(2n+1-\alpha)/\eta \geq 4(2n+1-\alpha)$, so that T is dominated by $n=0$ and $\alpha=1$.

By (A.20), the pair production rate by the forcing η in (A.23) can be derived from the analogous results for the pair production rate produced by an electric field E along B. The results from the latter[146, 144] imply a production rate $\dot{N}$ of particles given by

$$\dot{N} = \frac{e}{4\pi^2} \int \frac{\eta B e^{-\pi m_e^2/\eta}}{\tanh(\pi eB/\eta)} \sqrt{-g} d^3 x \sim \frac{e^2 B^2 Ma}{2\pi} \int \frac{r^2 - a^2 \cos^2 \theta}{r^2 + a^2 \cos^2 \theta} e^{-\pi m_e^2/\eta} \sin^3 \theta dr d\theta. \tag{A.25}$$

Here $1/\eta \sim (eBaM \sin^2 \theta)^{-1}(8a^2 + 12(r - \sqrt{3}a\cos\theta)^2)$ about $r = \sqrt{3}a\cos\theta$. For a rapidly spinning black hole, $\sqrt{3}a\cos\theta$ is outside the horizon in the small angle approximation, whereby after r-integration of (A.25) we are left with

$$\dot{N} \sim \frac{e^2 B^2 a^2 M}{8\pi\sqrt{3}c} \int e^{-8\pi c/\sin^2\theta} \sin^4\theta d\theta \sim \frac{N_H^2}{128\sqrt{3}\pi^2 M} \left(\frac{a}{M}\right)^4 c^{-7/2} e^{-8\pi c/\theta^2} \theta^7$$

(A.26)

asymptotically as $8\pi c/\theta^2 \gg 1$. Here, $c = m_e^2 a/eBM$, $N_H = m_e^2 M^2$ is a characteristic number of particles on the horizon, and θ is the half-opening angle of the outflow. The right-hand side of (A.26) forms a lower limit in case of $8\pi c/\theta \le 1$. When $a \sim M$, N_H/c is characteristic for the total number of flux surfaces ν_* which penetrate the horizon and $c \sim B_c/B$, where $B_c = 4.4 \times 10^{13}$ G is the field strength which sets the first Landau level at the rest mass energy. By (A.20) and (A.25), a similar calculation obtains for the luminosity in particles L_p normalized to isotropic emission the asymptotic expression valid for small opening angles, given by

$$L_p' = \frac{L_p}{\theta^2/2} \sim \frac{\sqrt{3}}{2} eBM\dot{N}.$$

(A.27)

This calculation shows that black-hole spin initiates pair production spontaneously for superstrong magnetic fields. An open magnetic flux tube hereby is continuously replenished with charged particles which, subsequently, will pair-produce through canonical cascade processes such as curvature radiation.

A saturation of (A.29) follows by nondissipative and dissipative backreactions. The magnetic field diminishes by azimuthal currents from charged particles, and the horizon potential V_F diminishes due to a finite impedance of 4π of the horizon surface[534]. This backreaction goes beyond the zero-current approximation in the Wald field solution. The resulting bound on the outflow satisfies

$$4\pi e\dot{N} < \nu\omega_H,$$

(A.28)

up to a logarithmic factor of order $\ln(\pi/2\theta)$, where ν is taken at the half-opening angle θ of the outflow.

Note that this bound holds true regardless of the state of the ergotube, whether perturbative about the vacuum Wald-field or approximately force-free.

The saturated isotropic luminosity (A.29) hereby satisfies

$$L_p' \simeq \left(10^{48}\frac{\text{erg}}{\text{sec}}\right)\left(\frac{B}{B_c}\right)^2\left(\frac{M}{M_\odot}\right)^2 \sin^2\theta.$$

(A.29)

This holds for a broad range of values of θ, upon appealing to canonical pair creation processes to circumvent the minimum angle $\theta_0 \sim \sqrt{B_c/3B}$, that arises from vacuum breakdown alone.

For closely related discussions on pair-creation around rotating black holes, the reader is referred to[261, 294]. At the classical field level, the results are a manifestation of the energetic coupling $E = \omega J$ of frame-dragging ω to the angular momentum $J = eA_\phi$ of charged particles, as discussed in Chapter 12.

Appendix E. Black hole spacetimes in the complex plane

The known solutions of black holes in asymptotically flat spacetimes are analytic at infinity (a function $f(z)$ is analytic at infinity iff $f(1/z)$ is analytic at $z = 0$). The singularities in these spacetimes may be viewed to be a consequence of Liouville's theorem. The cosmic censorship conjecture poses that these singularities are located within an event horizon.

Spacetimes that are analytic at infinity allow for an expansion

$$g_{ab}(x^a + \xi^a/s) = \eta_{ab} + s g_{ab}^{(1)}(x^a) + s^2 g_{ab}^{(2)}(x^a) + \cdots \qquad (\text{A.30})$$

for any choice of spacelike ξ^a for any choice of complex number s.

Schwarzschild black holes of mass M can be described in spherical coordinates by the line element

$$ds^2 = -\left(1 - \frac{2M}{r}\right)dt^2 + \frac{r}{r - 2M}dr^2 + r^2 d\Omega, \qquad (\text{A.31})$$

where $d\Omega = (d\theta^2 + \sin^2\theta d\phi^2)$ denote the surface element on the unit sphere. Based on

$$\tilde{r} = \frac{1}{2}\left[-M + r\left(1 + \sqrt{1 - \frac{2M}{r}}\right)\right], \quad r = \tilde{r}\left(1 + \frac{M}{2\tilde{r}}\right)^2, \qquad (\text{A.32})$$

the equivalent line element in isotropic coordinates is[577]

$$ds^2 = g_{ab}dx^a x^b = -\frac{(1 - M/2\tilde{r})^2}{(1 + M/2\tilde{r})^2}dt^2 + \left(1 + \frac{M}{2\tilde{r}}\right)^2 (d\tilde{x}^2 + d\tilde{y}^2 + d\tilde{z}^2), \quad (\text{A.33})$$

where $\tilde{r}^2 = \tilde{x}^2 + \tilde{y}^2 + \tilde{z}^2$. At large distances, (A.33) explicitly recovers the Minkowski metric in $(t, \tilde{x}, \tilde{y}, \tilde{z})$ at large distances,

$$ds^2 = -dt^2 + d\tilde{x}^2 + d\tilde{y}^2 + d\tilde{z}^2 + O(1/\tilde{r}). \qquad (\text{A.34})$$

This provides a starting point for the s-expansion about infinity.

According to the above, we consider a shift $g_{ab}(\tilde{x}+s^{-1}, \tilde{y}, \tilde{z})$, where s is a complex number. Explicitly, we have

$$
g_{ab} = \begin{pmatrix} -1 & 0 & 0 & 0 \\ 0 & 1 & 0 & 0 \\ 0 & 0 & 1 & 0 \\ 0 & 0 & 0 & 1 \end{pmatrix} + s \begin{pmatrix} 2M & 0 & 0 & 0 \\ 0 & -2M & 0 & 0 \\ 0 & 0 & -2M & 0 \\ 0 & 0 & 0 & -2M \end{pmatrix} \tag{A.35}
$$

$$
+ s^2 \begin{pmatrix} -2M\tilde{x}-2M^2 & 0 & 0 & 0 \\ 0 & -2M\tilde{x}+\frac{3}{2}M^2 & 0 & 0 \\ 0 & 0 & -2M\tilde{x}+\frac{3}{2}M^2 & 0 \\ 0 & 0 & 0 & -2M\tilde{x}+\frac{3}{2}M^2 \end{pmatrix}
$$

$$
+ s^3 g_{ab}^{(3)} + O(s^4) \tag{A.36}
$$

where

$$
g_{ab}^{(3)} = \begin{pmatrix} g_{tt}^{(3)} & 0 & 0 & 0 \\ 0 & g_{\tilde{r}\tilde{r}}^{(3)} & 0 & 0 \\ 0 & 0 & g_{\tilde{r}\tilde{r}}^{(3)} & 0 \\ 0 & 0 & 0 & g_{\tilde{r}\tilde{r}}^{(3)} \end{pmatrix}, \tag{A.37}
$$

$$
g_{tt}^{(3)} = 2M(\tilde{x}^2 - \frac{1}{2}\tilde{y}^2 - \frac{1}{2}\tilde{z}^2) - M(-M\tilde{x}+\frac{1}{4}M^2)
$$

$$
+ (M\tilde{x}+\frac{3}{4}M^2)M + M(M\tilde{x}+\frac{3}{4}M^2) +
$$

$$
g_{\tilde{r}\tilde{r}}^{(3)} = 2M(\tilde{x}^2 - \frac{1}{2}\tilde{y}^2 - \frac{1}{2}\tilde{z}^2) - M^2\tilde{x} + 2M(-M\tilde{x}+\frac{1}{4}M^2). \tag{A.38}
$$

The s-expansion (A.30) is a consequence of the analytic structure of general relativity: the Einstein equations are quadratic functions of the metric and its derivatives with constant coefficients. The singularities that spacetimes do have are concentrated near the real axis of the coordinates, representing a finite amount of mass M. We may extend the cosmic censorship conjecture to the complex plane to entail that all singularities are confined to a strip about the real axis ($s = \infty$) of width $2M$. In contrast, an essential singularity at infinity appear only in the approximation of a continuous radiation to infinity.

The s-expansion further shows that strongly nonlinear general relativity ($s \to \infty$) is analytically connected to weakly nonlinear relativity ($s \to 0$). It would be of interest to consider numerical relativity for $\mathrm{Im}(s) > 0$ in the weakly nonlinear regime as a means of studying the problem of black hole–black hole coalescence. Notice that $\mathrm{Im}(s) \neq 0$ suffices to avoid coordinate singularities with horizon surfaces.

It will be appreciated that the s-expansion (Eqn (A. 34)) is not uniformly valid for all $\tilde{x}$. An alternative expansion can be written in the small parameter M/L, where L is the box size corresponding to the distance between the source and the observer. More generally, a globally valid *non-singular* formulation for the initial value problem of black-hole spacetimes (e.g., for calculating gravitational radiation produced by a binary of two black holes) obtains in the form of the vacuum Einstein equations on a four-volume in the complex plane:

$$G_{ab} = 0 \quad \text{on} \quad z^a = x^a + iy^a, \quad y^2 > M^2, \tag{A.39}$$

where M denotes the total mass-energy of the spacetime. The initial data for this problem follow by analytic continuation of physical initial data on the real line $y_a = 0$ to $y_2 > M^2$ and, at the end of the computation, the desired gravitational waves follow from analytic continuation of the results on $y_2 > M^2$ back to the real line $y_a = 0$.

Appendix F. Some units, constants and numbers

Table A.1 *Physical constants*

Black body constant	$a = \pi^2 k^4 / 15 c^3 h^3 = 7.56 \times 10^{-15}$ erg cm^{-3} K^{-4}
Stefan–Boltzmann constant	$\sigma = \pi^2 k^4 / 60 \hbar^3 c^2 = 5.67 \times 10^{-5}$ g s^{-3} K^{-4}
Bekenstein–Hawking entropy	$S_H / A = kc^3 / 4G\hbar = 1.397 \times 10^{49}$ cm^{-2}
Bohr radius	$a_0 = \hbar^2 / m_e e^2 = 0.529 \times 10^{-8}$ cm
Boltzman constant	$k = 1.38 \times 10^{-16}$ erg K^{-1}
	$1/k = 1160$ K eV^{-1}
Critical magnetic field	$B_c = m_e^2 c^3 / e\hbar = 4.43 \times 10^{13}$ G
Compton wavelength	$\lambda_c / 2\pi = \hbar / m_e c = 3.86 \times 10^{-11}$ cm
Velocity of light	$c = 2.99792458 \times 10^{10}$ cm s^{-1}
Newton's constant	$G = 6.67 \times 10^{-8}$ cm^{-3} g^{-1} s^{-2}
	$\kappa = (16\pi G / c^4) = 2.04 \times 10^{-24}$ s cm$^{-1/2}$g$^{-1/2}$
Planck's constant	$\hbar = 1.05 \times 10^{-27}$ erg s^{-1}
Planck energy	$E_p = l_p c^4 / G = 2.0 \times 10^{16}$ erg $= 1.3 \times 10^{19}$ GeV
Planck density	$\rho_p = l_p^{-2} c^2 / G = 5.2 \times 10^{93}$ g cm^{-3}
Planck length	$l_p = (G\hbar / c^3)^{1/2} = 1.6 \times 10^{-33}$ cm
Planck mass	$m_p = l_p c^2 / G = 2.2 \times 10^{-5}$ g
Planck temperature	$T_p = E_p / k = 1.4 \times 10^{32}$ K
Planck time	$t_p = l_p / c = 5.4 \times 10^{-44}$ s
Electron charge	$e = 4.80 \times 10^{-10}$ esu
Electron volt	$1\,\mathrm{eV} = 1.60 \times 10^{-12}$ erg
Electron mass	$m_e = 9.11 \times 10^{-28}$ g
	$m_e c^2 = 0.511$ MeV
Fine structure constant	$\alpha = e^2 / \hbar c \simeq 1/137$
Proton mass	$m_p = 1.67 \times 10^{-24}$ g
	$m_p c^2 = 938.2592(52)$ MeV
Neutron mass	$m_n c^2 = 939.5527(52)$ MeV
	$= m_p c^2 + 2.31 \times 10^{-27}$ g
	$= m_p c^2 + 1.29$ MeV$/c^2$
Rydberg constant	$m_e e^4 / 2\hbar^2 = 13.6$ eV
Thomson cross-section	$8\pi e^4 / 3 m_e^2 c^4 = 0.665 \times 10^{-24}$ cm^2

Table A.2 *Astronomical constants*

1 second of arc (')	$= 4.85 \times 10^{-6}$ rad.
1 astronomical unit (AU)	$= 1.50 \times 10^{13}$ cm
1 light year (ly)	$= 0.946 \times 10^{18}$ cm
1 parsec (pc)	$= 3.26$ ly $= 3.09 \times 10^{18}$ cm

(A.37)

Table A.3 *Selected supernovae*

Supernova	Type	Reference
SN1983N	Ib	[125]
SN1984L	Ib	[186]
SN1987A	II	[422, 266, 268]
SN1987K	Ib/c	[186]
SN1987L	Ia	[186]
SN1987N	Ia	[186]
SN1990B	Ib/c	[546]
SN1990I	Ib	[539]
SN1991T	Ia	[186]
SN1993N	II	[125]
SN1993J	IIb	[9, 186, 367]
SN1994I	Ic	[125]
SN1996X	Ia	[471]
SN1997B	Ic	[539]
SN1998bw	Ic	[224, 539, 341]
SN1998L	Ib/c	[186]
SN1999dn	Ib	[539]
SN1999em	IIP	[248]
SN1999gi	II	[539]
SN2002lt	Ic	[154]
SN2003lw	Ic	[512]
SN2003dh	Ib/c	[506]

References

[1] Abbott, B. & the LIGO/GEO collaboration, 2004, *Nucl. Instrum. Meth. Phys. Res.* A, **517**, 154.

[2] Abbott, B., Abbott, R. & Adhikara, R. *et al.* (1992) *Science*, **292**, 325.

[3] Abrahams, A., Anderson, A., Choquet-Bruhat, Y., & York, Jr., J. W. (1995) *Phys. Rev. Lett.*, **75**, 3377.

[4] Acernese, F., *et al.* (2002) *Class. Quant. Grav.*, **19**, 1421.

[5] Adams, S. W. (1915) *Pub. Astron. Soc. Pac.*, **27**, 236.

[6] Aguirre, A. (2000) *ApJ*, **529**, L9.

[7] Akiyama, S., Wheeler, J. C. Meier, D. L. & Lichtenstadt, I. (2003) *ApJ*, **584**, 954.

[8] Alcubierre, M., Brügmann, B., Miller, M. & Suen, W.-M. (1999) *Phys. Rev. D.*, **60**, 4017.

[9] Aldering, G., Humphreys, R. M. & Richmond, M. (1994) AJ, **107**, 662.

[10] Allen, B., Romano, J. D. (1999) *Phys. Rev. D.*, **59**, 102001.

[11] Aller, M. F., Aller, H. D. & Hughes, P. A. (2003) *ApJ*, **586**, 33.

[12] Amati, L. (1999) Ph.D. thesis (unpublished).

[13] Amati, L., Frontera, F. & Tavani, M., *et al.* (2002) *A & A*, **390**, 81.

[14] Amati, L., Piro, L. & Antonelli, L.A., *et al.* (1998) *Nucl. Phys. B.*, **69**, 656.

[15] Ando, M. & TAMA Collaboration (2002) *Class. Quant. Grav.*, **19**, 1409.

[16] Anile, A. M. (1989) *Relativistic Fluids and Magneto-fluids*. Cambridge: Cambridge University Press.

[17] Antonelli, L. A., Piro, L. & Vietri, M., *et al.* (2000) *ApJ*, **545**, L39.

[18] Apostolatos, T. A., Cutler, C., Sussman, G. J. & Thorne, K. S. (1994) *Phys. Rev. D*, **49**, 6274.

[19] Arnowitt, R., Deser, R. & Misner, C. W. (1962) in *Gravitation: an introduction to current research*, ed. L. Witten. New York, Wiley, p. 227.

[20] Ashtekar, A. (1986) *Phys. Rev. Lett.*, **57**, 2244.

[21] Ashtekar, A. (1987) *Phys. Rev. D.*, **36**, 1587.

[22] Ashtekar, A. (1991) *Lectures on Non-Perturbative Canonical Gravity* Singapore: World Scientific.

[23] Ashtekar, A., Romano J. D. & Tate R. S. (1989) *Phys. Rev. D.*, **40**, 2572.

[24] Astone, P., Bassan, M. & Bonifazi, P., *et al.* (2002) *Phys. Rev. D.* (2002) **66**, 102002.

[25] Ayal, S., & Piran, T. (2001) *ApJ*, **555**, 23.

[26] Bahcall, J. N., Kirhakos, S. & Schneider, D. P., *et al.*, (1995) *ApJ*, **452**, L91.

[27] Band, D. L., Matteson, J. & Ford, L., *et al.* (1993) *ApJ*, **413**, 218.

[28] Barbero, G. J. F. (1985) *Class. Quantum Grav.*, **5**, L143.

[29] Barbero, G. J. F. (1994) *Phys. Rev. D.*, **49**, 6935.

[30] Bardeen, J. M. (1970) *Nature*, **226**, 64.

[31] Bardeen, J. M. & Buchman, L. T. (2002) *Phys. Rev. D.*, **65**, 064037.

[32] Bardeen, J. M., Carter, B. & Hawking, S. W. (1973) *Commun. Math. Phys.*, **31**, 181.

[33] Bardeen, J. M., Press, W. H. & Teukolsky, S. A. (1972) *ApJ* **178**, 347.

[34] Barish, B. & Weiss, R. (1999) *Phys. Today*, **52**, 44.

[35] Barker, B. M. & O' Connell, R. F. (1975) *Phys. Rev. D.*, **12**, 329.

[36] Baumgarte, T. W. & Shapiro, S. L. (1999) *Phys. Rev. D.*, **59**, 024007.

[37] Bazer, J. & Ericson, W. B. (1959) *ApJ*, **129**, 758.

[38] Bechtold, J., Siemiginowska, A., Shields, J. *et al.* (2002) *APJ*, **588**, 119.

[39] Beccaria, M., Bernardini, M., Braccini, S. *et al.* (1998) *Class. Quant. Grav.*, **15**, 3339.

[40] Begleman, M., Blandford, R. D. & Rees, M. J. (1984) *Rev. Mod. Phys.*, **56**, 225.

[41] Bekenstein, J. D. (1973) *ApJ*, **183**, 657.

[42] Bekenstein, J. D. (1973) *Phys. Rev. D.*, **7**, 2333.

[43] Bekenstein, J. D. (1974) *Phys. Rev. D.*, **9**, 3292.

[44] Belczynski, K., Kalogera, V. & Bulik, T. (2002) *ApJ*, **572**, to appear.

[45] Bennett, C. L., Halpern, M. & Hinshaw, G., *et al.* (2003) *ApJS*, **148**, 1.

[46] Bennetti, S., Cappellaro, E. & Turatto, M. (1991) *A & A*, **247**, 410B.

[47] Berger, E., Kulkarni, S. R. & Frail, D. A., *et al.* (2003) *ApJ*, **599**, 408.

[48] Beskin, V. S. (1997) Phys.-Uspekhi, **40**, 659.

[49] Beskin, V. S. (1997) *Usp. Fiz. Nauk*, **167**, 689 (Trans. in Physics – Uspekhi, **40**, 659).

[50] Beskin, V. S. & Kuznetsova, I. V. (2000) *ApJ*, **541**, 257.

[51] Beskin, V. S. & Kuznetsova, I. V. (2000) *Nuovo Cimento*, **115**, 795.

[52] Bethe, H. A. & Brown, G. E. (1998) *ApJ*, **506**, 780.

[53] Bethe, H. A., Brown, G. E. & Lee, C.-H. (2003) *Selected papers: formation and evolution of black holes in the galaxy*. World Scientific, p. 262.

[54] Bildsten, L. (1998) *ApJ*, **501**, L89.

[55] Bionta, R. M. Blewitt, G. & Bratton, C. B., *et al.* (1987) *Phys. Rev. Lett.*, **58**, 1494.

[56] Birell, N. D. & Davies, P. C. W. (1982) *Quantum fields in curved space*. Cambridge: Cambridge University Press.

[57] Biretta, J. A., Zhou, F. & Owen, F. N. (1995) *ApJ*, **447**, 582.

[58] Bisnovatyi-Kogan, G. S. (1970) *Astron. Zh.*, **47**, 813.

[59] Bisnovatyi-Kogan, G. S., Popov, Yu. P., & Samochin, A. A. (1976) *Astrophys. Space Sc* **41**, 321.

[60] Blanchet, L. (2002) *Living Rev. Rel.*, **5**, 3; gr-qc/0202016.

[61] Blandford, R. D. (1976) *MNRAS*, **176**, 465.

[62] Blandford, R. D. & Königl, A. (1979) *ApJ*, **232**, 34.

[63] Blandford, R. D. & Payne, D. G. (1982) *MNRAS*, **199**, 883.

[64] Blandford, R. D. & Znajek, R. L. (1977) *MNRAS*, **179**, 433.

[65] Blandford, R. D., McKee, C. F. & Rees, M. J. (1977) *Nature*, **267**, 211.

[66] Bloom, J. S., Djorgorski, S. G. & Kulkarni, S. R., *et al.* (1998) *ApJ*, **507**, L25.

[67] Bloom, J. S., Djorgovski, S. G. & Kulkarni, S. R. (2001) *ApJ*, **554**, 678.

[68] Bloom, J. S., Kulkarni, S. R. & Djorgovski, S. G. (2002) *Astron. J.*, **123**, 1111.

[69] Bloom, J. S., Kulkarni, S. R. & Djorgovski, S. G., *et al.* (1999) *Nature*, **401**, 453.

[70] Bloom, J. S., Kulkarni, S. R. & Harrison, F., *et al.* (1998) *ApJ*, **506**, L105.

[71] Bona, C. & Massó, J. (1992) *Phys. Rev. Lett.*, **68**, 1097.

[72] Bona, C., Massó, J., Seidel, E. & Stela, J. (1995) *Phys. Rev. Lett.*, **75**, 600.

[73] Bond, H. E. (1997) IAU Circ. No. 6664.

[74] Bonnell, I. A. & Pringle, J. E. (1995) *MNRAS*, **273**, L12.

[75] Borra, E. F., Landstreet, J. D. & Mestel, L. (1982) *ARA & A*, **20**, 191.

[76] Boyer, R. H. & Lindquist, R. W. (1967) *J. Math. Phys.*, **8**, 265.

[77] Brügmann, B. (2000) *Ann. Phys.*, **9**, 227.

[78] Bradaschia, C., del Fabbro & di Virgilio, A., *et al.* (1992) *Phys. Lett. A*, **163**, 15.

[79] Braginsky, V. B., *et al.* (1999) *Phys. Lett. A*, **264**, 1.

[80] Branch, D., Dogget, J. B. & Nomoto, K., *et al.* (1985) *ApJ*, **294**, 619.

[81] Branch, D., *et al.* (2001) in *SNe and GRBs*, eds. M. Livio, N. Panagia and K. Sahu. Cambridge: Cambridge University Press, p. 96.

[82] Bridle, A. H. & Perley, R. A. (1984) *ARA & A*, **22**, 319.

[83] Brodbeck, O., Frittelli, S. & Hübner, P., *et al.* (1999) *J. Math. Phys.*, **909**.

[84] Bromm, J. S. & Loeb, A. (2002) *ApJ*, **575**, 111.

[85] Brown, G. E. & Bethe, H. A. (1994) *ApJ*, **423**, 659.

[86] Brown, G. E., Bethe, H. A. & Lee, C.-H. (2003) *Selected papers: formation and evolution of black holes in the galaxy.* World Scientific, p. 262.

[87] Brown, G. E., Lee, C.-H. & Wijers R. A. M. J., *et al.* (2000) *NewA*, **5**, 191.

[88] Brown, G. E., Weingartner, J. C. & Wijers, R. A. M. J. (1996) *ApJ*, **463**, 297.

[89] Brown, L. F. (1990) *BAAS*, **22**, 1337.

[90] Brown, L. F., Roberts, D. H. & Wardle, J. F. C. (1994) *ApJ*, **437**, 108.

[91] Buchman, L. T. & Bardeen, J. M. (2003) *Phys. Rev. D.*, **67**, 084017.

[92] Burbidge, E. M. (1967) *ARA & A*, **5**, 399.

[93] Burgay, M., D' Amico, N. & Possenti, A., *et al.* (2003) *Nature*, **426**, 531.

[94] Burns, J. O., Norman, M. L. & Clarke, D. A. (1991) *Science*, **253**, 522.

[95] Burrows, A. & Lattimer, J. M. (1987) *ApJ*, **318**, L63.

[96] Butcher, H. R., van Breugel, W. & Miley, G. K. (1980) *ApJ*, **235**, 749.

[97] Camenzind, M. (1990) *Rev. Mod. Astron.*, **3**, 234.

[98] Canuto, C., Hussaini, M. Y., Quarteroni, A. & Zang, T. A. (1988) *Spectral Methods in Fluid Mechanics.* Berlin: Springer-Verlag.

[99] Cappellaro, E. (2004) *Mem. Soc. Astron. Italiana*, **75**, 206.

[100] Cappellaro, E., Barbon, R. & Turatto, M. (2003) IAU 192, *Supernovae: 10 years of 1993J*, Valencia, spain, April 22–6, eds. J. M. Marcaide and K. W. Weiler.

[101] Cappellaro, E., Mazzali, P. A. & Benetti, S., *et al.*, 1997, *MNRAS*, **328**, 203.

[102] Cappellaro, E., Turatto, M. & Tsvetkov, D. Yu. *et al.*, (1997) *A & A*, **322**, 431.

[103] Cardoso, V., Dias, O. J. C., Lemos, J. P. S. & Yoshida, S. (2004) *Phys. Rev. D.*, **70**, 044039.

[104] Carter, B. (1968) *Phys. Rev.*, **174**, 1559.

[105] Carter, B. (1973) in C. DeWitt & B. S. DeWitt eds., *Black holes.* New York: Gordon & Breach, p. 57.

[106] Castro-Tirado, A. J., Gorosabel, J. & Benitez, N., *et al.* (1998) *Science*, **279**, 1011.

[107] Cavallo, G. & Rees, M. J. (1978) *MNRAS*, **183**, 359.

[108] Chan, K. L. & Hendriksen, R. N. (1980) *ApJ*, **241**, 534.

[109] Chandrasekhar, S. (1981) *Hydrodynamic and hydromagnetic stability.* Dover Publications.

[110] Chandrasekhar, S. (1983) *The Mathematical theory of black holes.* Oxford: Oxford University Press.

[111] Chandrasekhar, S. & Esposito, F. P. (1970) *ApJ*, **160**, 153.

[112] Chevalier, R. (1998) *ApJ*, **499**, 810.

[113] Chevalier, R. A. & Li, Z.-Y. (1999) *ApJ*, **520**, L29.

[114] Chiaberge, M., Capetti, A. & Celotti, (2000) *A & A*, **355**, 837.

[115] Choquet-Bruhat Y. (1960) *Acta Astron*, **6**, 354.

[116] Choquet-Bruhat Y. (1966) *Commun. Math. Phys*, **3**, 334.

[117] Choquet-Bruhat, Y. (1994) in T. Ruggeri (ed.) Proc. VIIth Conf. *Waves and stability in continuous media*, Bologna, 1993, World Scientific.

[118] Choquet-Bruhat, Y. (1994) *C. R. Acad. Sci. Paris, Sér. I Math.*, **318**, 775.
[119] Choquet-Bruhat, Y. & York, J. W. (1995) gr-qc/9506071, IEP-UNC-509, TAR-UNC-047 (unpublished).
[120] Choquet-Bruhat, Y., DeWitt-Morette, C. & Dillard-Bleick, M. (1977) *Analysis, manifolds and physics*. Dordrecht: North-Holland.
[121] Chu, Y.-H., Kim, S. & Points, S. D., *et al.* (2000) *ApJ*, **119**, 2242.
[122] Ciufolini, I., Pavils, E. C., Chieppa, F. *et al.* (1998) *Science*, **279**, 2100.
[123] Ciulini, I. & Pavils, E. C. (2004) *Nature*, **431**, 958.
[124] Clarke, D. A., Norman, M. L. & Burns, J. O. (1986) *ApJ*, **311**, L63.
[125] Clocchiatti, A., Wheeler, J. C., & Brotherton, M. S., *et al.*, (1996) *ApJ*, **462**, 462.
[126] Coburn, W. & Boggs, S. E. (2003) *Nature*, **423**, 415.
[127] Cohen, E., Piran, T., & Sari, R., 1998, *ApJ*, 509, 717
[128] Cohen, J. M. & Wald, R. M. (1971) *J. Math. Phys.*, **12**, 1845.
[129] Cohen, J. M., Tiomno, J. & Wald, R. M. (1973) *Phys. Rev. D.*, **7**, 998.
[130] Colgate, S. A. & McKee, C. (1969) *ApJ*, **157**, 623.
[131] Colgate, S. A., Petschek, A. G. & Kriese, J. T. *et al.* (1980) *ApJ*, **237**, L81.
[132] Conway, R. G., Garrington, S. T., Perley, R. A. & Biretta, J. A. (1993) *A & A*, **267**, 347.
[133] Cook, G. B., *et al.* (1998) *Phys. Rev. Lett.*, **80**, 2512.
[134] Copson, E. T. (1928) *Proc. Roy. Soc. London*, **A118**, 184.
[135] Costa, E., *et al.* (1997) *Nature*, **387**, 878.
[136] Cottam, J., Paerels, F. & Mendez, M. (2002) *Nature*, **420**, 51.
[137] Coward, D. M., Burman, R. R. & Blair, D. (2001) *MNRAS*, **324**, 1015.
[138] Coward, D. M., van Putten, M. H. P. M. & Burman, R. R. (2002) *ApJ*, **580**, 1024.
[139] Cutler, C., Apostalatos, T. A. & Bildsten, L. *et al.* (1993) *Phys. Rev. Lett.*, **70**, 2984.
[140] Cutler, C., & Thorne, K. S. (2002) in *Proc. GR16*, Durban, South Africa.
[141] Dado, S., Dar, A. & De Rújula, A. (2002) *A & A*, **388**, 1079.
[142] Dado, S., Dar, A. & De Rújula, A. (2003) *A & A*, **401**, 243.
[143] Daigne, F. & Mochkovitch, R. (1998) *MNRAS*, **296**, 275.
[144] Damour, T. (1976) in R. Ruffini, (ed.) *Proc. 1st Marcel Grossman Meeting on General Relativity*, Amsterdam: North-Holland, p. 459.
[145] Damour, T. (1979) in R. Ruffini, (ed.) *Proc. 2nd Marcel Grossman Meeting on General Relativity*, Amsterdam: North-Holland, p. 587.
[146] Damour, T. & Ruffini, R. (1975) *Phys. Rev. Lett.*, **35**(7), 463.
[147] Danzmann, K., in *First Edoardo Amaldi Conference Gravitation Wave Experiments*. E. Coccia, G. Pizella, F. Ronga (eds.). Singapore: World Scientific, p. 100.
[148] Dar, A. & de Rújula, A. (2003) astro-ph/0308248.
[149] Davies, M. B., King, A. & Rosswog, S. *et al.*, (2002) *ApJ*, **579**, L63.
[150] Davies, P. C. W. & Fulling, S. A. (1977) *Proc. R. Soc. London*, **A356**, 237.
[151] de Bernardis, P., Ade, P. A. R. & Bock, J. J. *et al.* (2000) *Nature*, **404**, 995.
[152] DeWitt, B. S. (1962) in Witten, L. ed., *Gravitation: an introduction to current research*. New York: Wiley & Sons, p. 266.
[153] DeWitt, B. S. (1975) *Phys. Rep.*, **C19**, 297.
[154] Della Valle, M., Maleseni, D. & Benetti, S. *et al.*, (2003) *IAU Circ.* No. 8197.
[155] Della Valle, M., *et al.*, 2003, *A & A*, **406**, 33.
[156] Dendy, R. (ed.), 1993, *Plasma physics: an introductory course*. Cambridge University Press.
[157] Deruelle, N. in R. Ruffini (ed.), *Proc. First Marcel Grossmann Meeting on General Relativity*, edited by Amsterdam: North Holland, 1977, pp. 483–8.
[158] Deruelle, N. & Ruffini, R. (1974) *Phys. Lett.*, **52B**, 437.

[159] Dey, A. & van Breugel, W. J. M. (1994) *AJ*, **107**(6), 1977.
[160] Djorgovski, S. G., Kulkarni, S. R. & Bloom, J. S. *et al.* (1998) *ApJ*, **508**, L17.
[161] Djorgovski, S. G., Metzget, M. R. & Kulkarni, S. R. *et al.* (1997) *Nature*, **387**, 876.
[162] Dokuchaev, V. I. (1987) *Sov. Phys. JETP*, **65**, 1079.
[163] Dubal, M. R. & Pantano, O. (1993) *MNRAS*, **261**, 203.
[164] Duez, M. D., Shapiro, S. L., & Yo, H.-J. (2004) gr-qc/0401076.
[165] Duncan, G. C. & Hughers, P. A. (1994) *ApJ*, **241**, 534.
[166] Duncan, R. C. (2000) astro-ph/0002442.
[167] Eichler, D. & Levinson, A. (2000) *ApJ*, **529**, 146.
[168] Eichler, D. & Levinson, A. (2003) *ApJ*, **596**, L147.
[169] Eichler, D., & Levinson, A. (1999) *ApJ*, **521**, L117.
[170] Eikenberry, S. & van Putten, M. H. P. M. (2003) *ApJ*, submitted.
[171] Ernst, J. F. (1976) *J. Math. Phys.*, **17**, 54.
[172] Ernst, J. F. & Wild, W. J. (1976) *J. Math. Phys.*, **17**, 182.
[173] Estabrook, F. B., Robinson, R. S. & Wahlquist, H. D. (1997) *Class. Quant. Grav.*, **14**, 1237.
[174] Fabian, A. C. (2004) *From X-ray binaries to quasars:black hole accretion on all mass scales*, ed. T. J. Maccarone, R. P. Fender and G. C. Ho. Dordrecht: Kluwer.
[175] Fabian, A. C., Vaughan, S. & Nandra, K. *et al.* (1997) *Nature*, **389**, 261.
[176] Fabian, A. C., Vaughan, S. & Nandra, K. *et al.* (2002) *MNRAS*, **335**, L1.
[177] Fabian, A. C., Rees, M. J., Stella, L. & White, N. E. (1989) *MNRAS*, **238**, 729.
[178] Fanaroff, B. L. & Riley, J. M. (1974) *MNRAS*, **167**, 31*.
[179] Feroci, M., Hurley, K. & Duncan, R. C., *et al.* (2001) *ApJ*, **549**, 1021.
[180] Ferrari, V., Matarrese, S. & Schneider, R. (1999) *MNRAS*, **303**, 247.
[181] Ferrari, V., Matarrese, S. & Schneider, R. (1999) *MNRAS*, **303**, 258.
[182] Ferrari, V., Miniutti, G. & Pons, J. A. (2003) *MNRAS*, submitted; astro-ph/02 10581(v2).
[183] Ferrari A., Trussoni E. & Rosner R., *et al.* (1986) *ApJ*, **300**, 577.
[184] Fierz, M. & Pauli, W. (1939) *Proc. Roy. Soc. Lond.*, **A173**, 211.
[185] Fijimoto, M. K. (2002) priv. comm.
[186] Filippenko, A. V. (1997) *Ann. Rev. Astron. Astrophys.*, **35**, 309.
[187] Filippenko, A. V. (2001) in S. S. Holt & U. Hwang, eds., *Young Supernova remnants*. Conference Proceedings 565, New York: AIP, p. 40.
[188] Finn, L. S., Mohanty, S. D. & Romano, J. D. (1999) *PRD*, **60**, 121101.
[189] Fischer, A. E. & Marsden, J. E. (1972) *Commun. Math. Phys.*, **28**, 1.
[190] Fishman, G. J., Meegan, C. A. & Wilson, R. B. *et al.* (1994) *ApJS*, **92**, 229.
[191] Flanagan, E. & Hughes, S. A. (1998) *Phys. Rev. D.*, **57**, 4535.
[192] Flatters, C. & Conway, R. G. (1985) *Nature*, **314**, 425.
[193] Ford, L. A. (1995) *ApJ*, **439**, 307.
[194] Frail, D. A. Berger, E. & Galama, T., *et al.* (2000) *ApJ*, **538**, L129.
[195] Frail, D. A., Kulkarni, S. R. & Nicastro S. R., *et al.* (1997) *Nature*, **389**, 261.
[196] Frail, D. A., Kulkarni, S. R. & Sari, R., *et al.* (2001) *ApJ*, **562**, L55.
[197] Frail, D. A., Kulkarni, S. R. & Shepherd, D. S., *et al.* (1998) *ApJ*, **502**, L119.
[198] Frail, D. A., *et al.* (1999) in *Proc. Fifth Huntsville meeting on gamma-ray burst symposium*.
[199] Fraix-Burnet, D., Nieto J. L. & Lelièvre G., *et al.*, (1989) *ApJ*, **336**, 121.
[200] Fraix-Burnet, D., Nieto, J.-L. & Poulain, P. (1989), *A & A*, **221**, L1.
[201] Freedman, W. L., Madore, B. F. & Gibson, B. K. *et al.* (2001) *ApJ*, **553**, 47.
[202] Friedman, A. (1922) *Z. Phys.*, **10**, 377.
[203] Friedrichs, K. O. (1974) *Commun. Pure Appl. Math.*, **28**, 749.

[204] Friedrichs, K. O. & Lax P. D. (1971) *Proc. Natl. Acad. Sc. USA*, **68**, 1686.
[205] Fritelli, S. & Reula, O. A. (1994) *Commun. Math. Phys.* **166**, 221.
[206] Fritelli, S. & Reula, O. A. (1996) *Phys. Rev. Lett.*, **76**, 4667.
[207] Fritelli, S. & Reula, O. A. (1999) *J. Math. Phys.*, **40**, 5143.
[208] Frolov V. & Novikov, I. D., (1989) *Black Hole Physics*. Dondrecht: Kluwer.
[209] Frontera, F., Costa, E. & Piro, L. *et al.* (1998) *ApJ*, **493**, L67.
[210] Fruchter, A. S., Pian, E. & Thorsett, S. E. *et al.* (1999) *ApJ*, **516**, 683.
[211] Fruchter, A. S., Thorsett, S. E. & Metzger, M. R. *et al.* (1999) *ApJ*, **519**, L13.
[212] Fryer, C. L., Holz, D. E. & Hughes, S. A. (2002) *ApJ*, **565**, 430.
[213] Fryer, C. L., Holz, D. E. & Hughes, S. A. (2004) astro-ph/0403188.
[214] Fryer, C. L., Woosley, S. E. & Hartman, D. H. (1999) *ApJ* **526**, 152.
[215] Fryer, C. L., Woosley, S. E. & Heger, A. (2001) *ApJ*, **550**, 372.
[216] Gómez R., Lehner, R., Marsa, R. L. *et al.* (1998) *Phys. Rev. Lett.*, **80**, 3915.
[217] Gómez J.-L., Marscher A. P. & Ibáñez, J. M., & Marcaide J. M. (1995) *ApJ*, **449**, L19.
[218] Gómez, J.-L., Mueller E. & Font J. A. *et al.* (1997) *ApJ* **479**, 151.
[219] Gómez, J.-L., Marscher, A. P. & Alberdi, A. *et al.* (2000) *Science*, **289**, 2317.
[220] Gómez, J. L. (2001) in Georganopoulos *et al.*, (eds.) *Proceedings, Mykonos Conference on Relativistic flows in astrophysics*. Springer-Verlag Lecture Notes in Physics; astro-ph/0109338.
[221] Gal-Yam, A., Moon, D.-S. & Fox, D. B. *et al.* (2004) *ApJ*, **609**, 59.
[222] Gal-Yam, A., Poznanski, D. & Maoz, D., *et al.*, (2004) astro-ph/0403296.
[223] Galama, T. J., Tanvir, N. & Vreeswijk, P. M. *et al.* (2000) *ApJ*, **536**, 185.
[224] Galama T. J., Vreeswijk P. M. & van Paradijs, J., *et al.* (1998) *Nature*, **395**, 670.
[225] Garabedian, P. (1986) *Partial Differential Equations*. New York: Chelsea.
[226] Garcia, M. R., Callanan, P. J. & Moraru, D. *et al.* (1998) *ApJ*, **500**, L105.
[227] Gavriil, F. P., Kaspi, V. M. & Woods, P. M. (2002) *Nature*, **419**, 142.
[228] Gertsenshtein, M. E. & Pustovoit, V. I. (1962) *Sov. Phys. – JETP* 14, 433.
[229] Ghisellini, G., Lazatti, D. Rossi, E., & Rees, M. J. (2002) *A & A*, **389**, L33.
[230] Ghisellini, G., Padovani P. Celotti A. & Maraschi, L. (1993) *ApJ*, 407, 65.
[231] Gibbons, G. W. (1975) *Commun. Math. Phys.*, **44**, 245.
[232] Gibbons, G. W. (1976) *MNRAS*, **177**, 37P.
[233] Goedbloed, H., & Keppens, R., 2004, in 12th Int. Congress Plasma Physics, Nice; physics/0411180).
[234] Goldreich, P. & Julian, W. H. (1969) *ApJ*, **157**, 869.
[235] Goldreich, P., Goodman, J. & Narayan, R. (1986) *MNRAS*, **221**, 339.
[236] Goodman, J. (1986) *ApJ*, **308**, L47.
[237] Gotzes, S. (1992) *Acta Phys. Pol. B*, **23**, 433.
[238] Granot, J., Miller, M. & Piran, T., *et al.* (1999) in R. M. Kippen, R. S. Mallozi & G. J. Fishman (eds.) *Gamma-ray Bursts*. Fiftieth Huntsville Symposium (Conference Proceedings 526). New York: AIP (2000), p. 540.
[239] Granot, J., Miller, M. & Piran, T. *et al.* (2001) in *Gamma-ray bursts in the afterglow era*, p. 312.
[240] Groot, P. J., Galama, T. J. & Vreeswijk, P. M., *et al.* (1998) *ApJ*, **502**, L123.
[241] Gruziuov, A. (1999) *A & A* astro-ph/0301536.
[242] Guetta, D. & Piran, T. (2004) *A & A*, subm., astro-ph/0407429.
[243] Guetta, D., Spada, M. & Waxman, E. (2001) *ApJ*, **559**, 101.
[244] Halpern, J. P., Uglesich, R. & Mirabal, N. *et al.* (2000) *ApJ*, **543**, 697.
[245] Hamuy, M. (2003) in C. L. Fryer, ed., *Core collapse of massive stars*. Conference Proceedings 302. Dordrecht: Kluwer Academic Publishers.
[246] Hamuy, M., Phillips, M. M. & Maza, J. *et al.* (1995) *Astron. J.*, **109**, 1.

[247] Hamuy, M., Phillips, M. M. & Schommer, R. A. *et al.* (1996) *Astron. J.*, **112**, 2391.

[248] Hamuy, M., Pinto, P. A. & Maza, J. *et al.* (2001) *ApJ*, **558**, 615.

[249] Hanany, S., Ade, P. & Balbi, A. *et al.* (2000) *ApJ*, **545**, 5.

[250] Hanni, R. S. & Ruffini, R. (1973) *Phys. Rev. D.*, **8**, 3259.

[251] Hardcastle, M. J., Alexander, P., Pooley, C. G. & Riley, J. M. (1996) *MNRAS*, **278**, 273.

[252] Harkness, R. P. & Wheeler, J. C. (1990) in A. G. Petschek, *Supernovae.* (ed.) (New York: Springer-Verlag), p. 1.

[253] Harrison, F. A., Bloom, J. S. & Frail, D. A. *et al.* (1999) *ApJ*, **523**, L121.

[254] Hawking, S. W. (1975) *Commun. Math. Phys.*, **43**, 199.

[255] Hawking, S. W. (1976) *Phys. Rev. D.*, **13**, 191.

[256] Hawley, J. F. (2000) *ApJ*, **528**, 462.

[257] Heise, J., Zand, J., Kippen, R. M. & Woods, P. M. (2000) in E. Costa, F. Frontera and J Hjorth (eds.) *Gamma-ray Bursts in the Afterglow Era*, Rome: CNR (2000) and Berlin/Heidelberg: Springer, p. 16.

[258] Hello, P. (1997) in M. Davier & P. Hello (eds.) *Second Workshop on Gravitational Wave Analysis*. Orsay, France, p. 87.

[259] Hern, S. D. (2000) Ph.D. thesis, Cambridge University, gr-qc/0004036.

[260] HETE-II (2000) http://space.mit.edu/HETE.

[261] Heyl, J. S (2001) *Phys. Rev. D*, **63**, 064028.

[262] Higdorn, J. & Lingenfelter, R. E. (1990) *Ann. Rev. Astron. & Astroph.*, **28**, 401.

[263] Hirata, K., Kajita, T. & Koshiba, M. *et al.* (1987) *Phys. Rev. Lett.*, **58**, 1490.

[264] Hjelming, R. M. & Rupen, M. P. (1995) *Nature*, **375**(8), 464.

[265] Hjorth, J., *et al.*, (2003) *ApJ*, **423**, 847.

[266] Höflich, P. J. (1991) *A & A*, **246**, 481.

[267] Höflich, P., Khokhlov, A. & Wang, L. (2001) in *20th Texas Symposium on Relative Astrophysics*, ed. J. C. Wheeler & H. Martel, Melville, NY: AIP 2001 (Conference Proceedings 586), p. 459; astro-ph/0104025.

[268] Höflich, P., Wheeler, J. C. & Wang, L. (1999) *ApJ*, **521**, 179.

[269] Howell, E., Coward, D. & Burman, R. *et al.* (2004) MNRAS, **351**, 1237.

[270] Huang, K. (1998) Quantum field theory: from operators to path intervals. New York: John Wiley.

[271] Hulse, R. A., & Taylor, J. H. (1975) *ApJ*, **195**, L51.

[272] Hunter, C. (1972) *Ann. Rev. Fl. Dynam.*, 219.

[273] Hurley, K. (2003) priv. comm.

[274] Hurley, K., Costa, E. & Feroci, M. *et al.* (1997) *ApJ*, **485**, L1.

[275] Ibrahim, A. I., Strohmayer, T. E. & Woods, P. M. *et al.* (2001) *ApJ*, **558**, 237.

[276] Iorio, L. (2001), *Class. Quant. Grav.*, **19**, 5473.

[277] Israelian, G., Rebolo, R. & Basri, G. *et al.* (1999) *Nature*, **401**, 142.

[278] Itzykson, C. & Zuber J.-B. (1980) *Quantum Field Theory*. Maidenhead: McGraw-Hill Book Company.

[279] Iwamoto, K. (1999) *ApJ*, **512**, L47.

[280] Iwamoto, K., Mazzali, P. A. & Nomoto, K. *et al.* (1998) *Nature*, **395**, 672.

[281] Iwasawa, K., Fabian, A. C. & Reynolds, C. S. *et al.* (1996) *MNRAS*, **282**, 1038.

[282] Jackson, J. D. (1975) *Classical Electrodynamics*. New York: Wiley.

[283] Jackson, N., Browne I. W. A., Shone, D. L. & Lind K., (1990) *MNRAS*, **244**, 750.

[284] Jackson, N., Sparks, W. B. & Miley, G. K. & Machetto, F., (1993) *A & A*, **269**, 128.

[285] Jacob, S. Barrigá, P. & Blair, D. G., *et al.* (2003) *Publ. Astron. Soc. Aust.*, **20**, 223.

[286] Junor, W., Biretta, J. A. & Livio, M. (1999) *Nature*, **401**, 891.

[287] Kalogera, V., Narayan, R., Spergel, D. N. & Taylor, J. H. (2001) *ApJ*, 556, 340.

[288] Katz, J. I. (1994) *ApJ*, **422**, 248.

[289] Katz, J. I. (1994) *ApJ*, **432**, L107.

[290] Katz, J. I. & Canel, L.M. (1996) *ApJ*, **471**, 915.

[291] Kawabata, K. S., Deng, J. & Wang, L. *et al.* (2003) *ApJ*, **593**, L19.

[292] Keller, H. B. (1987) *Numerical methods in bifurcation problems*. Berlin: Springer-Verlag/Institute for Fundamental Research.

[293] Kerr, R. P. (1963) *Phys. Rev. Lett.*, **11**, 237.

[294] Kim, S. P. & Page, D. N. (2003) Arxiv Preprint hep-th/0301132.

[295] Kippenhahn, R. & Weigert, A. (1990) *Stellar structure and evolution*. New York: Springer-Verlag. p. 178.

[296] Klebesadel, R., Strong I. & Olson R. (1973) *ApJ*, **182**, L85.

[297] Kobayashi, S. & Mészáros, P. (2002) *ApJ*, **585**, L89.

[298] Kobayashi, S., Piran, T. & Sari, R. (1997) *ApJ*, **490**, 92.

[299] Koide, S., Meier, D. L. & Shibata, K. *et al.* (2000) *ApJ*, **536**, 668.

[300] Koide, S., Nishikawa, K. & Mutel, R. L. (1996) *ApJ*, **463**, L71.

[301] Koide, S., Shibata, K. & Kudoh, T. (1998) *ApJ*, **495**, L63.

[302] Koide, S., Shibata, K. & Kudoh, T. *et al.* (2002) *Science*, **295**, 1688.

[303] Koldova, A. V., Kuznetsov, O. A. & Ustyugova, G. V. (2002) *MNRAS*, **333**, 932.

[304] Kommissarov, S. S. (1997) *Phys. Lett. A.*, **232**, 435.

[305] Kouveliotou, C., Meegan, C. A. & Fishman, G. J. *et al.* (1993) *ApJ*, **413**, L101.

[306] Kouveliotou, C., Strohmayer, T. & Hurley, K. *et al.* (1999) *ApJ*, **510**, L115.

[307] Kozai, Y., 1962, *AJ*, **67**, 9

[308] Kraus, L. M. (2004) in *XIV Canary Island Winter School in Astrophysics 2002: Dark matter and dark energy in the universe* (to appear).

[309] Kudritzki, R. (2003) priv. comm.; see http://www.ifa.hawaii.edu/pan-starrs.

[310] Kulkarni, S. R., Berger, E. & Bloom, J. S. *et al.* (2000) in Proc. SPIE, **4005**, 9.

[311] Kulkarni, S. R., Djorgovski, S. G. & Odewahn, S. C. *et al.* (1999) *Nature*, **398**, 389.

[312] Kulkarni, S. R., Frail, D. A. & Sari, R. *et al.* (1999) *ApJ*, **522**, L97.

[313] Kulkarni, S. R., Frail, D. A. & Wieringa, M. H. *et al.* (1998) *Nature*, **395**, 663.

[314] Kundt, W. (1976) *Nature*, **261**, 673.

[315] Lamb, D. Q., Donaghy, T. Q. & Graziani, C. (2003) *ApJ*, astroph/0312634, astro-ph/0312504.

[316] Landau, L. D. (1932) *Phys. z. Sowjetunion*, **1**, 285.

[317] Landau, L.D., & Lifshitz, E.M. (1987), *Fluid mechanics* (New York: Pergamon).

[318] Landau, L. D., & Lifshitz, E. M. (1995) *The classical theory of fields*. Oxford: Butterworth-Heinemann.

[319] Laor, A. (1991) *ApJ*, **376**, 90.

[320] Lawden, D. F. (1989) *Elliptic functions and applications*. New York: Springer-Verlag.

[321] Layzer, D. (1965) *ApJ*, **141**, 837.

[322] Lazzatti, D., Campana, S. & Ghisellini, G. (1999) *MNRAS*, **304**, L31.

[323] Lazzati, D. (2003) 30 Years of Discovery in E. E. Fenimore & M. Galassi (eds.) *Gamma-ray Burst Symposium* (Conference Proceedings 727). New York: AIP (2004), p. 251.

[324] Lazzati, D. (2004) INT workshop on the supernova association to GRBs, Seattle, 12–14 July (online talks).

[325] Lazzati, D., Ramirez-Ruiz, E. & Rees, M. J. (2002) *ApJ*, **572**, L57.

[326] Lazzati, D., Rossi, E. & Ghisellini, G. *et al.* (2004) *MNRAS*, **347**, L1.

[327] LeBlanc, J. M., & Wilson, J. R. (1970) *ApJ*, **161**, 541.

[328] Le Brun, V., Bergeron J. & Boissé P. (1996) *A & A*, **306**, 691.

[329] Le Brun, V., Bergeron J. & Boissé P. & Deharveng, J. M. (1997) *A & A*, **321**, 733.
[330] Lee, C.-H., Brown, G. E. & Wijers, R. A. M. J. (2002) *ApJ*, **575**, 996.
[331] Lehner, L. (2001) *Class. Quant. Grav.*, **R25**.
[332] Lense, J., & Thirring, H. (1918) *Phys. Z.*, **19**, 156.
[333] Levinson, A. (2004) *ApJ*, **608**, 411.
[334] Levinson, A. & Blandford, R. D. (1996) *ApJ*, **456**(1), L29.
[335] Levinson, A. & Eichler, D. (2000) *Phys. Rev. Lett.*, **85**, 236.
[336] Levinson, A. & Eichler, D. (2003) *ApJ*, **594**, L19.
[337] Levinson, A. & van Putten, M. H. P. M. (1997) *ApJ*, **488**, 69.
[338] Levinson, A., Ofek, E. & Waxman, E. *et al.* (2002) *ApJ*, **576**, 923.
[339] Lewandowski, J., Tafel, J. & Trautman, A. (1983) *Lett. Math. Phys.*, **7**, 347.
[340] Lewin, W. H. G., van Paradijs, J. & van den Heuvel E. P. J. eds. (1995) *X-ray binaries*. Cambridge: Cambridge University Press.
[341] Li, Z.-Y. & Chevalier, R. A. (1999) *ApJ*, **526**, 716.
[342] Li, Z.-Y. & Chevalier, R. A. (2000) *ApJ*, submitted; astro-ph/0010288.
[343] Lichnerowicz, A. (1967) *Relativistic hydrodynamics and magnetohydrodynamics*. New York: W. A. Benjamin Inc.
[344] Linet, B. (1976) *J. Phys. A.*, **9**, 1081.
[345] Liu, Y. T. & Thorne, K. S. (2000) *Phys. Rev. D.*, 122002.
[346] Livio, K., Ogilvie, G. I. & Pringle, J. E. (1999), *ApJ*, **512**, 100.
[347] Lovelace, R. V. (1976) *Nature*, **262**, 649.
[348] Lynden-Bell, D. (1969) *Nature*, **233**, 690.
[349] Lynden-Bell, D. & Rees, J. M. (1971) *MNRAS*, **152**, 461.
[350] Lyne, A. G., Burgay, M. & Kramer, M. *et al.* (2004) *Science*, **303**, 1153.
[351] Lyne, A. G. & Kramer, M. (2004) priv. comm.
[352] Mönchmeyer, R., Schäfer, G. & Müller, E. *et al.* (1991) *A & A*, **246**, 417.
[353] Mészáros, P. (2002) *ARA & A*, **40**, 137.
[354] Mészáros, P. & Rees, M. J. (1997) *ApJ*, **476**, 232.
[355] Mészáros, P. & Rees, M. J. (1999) *MNRAS*, **306L**, 39.
[356] MacFadyen, A. I. (2003) astro-ph/0301425.
[357] MacFadyen, A. I. (2003) in *Proceedings From twilight to highlight – the physics of supernovae workshop*, Garching, 2002.
[358] MacFadyen, A. I. & Woosley, S. E. (1999) *ApJ*, **524**, 262.
[359] Madau, P. & Pozzetti, L. (2000) *MNRAS*, **312**, L9.
[360] Maggiore, M. (2000) *Phys. Rep.*, **331**, 283.
[361] Malesani, D., Tagliaferri, G. & Chinearini, G. *et al.* (2004) *ApJ*, **609**, L5.
[362] Martí, J. M. & Müller, E. (1999) *Living Revs*, **2**, 3.
[363] Martí, J. M., Müller, E. & Font, J. A. *et al.* (1995) *ApJ*, **448**, L105.
[364] Mashhoon, B. (2000) *Class. Quant. Grav.*, **31**, 681.
[365] Mashhoon, B. & Muench, U. (2002) *Ann. Phys.*, **7**, 532.
[366] Matz, S. M. & Share, G. H. (1990) *ApJ*, **362**, 235.
[367] Maund, J. R., Smartt, S. J. & Kudritzki, R. P. *et al.* (2004) *Nature*, **427**, 129.
[368] Mazets, E. P., Golenetskii, S. V. & Ilinskii, V. N. (1974) *JETP*, **19**, L77.
[369] Meegan, C. A., Fishman, G. J. & Wilson, R. B., *et al.* (1992) *Nature*, **355**, 143.
[370] Meier, D. L., Koide, S. & Uchida, Y. (2001) *Science*, **291**, 84.
[371] Metzger, M., Djorgovski, S. G. & Kulkarni, S. R., *et al.* (1997) *Nature*, **387**, 879.
[372] Michelson, A. A., & Morley, E. W. (1887) *Am. J. Sci.*, **34**, 333.
[373] Miller, J. M., Fabian, A. C. & Reynolds, C. S. *et al.* (2004) *ApJ*, **606**, L131.
[374] Miller, J. M., Fabian, A. C. & Wijnands, R. *et al.* (2002) *ApJ*, **570**, L69.
[375] Milne, P. A., The, L.-S. & Leising, M. D. (2001) *ApJ*, **559**, 1019.
[376] Mineshige, S., Hosokawa, T. & Machida, M. *et al.* (2002) *PASJ*, **54**, 655.

[377] Miniutti, G., Fabian, A. C. & Miller, J. M. (2004) *MNRAS*, **351**, 466.
[378] Mirabel, I. F. & Rodríguez, L. F. (1994) *Nature*, **371**, 46.
[379] Mirabel, I. F. & Rodríguez, L. F. (1995) in H. Böringer, G. E. Morfill J.E. Trümper (eds.), Seventh Texas Symposium on Relativistic Astrophysics, *Ann. NY Acad. Sci.*, **759**, 1 p. 21.
[380] Mirabel, I. F. & Rodríguez, L. F. (1999) *ARA & A*, **37**, 409.
[381] Mirabel, I. F., & Rodríguez, L. F. (1996) in H. Böringer, G. E. Morfill & J. E. Trümper (eds.) *Ann. NY Acad. Sc.* **759**. New York: New York Academy of Science.
[382] Misner, C. W., Thorne, K. S. & Wheeler, A. (1974) *Gravitation*. San Francisco:
[383] Miyoshi, M., Moran, J. & Herrnstein, J. *et al.* (1995) *Nature*, **373**, 127.
[384] Modestino, G. & Moleti, A. (2002) *PRD*, **65**, 022005.
[385] Nakamura, T. & Fukugita, M. (1989) *ApJ*, **337**, 466.
[386] Namiki, M. & Otani, C., *et al.* (1999) *A & A Suppl.*, **138**, 433.
[387] Narayan, R., Piran, T. & Shemi, A. (1991) *ApJ*, **379**, L17.
[388] Nikishov, A. I. (1969) *Zh. ETF*, **57**, 1210 [*Sov. Phys. JETP*, **30**(4), 660 (1970)].
[389] Nishikawa, K, Koide, S. & Sakai, J. *et al.* (1997) *ApJ*, L45.
[390] Nomoto, K., Iwamoto, K. & Suzuki, T. (1995) *Phys. Rep.*, **256**, 173.
[391] Nomoto, K., Mazzali, P. A. & Nakamura, T. *et al.* (2000) in M. Livio, N. Panagia & K. Sahu, eds., *The greatest explosions since the big bang: supernovae and gamma-ray bursts*. Cambrige: Cambridge University Press, astro-ph/0003077.
[392] Nomoto, K., Thielemann, F. K. & Yokoi, K. (1984) *ApJ*, **286**, 644.
[393] Nomoto, K., Yamaoka, H. & Pols, O. R. *et al.* (1994) *Nature*, **371**, 227.
[394] Nomoto, K., Mazzali, P. A. & Nakamura, T. *et al.* (2001) in M. Livio, N. Panagia & K. Sahu eds., *Supernovae and gamma-ray bursts*. Cambridge University Press, p. 144.
[395] Norman, M. L. & Winkler, K.-H. (1986) *Astrophysical Radiative Hydrodynamics*. Dordrecht: D. Reidel.
[396] Norman, M. L., Smarr, L., Winkler, K. H. A. & Smith, M. D. (1982) *A & A*, **113**, 285.
[397] Norris, J. P., Share, G. H. & Messina, D. C. *et al.* (1986) *ApJ*, **301**, 213.
[398] Ohanian, H. C. & Ruffini, R. (1994) *Gravitation and spacetime*. New York: W. W. Norton & Company.
[399] Okamoto, I. (1992) *MNRAS*, **253**, 192.
[400] Ostriker, J. P. & Gunn, J. E. (1971) *ApJ*, **164**, L95.
[401] Paciesas W.S., Meegan, C. A. & Pendleton, G. N. *et al.* (1999) *ApJ Suppl.*, **122**, 465.
[402] Paczyński, B. P. (1986) *ApJ*, **308**, L43.
[403] Paczyński, B. P. (1991) *Acta. Astron.*, **41**, 257.
[404] Paczyński, B. P. (1998) *ApJ*, **494**, L45.
[405] Paczyński, B. P. (2001) *Acta Astron.*, **51**, 81.
[406] Paczyński, B. & Rhoads, J. E. (1993) *ApJ*, **418**, L5.
[407] Page, D. N. (1976) *Phys. Rev. D.*, **14**, 3260.
[408] Panaitescu, A. & Kumar, P. (2000) *ApJ*, **543**, 66.
[409] Papaloizou, J. C. B. & Pringle, J. E. (1984) *MNRAS*, **208**, 721.
[410] Papapetrou, A. (1951) *Proc. Roy. Soc.*, **209**, 248.
[411] Parna, R., Sari, R. & Frail, D. A. (2003) *ApJ*, **594**, 379.
[412] Pearson, T. J., Unwin, S. C. & Cohen, M. H., *et al.* (1981) *Nature*, **290**, 365.
[413] Pederson, H., Jaunsen, A. O. & Grav, T. *et al.* (1998) *ApJ*, **496**, 311.
[414] Peebles, P. J. E. (1993) Principles of physical cosmology. Princeton University Press.
[415] Pendleton, G. N., Mallozzi, R. S. & Paciesas, W. S. *et al.* (1996) *ApJ*, **464**, 606.

[416] Penrose, R. (1969) *Rev. del Nuovo Cimento*, **1**, 252.

[417] Penrose, R. & Floyd, R. M. (1971) *Nature Phys. Sci.*, **229**, 177.

[418] Perlmutter, S., Aldering, G. & Goldhaber, G. *et al.* (1999) *ApJ*, **517**, 565.

[419] Peters, P. C., & Mathews, J. (1963) *Phys. Rev.*, **131**, 435.

[420] Peyret, R. & Taylor, T. D. (1983) *Computational Methods for Fluid Flow*. New York: Springer-Verlag.

[421] Phillips, M. M. (1993) *ApJ*, **413**, L105.

[422] Phillips, M. M., Heatcote, S. R. & Hamuy, M., *et al.* (1988) *Astron. J.*, **95**, 1087.

[423] Phinney, E. S. (1983) in *Proc. Astrophys. Jets.* (Dordrecht: Reidel), p. 201.

[424] Phinney, E. S. (1991) *ApJ*, **380**, L17.

[425] Phinney, E. S. (2001) astro-ph/0108028.

[426] Piran, T. (1998) *Phys. Rep.*, **314**, 575.

[427] Piran, T. (1999) *Phys. Rep.*, **314**, 575; *ibid.* (2000) *Phys. Rep.* **333**, 529.

[428] Piran, T. (2004) *Rev. Mod. Phys.*, to appear; astro-ph/0405503.

[429] Piran, T. & Sari, R. (1998) A. V. Olinto, J. A. Friedman & D. N. Schramm (eds.) in 18th Texas Symposium Relativity, Astrophysics and Cosmology. Singapore: World Scientific, p. 34.

[430] Pirani, F. A. E. (1956) *Act. Phys. Pol.*, **XV**, 389.

[431] Pirani, F. A. E. (1957) *Phys. Rev. D*, **105**, 1089.

[432] Piro, L., Costa, E. & Feroci, M. *et al.* (1999) *ApJ*, **514**, L73.

[433] Piro, L., Feroci, M. & Costa, E. *et al.* (1997) IAU Circ. No. 6656.

[434] Piro, L., Garmire, G. & Garcia, M., *et al.* (2000) *Science*, **290**, 955.

[435] Piro, L., Heise, J. & Jager, R. *et al.* (1998) *A & A*, **329**, 906.

[436] Piro, L., Scarsi L. & Butler L. C. (1995) *Proc. SPIE*, **2517**, 169.

[437] Podsiadlowski, Ph., Mazzali, P. A. & Nomoto, K. *et al.* (2004) *ApJ*, **607**, L17.

[438] Popov, S. B. (2004) astro-ph/0403710.

[439] Porciani, C. & Madau, P. (2001) *ApJ*, **548**, 522.

[440] Portegies Zwart, S. F. & McMillan, S. F. W. (2000) *ApJ*, **528**, L17.

[441] Press, W. H. & Teukolsky, S. A. (1972) *ApJ.*, **178**, 347.

[442] Price, P. A., Fox, D. W. & Kulkarni, S. R. *et al.* (2003) *Nature*, **423**, 844.

[443] Pruet, J., Surman, R. & McLaughlin, G. C. (2004) *ApJ*, L101.

[444] Punsly, B. & Coronity, F. V. (1990) *ApJ*, **550**, 518.

[445] Punturo, M. (2003) http://www.virgo.infn.it/senscurve/VIR-NOT-PER-1390-51.pdf.

[446] O'Raifeartaigh, L. (1997) *The dawning of gauge theory*. Princeton Series in Physics. Princeton: Princeton University Press.

[447] Ramirez-Ruiz, E., MacFadyen, A.I., Lazzati, D., 2002, *MNRAS*, 331, 197

[448] Ramirez-Ruiz, E. (2004) *MNRAS*, **349**, L38.

[449] Rees, J. M. & Mészáros P. (1993) *ApJ*, **418**, L59.

[450] Rees, M. J. (1978) *MNRAS*, **184**, 61P.

[451] Rees, M. J. & Mészáros, P. (1992) *MNRAS*, **258**, 41P.

[452] Rees, M. J. & Mészáros, P. (1994) *ApJ*, **430**, L93.

[453] Rees, M. J., Ruffini, R. & Wheeler, J. A. (1974) *Black holes, gravitational waves and cosmology: an introduction to current research*. New York: Gordon & Breach, Section 7.

[454] Reeves, J. N. Watson, D. & Osbourne, J. P. *et al.* (2002) *Nature*, **416**, 512.

[455] Reichart, D. E. (1997) *ApJ*, **485**, L57.

[456] Reichart, D. E. (1999) *ApJ*, **521**, L111.

[457] Reichart, D. E. (2001) *ApJ*, **554**, 643.

[458] Reula, O. (1998) *Living reviews in relativity*. (http://www.livingreviews.org).

[459] Rhoads, J. E. (1997) *ApJ*, **487**, L1.

[460] Rhoads, J. E. (1999) *ApJ*, **525**, 737.

[461] Ries, A. G., Press, W. H. & Kirshner, R. P. (1995) *ApJ*, **438**, L17.

[462] Ries, A. G., Press, W. H. & Kirshner, R. P. (1996) *ApJ*, **473**, 88.

[463] Robertson, H. P. (1936) *ApJ*, **83**, 187.

[464] Rodríguez, L. F., & Mirabel, I. F. (1999) *ApJ.*, **511**, 398.

[465] Rossi, E., Lazzati, D. & Rees, M. J. (2002) *MNRAS*, **332**, 945.

[466] Ruffini, R. & Wilson, J. R. (1975) *Phys. Rev. D.*, **12**, 2959.

[467] Ruggeri, T. & Strumia, A. (1981) *J. Math. Phys.*, **22**, 1824.

[468] Rybicki, G. B., & Lightman, A. P. (1979) *Radiative processes in astrophysics.* New York: Wiley & Sons.

[469] Sahu, K., Livio, M. & Petro, L. *et al.* (1997) *Nature*, **387**, 476.

[470] Sakamoto, T., Lamb, D. Q. & Graziani, C. *et al.* (2004) *ApJ*, **602**, 875.

[471] Salmonson, J. D. (2001) *ApJ*, **546**, L29.

[472] Sari, R. (2000) in R. M. Kippen, R. S. Mallozi & G. J. Fishman (eds.) *Gamma-ray bursts.* Fiftieth Huntsville Symposium (Conference Proceedings 526). New York: AIP (2000), p. 504.

[473] Sari, R., Piran, T. & Haplern, J. P. (1999) *ApJ*, **519**, L17.

[474] Saulson, P. R. (1984) *Phys. Rev. D.*, **30**, 732.

[475] Saulson, P. R. (1990) *Phys. Rev. D.*, **42**, 2437.

[476] Saulson, P. R. (1994) *Fundamentals of interferometric wave detectors.* World Scientific.

[477] Sazonov, A. U., Lutocinov, A. A. & Sunyaev, R. A. (2004) *Nature*, **430**, 646.

[478] Schaefer, B. E., Deng, M. & Band, D. L. (2001) *ApJ*, **563**, L123.

[479] Scheel, M. A., Baumgarte T. W. & Cook G. B. (1998) *Phys. Rev. D.*, **58**, 044020.

[480] Schmidt, M. (1963) *Nature*, **197**, 1040.

[481] Schmidt, M. (1999) *A & A Suppl.*, **138**, 409.

[482] Schmidt, M., Higdon J. C. & Heuter, G. (1988) *ApJ*, **329**, L85.

[483] Schödel, R., Ott, T. & Genzel, R. *et al.* (2002) *Nature*, **419**, 694.

[484] Schutz, B. F. (1980) *Phys. Rev. D.*, **22**, 249.

[485] Schutz, B. F. (1985) *A first course in general relativity.* Cambridge: Cambridge University Press.

[486] Schutz, B. F. (1997) in M. Davier & P. Hello (eds.), *Proceedings, Second workshop on gravitational wave analysis.* Orsay, France, p. 133.

[487] Schutz, B. F. & Verdaguer, E. (1983) *MNRAS*, **202**, 881.

[488] Sethi, S. & Bhargavi, S. G. (2001) *A & A*, **376**, 10.

[489] Shapiro I. I. (1964) *Phys. Rev. Lett.* **13**, 789.

[490] Shapiro, S. L., & Teukolsky, S. A. (1983) *Black holes, white dwarfs, and neutron stars.* New York: Wiley.

[491] Shatskiy, A. A. (2003) *Astron. Lett.*, **29**, 155; astro-ph/0301536.

[492] Shaviv, N. & Dar, A. (1995) *ApJ*, **447**, 863.

[493] Shemi, A. & Piran, T. (1990) *ApJ*, **365**, L55.

[494] Shigeyama, T., Nomoto, K. & Tsujimoto, T., *et al.* (1990) *ApJ*, **361**, L23.

[495] Sigg, D. (1998) in *Proceedings, TASI*, Boulder, Colorado.

[496] Smith, I. A., Gruendl, R. A. & Liang, E. P. *et al.* (1997) *ApJ*, **487**, L5.

[497] Soderberg, A. M., Kulkarni, S. R. & Berger, E. *et al.* (2004) *ApJ*, **606**, 994.

[498] Soderberg, A. M., Kulkarni, S. R. & Berger, E. *et al.* (2004) *Nature*, **430**, 648.

[499] Soderberg, A. M., Kulkarni, S. R. & Frail, D. A. (2003) *GCN Circ. No. 2483*.

[500] Soderberg, A. M., Price, P. A. & Fox, D. W. *et al.* (2002) *GCN Circ. No. 1554*.

[501] Sokolov, V. V., Kopylov, A. I. & Zharikov, S. V., *et al.* (1997) in C. A. Meegan, T. M. Koshut and R. D. Preece, eds., *Gamma-ray bursts.* Fourth Huntsville Symposium (Conference Proceedings 428). New York: AIP, p. 525.

[502] Sol, H. & Pelletier, G. (1989) *MNRAS*, **237**, 411.
[503] Spallicci, A. D. A. M., Abramovici, A., Althouse, W. E. & Drever, R. W. S., *et al.*, (2004), gr-qc/0406076.
[504] Spyromilio, J., Meikle, W. P. S. & Allen, D. A. (1992) *MNRAS*, **258**, p. 53.
[505] Stanek, K. Z., Garnavich, P. M. & Kaluzny, J., *et al.* (1999) *ApJ*, **522**, L39.
[506] Stanek, K. Z., Matheson, T. & Garnavich, P. M. *et al.* (2003) *ApJ*, **591**, L17.
[507] Starobinsky, A. A. (1972) *Zh. ETF*, **64**, 48 [*Sov. Phys. JETP*, **37**:28 (1973)].
[508] Stella, L. (2000) in *Proceedings, X-ray astronomy, 1999*: G. Malaguti, G. Palumbo & N. White(eds.) *Stellar endpoints, AGN and the diffuse background.* (Singapore and New York: Gordon and Breach.
[509] Stephani, H. (1990) *General relativity*, 2nd edn. Cambridge: Cambridge University Press.
[510] Swartz, D. A., Filippenko, A. V. & Nomoto, K., *et al.* (1993) *ApJ*, **411**, 313.
[511] Sweeney, M. A. (1976) *Astron. Astroph.* **49**, 375.
[512] Tagliaferri, G., Corvino, S. & Fugazza, D. *et al.* (2004) *IAU Circ. No. 8308.*
[513] Takahashi, M., Shinya, N., Tatematsu, Y. & Tomimatsu, A. (1990) *ApJ*, **363**, 206.
[514] Tanaka, Y. & Lewin, W. H. G. (1997) in W. H. G. Lewin, J. van Paradijs & E. P. J. van den Heuvel (eds.) Cambridge: *Black hole binaries*. Cambridge University Press, p. 126.
[515] Tanaka, Y., Nandra, K. & Fabian, A. C. (1995) *Nature*, **375**, 659.
[516] Taub, A. H. (1948) *Phys. Rev.*, **74**, 328.
[517] Taylor, B., Chambers, C. M. & Hiscock W. A. (1998) *Phys. Rev. D.*, **58**(4), 40121.
[518] Taylor, J. H. (1994) *Rev. Mod. Phys.*, **66**, 711.
[519] Taylor, J. H. & Weisberg, J. M. (1982) *ApJ*, **253**, 908.
[520] Taylor, J. H. & Weisberg, J. M. (1989) *ApJ*, **345**, 434.
[521] Ternov, I. M., Gaina, A. B. & Chizhov, G. A. (1986) *Yad. Fiz.*, **44**, 533 [*Sov. J. Nucl. Phys.*, **44**(2), 343 (1986)].
[522] Teukolsky, S. A. (1973) *ApJ.*, **185**, 635.
[523] Teukolsky, S. A. & Press, W. H. (1974) *ApJ.*, **193**, 443.
[524] Thompson, C. & Duncan, R. C. (2001) *ApJ*, **561**, 980.
[525] Thomsen, B., Hjorth, J. & Watson, D. *et al.* (2004) *A & A*, **419**, L21.
[526] Thomson, R. C., Mackay, C. D. & Wright, A. E. (1993) *Nature*, **365**, 133.
[527] 't Hooft, G. (2002) *Introduction to general relativity*. Princeton: Rinton Press.
[528] 't Hooft, G. (2003) priv. comm.
[529] Thorne, K. S. (1969) *ApJ*, **158**, 1.
[530] Thorne, K. S. (1969) *ApJ*, **158**, 997.
[531] Thorne, K. S. (1987) in S. W. Hawking & W. Israel (eds.) *300 Years of gravitation*. Cambridge: Cambridge University Press, pp. 330–458.
[532] Thorne, K. S. (1995) Arxiv Preprint gr-qc/9506086; *ibid.* 1995, gr-qc/9506084.
[533] Thorne, K. S. (1997) *Rev. Mod. Astron.*, **10**, 1.
[534] Thorne, K. S., Price, R. H. & MacDonald, D. A. (1986) *Black holes: the membrane paradigm*. New Haven: Yale University Press.
[535] Timmes, F. X., Woosley, S. E. & Hartman, D. H., *et al.* (1996) *ApJ*, **464**, 332.
[536] Tinney, C., Stathakis, R. & Cannon, R. *et al.* (1998) *IAU Circ. 6896.*
[537] Tricarico, P., Ortolan, A. & Solaroli, A. *et al.* (2001) *Phys. Rev. D.*, **63**, 082002.
[538] Trimble, V. & Weber, J. (1973) *Ann. N.Y. Acad. Sci.*, **224**, 93.
[539] Turatto, M. (2003) in K. W. Weiler (ed.) *Supernovae and Gamma-ray Bursters*. Heidelberg: Springer-Verlag, p. 21.
[540] Turatto, M. (2003) in K. W. Weiler ed., *Proceedings, supernovae and gamma-ray bursts*. astro-ph/0301107.
[541] Uemura, M., Kato, T. & Ishioka, R. *et al.* (2003) *Nature*, **423**, 843.

[542] Unruh, W. G. (1974) *Phys. Rev. D.*, **10**, 3194.

[543] Utiyama, R. (1956) *Phys. Rev.*, **101**, 1597.

[544] Utiyama, R. (1980) *Prog. Theor. Phys.*, **64**, 2207.

[545] Uzdensky, D. A. (2004) *ApJ*, **603**, 652.

[546] van Dyk, S. D., Sramek, R. A., Weiler, K. W. *et al.* & Panagia, N. (1993) *ApJ*, **409**, 162.

[547] van Paradijs, J., *et al.* (1997) *Nature*, **386**, 686.

[548] van Putten, M. H. P. M. (1991) *Commun. Math. Phys.*, **141**, 63.

[549] van Putten, M. H. P. M. (1993) *ApJ*, **408**, L21.

[550] van Putten, M. H. P. M. (1993) *J. Comput. Phys.*, **105**(2), 339.

[551] van Putten, M. H. P. M. (1994) in T. Ruggeri (ed.) *Proceedings, Seventh conference on waves and stability in continuous media*, Bologna, 1993. Italy, 1994: World Scientific.

[552] van Putten, M. H. P. M. (1994) *In. J. Bifurc. & Chaos*, **4**, 57.

[553] van Putten, M. H. P. M. (1994) *Phys. Rev. D.* **50**(10), 6640.

[554] van Putten, M. H. P. M. (1995) *SIAM J Numer. Anal.*, **32**(5), 1504.

[555] van Putten, M. H. P. M. (1996) *ApJ*, **467**, L57.

[556] van Putten, M. H. P. M. (1997) *Phys. Rev. D.*, **55**, 4705.

[557] van Putten, M. H. P. M. (1999) *Science*, **294**, 115.

[558] van Putten, M. H. P. M. (2000) *Phys. Rev. Lett.*, **84**, 3752; astro-ph/9911396.

[559] van Putten, M. H. P. M. (2001) *Phys. Rev. Lett.,* **87**, 091101.

[560] van Putten, M. H. P. M. (2001) *Phys. Rep.* **345**, 1.

[561] van Putten, M. H. P. M. (2002) *ApJ*, **575**, L71.

[562] van Putten, M. H. P. M. (2002) *J. Math. Phys.*, **43**, 6195.

[563] van Putten, M. H. P. M. (2003) *ApJ*, **583**, 374.

[564] van Putten, M. H. P. M. (2004) *ApJ Lett*, **611**, L81.

[565] van Putten, M. H. P. M. Lee, H.-K., Lee, C.-H. & Kim, H. (2004) *Phys. Rev. D.*, **69**, 104026.

[566] van Putten, M. H. P. M. & Eardley D. M. (1996) *Phys. Rev. D.*, **53**, 3056; gr-qc/9505023.

[567] van Putten, M. H. P. M. & Levinson, A. (2002) *Science*, **294**, 1837.

[568] van Putten, M. H. P. M. & Levinson, A. (2003) *ApJ*, **584**, 937; astro-ph/0212297.

[569] van Putten, M. H. P. M. & Ostriker, E. (2001) *ApJ*, **552**, L31.

[570] van Putten, M. H. P. M. & Regimbau, T. (2003) *ApJ*, **593**, L15.

[571] van Putten, M. H. P. M. & Sarkar, A. (2000) *Phys. Rev. D*, **62**, 041502(R).

[572] van Putten, M. H. P. M. & Wilson, A. (1999) in R. Narayan, R. Antonucci, N. Gehrels and S. M. Kahn, eds. (1999) *Theory confronts reality*. Proceedings, ITP Conference, 2–5 February.

[573] van Putten, M. H. P. M., Levinson, A. & Regimbau, T., *et al.* (2004) *Phys. Rev. D.*, **69**, 044007.

[574] Vietri, M. (1997) *ApJ*, **488**, L105.

[575] Vietri, M. (1997) *ApJ*, **478**, L9.

[576] Wald, R. M. (1974) *Phys. Rev. D.*, **10**, 1680.

[577] Wald, R. M. (1984) *General relativity*. London: University of Chicago Press.

[578] Walker, A. G. (1936) *Proc. London Math. Soc.*, **42**, 90.

[579] Walker, M. & Penrose, R. (1970) *Commun. Math. Phys.*, **18**, 265.

[580] Wang, L. & Wheeler, J. C. (1998) *ApJ*, **508**, L87.

[581] Wardle, J., Homan, D. C., Ojha, R. & Roberts, D. H. (1998) *Nature*, **395**, 457.

[582] Watson, D., Hjorth, J. & Levan, A. *et al.* (2004) **605**, L101.

[583] Waxman, E., Kulkarni, S. R. & Frail, D. A. (1998) *ApJ*, **487**, 288.

[584] Weber, J. (1973) *Phys. Rev. Lett.*, **31**, 779.

[585] Weiler, K. W., Panagia, N. & Montes, M. J. (2001) *ApJ*, **562**, 670.

[586] Weiler, K. W., van Dyk, S. D. & Montes, M. J. *et al.* (1998) *ApJ*, **500**, 51.

[587] Weinberg, S. (1972) *Gravitation and cosmology*. New York: Wiley & Sons.

[588] Weinberg, S. (1989) *Rev. Mod. Phys.*, **61**, 1.

[589] Weisberg, J. M., & Taylor, H., in Radio Pulsars, eds. M. Bailes, D. J. Nice & S. E. Thorsett, San Francisco: APS 2003 (Conf. Series CS 302), p. 93.

[590] Weiss, R. (1972) *Quar. Prog. Rep.*, **105**, Lincoln Research Laboratory. Massachusetts Institute of Technology.

[591] Wen, L., 2003, *ApJ*, 598, 419

[592] Wen, L., Panaitescu, A. & Laguna, P. (1997) *ApJ*, **486**, 919.

[593] Wheeler, J. C. (2002) *Am. J. Phys.*, **71**, 11.

[594] Wheeler, J. C., & Harkness, R. P. (1986) in B. F. Madore, R. B. Tully (ed.) *Galaxy distances and deviations from universal expansion*. Dordrecht: Reidel, p. 45.

[595] Wheeler, J. C. & Levreault, R. (1985) *ApJ*, **294**, L17.

[596] Wheeler, J. C., Yi, I. & Höflich, P., *et al.* (2000) *ApJ*, **537**, 810.

[597] Whitham, G. (1973) *Linear and nonlinear waves*. New York: Wiley-Interscience.

[598] Wijers, R. A. M. J., Rees, M. J. & Mészáros, P. (1997) *MNRAS*, **288**, L51.

[599] Wilkins, D. (1972) *Phys. Rev. D5*, 814.

[600] Willingale, R., Osborne, J. P. & O Brien, P. T., *et al.* (2004) *MNRAS*, **349**, 31.

[601] Willke, B. (2002) *Class. Quant. Grav.*, **19**, 1377.

[602] Wilms, J., *et al.* (2001) *MNRAS*, **328**, L27.

[603] Wilson, A. S., Young, A. J. & Shopbell, P. L. (2000) *ApJ*, **544**, L27.

[604] Wilson, M. J. (1987) *MNRAS*, **226**, 447.

[605] Witten, L. (ed.) (1962) *Gravitation: an introduction to current research*. New York: Wiley & Sons.

[606] Woosley, S. E. & Weaver, T. A. (1986) *ARA & A*, **386**, 181.

[607] Woosley, S. E. & Weaver, T. A. (1995) *ApJS*, **101**, 181.

[608] Woosley, S. E., Langer, N. & Weaver, T. A. (1993) *ApJ*, **411**, 823.

[609] Woosley, S. E., Langer, N. & Weaver, T. A. (1995) *ApJ*, **448**, 315.

[610] Yoshida, A. (2000) *ASpR*, **25**, 761.

[611] Yoshida, A., Namiki, M. & Otani, C., *et al.* (1997) in C. A. Meegan, T. M. Koshut & R. D. Preece, eds., *Gamma-ray bursts*. Fourth Huntsville Symposium (Conference Proceedings 428). New York: AIP, p. 441.

[612] Yoshida, A., Namiki, M. & Otani, C. *et al.* (1999) *A & A Suppl.*, **138**, 433.

[613] Yoshida, A., Namiki, M. & Otani, C., *et al.* (2000) *ASpR*, **25**, 761.

[614] Zel'dovich Ya, B. (1971) *Zh. Eks. Teor. Fiz.*, **14**, 270 [transl. JETP Lett. **14**, 180(1971)].

[615] Zensus, J. A. (1997) *ARA & A*, **35**, 607.

[616] Zhang, B. & Mészáros, P. (2002) *ApJ*, **571**, 876.

[617] Znajek, R. L. (1977) *MNRAS*, **179**, 457.

Index